Springer Series in Information Sciences 7

Springer Series in Information Sciences

Editors: Thomas S. Huang   Teuvo Kohonen   Manfred R. Schroeder

Managing Editor: H.K.V. Lotsch

M. R. Schroeder

Number Theory
in Science
and Communication

With Applications in Cryptography,
Physics, Digital Information, Computing,
and Self-Similarity

Second Enlarged Edition

With 81 Figures

Springer-Verlag Berlin Heidelberg New York
London Paris Tokyo Hong Kong

Professor Dr. Manfred R. Schroeder

Direktor, Drittes Physikalisches Institut, Universität Göttingen, Bürgerstrasse 42–44,
D-3400 Göttingen, Fed. Rep. of Germany and

Past Director, Acoustics, Speech and Mechanics Research, Bell Laboratories,
Murray Hill, NJ 07974, USA

Series Editors:

Professor Thomas S. Huang

Department of Electrical Engineering and Coordinated Science Laboratory,
University of Illinois, Urbana, IL 61801, USA

Professor Teuvo Kohonen

Laboratory of Computer and Information Sciences, Helsinki University of Technology,
SF-02150 Espoo 15, Finland

Professor Dr. Manfred R. Schroeder

Drittes Physikalisches Institut, Universität Göttingen, Bürgerstrasse 42–44,
D-3400 Göttingen, Fed. Rep. of Germany

Managing Editor: Helmut K. V. Lotsch

Springer-Verlag, Tiergartenstrasse 17,
D-6900 Heidelberg, Fed. Rep. of Germany

QA
241
S318
1990

Corrected Printing 1990

ISBN 3-540-15800-6 2. Aufl. Springer-Verlag Berlin Heidelberg New York
ISBN 0-387-15800-6 2nd ed. Springer-Verlag New York Berlin Heidelberg

ISBN 3-540-12164-1 1. Aufl. Springer-Verlag Berlin Heidelberg New York
ISBN 0-387-12164-1 1st ed. Springer-Verlag New York Berlin Heidelberg

Library of Congress Cataloging-in-Publication Data. Schroeder, M. R. (Manfred Robert), 1926– . Number theory in science and communication : with applications in cryptography, physics, digital information, computing, and self-similarity / M. R. Schroeder.–2nd enl. ed., corr. printing. p. cm.–(Springer series in information sciences ; 7) Includes bibliographical references (p.). 1. Number theory. I. Title. II. Series. QA241.S318 1990 512'.7–dc20   90-9435

Printing: Druckhaus Beltz, 6944 Hemsbach/Bergstr.
Binding: J. Schäffer GmbH & Co. KG, 6718 Grünstadt
2154/3150-543210 – Printed on acid-free paper

Dedicated to the Memory of
Hermann Minkowski

who added a fourth dimension to our World
and many more to Number Theory

Foreword

*"Beauty is the first test: there is no
permanent place in the world for
ugly mathematics."*
— *G. H. Hardy*

Number theory has been considered since time immemorial to be the very
paradigm of pure (some would say useless) mathematics. In fact, the Chinese
characters for mathematics are *Number Science.* "Mathematics is the queen
of sciences — and number theory is the queen of mathematics," according to
Carl Friedrich Gauss, the lifelong Wunderkind, who himself enjoyed the
epithet "Princeps Mathematicorum." What could be more beautiful than a
deep, satisfying relation between whole numbers. (One is almost tempted to
call them wholesome numbers.) In fact, it is hard to come up with a more
appropriate designation than their learned name: the integers — meaning the
"untouched ones". How high they rank, in the realms of pure thought and
aesthetics, above their lesser brethren: the real and complex numbers —
whose first names virtually exude unsavory involvement with the complex
realities of everyday life!

Yet, as we shall see in this book, the theory of integers can provide totally
unexpected answers to real-world problems. In fact, discrete mathematics is
taking on an ever more important role. If nothing else, the advent of the
digital computer and digital communication has seen to that. But even
earlier, in physics the emergence of quantum mechanics and discrete
elementary particles put a premium on the methods and, indeed, the spirit of
discrete mathematics.

And even in mathematics proper, Hermann Minkowski, in the preface to
his introductory book on number theory, *Diophantische Approximationen,*
published in 1907 (the year he gave special relativity its proper four-
dimensional clothing in preparation for its journey into general covariance and
cosmology) expressed his conviction that the "deepest interrelationships in
analysis are of an arithmetical nature."

Yet much of our schooling concentrates on analysis and other branches of
continuum mathematics to the virtual exclusion of number theory, group
theory, combinatorics and graph theory. As an illustration, at a recent
symposium on information theory, the author met several young women
formally trained in mathematics and working in the field of *primality testing,*
who — in all their studies up to the Ph.D. — had not heard a single lecture
on number theory!

Or, to give an earlier example, when Werner Heisenberg discovered
"matrix" mechanics in 1925, he didn't know what a matrix was (Max Born
had to tell him), and neither Heisenberg nor Born knew what to make of the
appearance of matrices in the context of the atom. (David Hilbert is reported

to have told them to go look for a differential equation with the same eigenvalues, if that would make them happier. — They did not follow Hilbert's well-meant advice and thereby may have missed discovering the Schrödinger wave equation.)

The present book seeks to fill this gap in basic literacy in number theory — not in any formal way, for numerous excellent texts are available — but in a manner that stresses intuition and interrelationships, as well as applications in physics, biology, computer science, digital communication, cryptography and more playful things, such as puzzles, teasers and artistic designs.

Among the numerous applications of number theory on which we will focus in the subsequent chapters are the following:

1) The division of the circle into equal parts (a classical Greek preoccupation) and the implications of this ancient art for modern fast computation and random number generation.

2) The Chinese Remainder Theorem (another classic, albeit *Far* Eastern) and how it allows us to do coin tossing over the telephone (and many things besides).

3) The design of concert hall ceilings to scatter sound into broad lateral patterns for improved acoustic quality (and wide-scatter diffraction gratings in general).

4) The precision measurement of delays of radar echoes from Venus and Mercury to confirm the general relativistic slowing of electromagnetic waves in gravitational fields (the "fourth" — and last to be confirmed — effect predicted by Einstein's theory of general relativity).

5) Error-correcting codes (giving us distortion-free pictures of Jupiter and Saturn and their satellites).

6) "Public-key" encryption and deciphering of secret messages. These methods also have important implications for computer security.

7) The creation of artistic graphic designs based on prime numbers.

8) How to win at certain parlor games by putting the Fibonacci number systems to work.

9) The relations between Fibonacci numbers and the regular pentagon, the Golden ratio, continued fractions, efficient approximations, electrical networks, the "squared" square, and so on — almost ad infinitum.

I dedicated this book to Hermann Minkowski because he epitomizes, in my mind, the belief in the usefulness outside mathematics of groups and number theory. He died young and never saw these concepts come to full fruition in general relativity, quantum mechanics and some of the topics touched upon here. I am therefore glad that the town of Göttingen is moving to honor its former resident on the occasion of the 100th anniversary of his doctorate on 30 July, 1885 (under F. Lindemann, in now transcendental Königsberg). The late Lilly Rüdenberg, née Minkowski (born in Zürich, while her father was teaching a still unknown Albert there), communicated valuable information in preparation for this late recognition.

Preface to the Second Edition

The first edition quickly shrank on Springer's shelves, giving me a welcome opportunity to augment this volume by some recent forays of number theory into new territory. The most exciting among these is perhaps the discovery, in 1984, of a *new state of matter*, sharing important properties with both perfect crystals and amorphous substances, without being either one of these. The atomic structure of this new state is intimately related to the Golden ratio and a certain *self-similar* (rab)bit sequence that can be derived from it. In fact, certain generalizations of the Golden ratio, the "Silver ratios" — numbers that can be expressed as periodic continued fractions with period length one — lead one to postulate quasicrystals with "forbidden" 8- and 12-fold symmetries and additional quasicrystals with 5-fold symmetry whose lattice parameters are generated by the Lucas numbers (Sect. 30.1).

The enormously fruitful concept of self-similarity, which pervades nature from the distribution of atoms in matter to that of the galaxies in the universe, also occurs in number theory. And because self-similarity is such a pretty subject, in which *Cantor* and *Julia sets* join *Weierstrass functions* to create a new form of art (distinguished by *fractal Hausdorff dimensions)*, a brief celebration of this strangely attractive union seems all but self set.

It is perhaps symptomatic that with *set theory* still another abstract branch of mathematics has entered the real world. Who would have thought that such utterly mathematical constructions as *Cantor sets,* invented solely to reassure the skeptics that sets could have both measure zero and still be uncountable, would make a difference in any practical arena, let alone become a pivotal concept? Yet this is precisely what happened for many natural phenomena from gelation, polymerization and coagulation in colloidal physics to nonlinear systems in many branches of science. Percolation, dendritic growth, electrical discharges (lightnings and *Lichtenberg* figures) and the composition of glasses are best described by set-theoretic *fractal dimensions.*

Or take the weird functions Weierstrass invented a hundred years ago purely to prove that a function could be both everywhere continuous and yet nowhere differentiable. The fact that such an analytic pathology describes something in the real world — nay, is *elemental* to understanding the basins of attraction of strange attractors, for one — gives one pause. These exciting new themes are sounded in Chap. 30.

Physicists working in *deterministic chaos* have been touting the Golden ratio *g* as the "most irrational" of irrational numbers; and now they gladden us with yet another kind of new number: the *noble numbers,* of which *g* (how aptly named!) is considered the noblest. *Nonlinear dynamical systems* governed by these "new" numbers (whose continued-fraction expansions end in infinitely many 1's) show the greatest resistance to the onslaught of chaotic motion (such as turbulence). The rings of Saturn and, indeed, the very stability of the solar system are affected by these numbers. This noble feat, too, merits honorable mention (in Sects. 5.3 and 30.1).

Another topic, newly treated and of considerable contemporary import, is *error-free computation,* based on Farey fractions and p-adic "Hensel codes" (Sect. 5.12).

Other recent advances recorded here are applications of the *Zech logarithm* to the design of optimum *ambiguity functions* for radar, new *phase-arrays* with unique radiation patterns, and *spread-spectrum communication* systems (Sects. 25.7-9).

A forthcoming Italian translation occasioned the inclusion of a banking puzzle (Sect. 5.11) and other new material on Fibonacci numbers (in Sect. 1.1 and Chap. 30).

Murray Hill and Göttingen   *Manfred R. Schroeder*
April 1985

Acknowledgements

I thank Martin Kneser, Don Zagier, Henry Pollak, Jeff Lagarias and Hans Werner Strube for their careful reviews of the manuscript.

Wilhelm Magnus, Armin Kohlrausch, Paul C. Spandikow, Herbert Taylor and Martin Gardner spotted several surviving slips.

Joseph L. Hall of AT&T Bell Laboratories computed the spectra of the self-similar sequences shown in Chap. 30. Peter Meyer, University of Göttingen, prepared the auditory paradox, described in Sect. 30.2, based on a "fractal" signal waveform.

Norma Culviner's red pencil and yellow paper slips have purged the syntax of original sin. Angelika Weinrich, Irena Schönke and Elvire Hung, with linguistic roots far from English soil, and Dorothy Caivano have typed pages and fed computers to near system overflow. Gisela Kirschmann-Schröder and Liane Liebe prepared the illustrations with professional care and love. Anny Schroeder piloted the ms. through the sea of corrections inundating her desk from two continents.

Dr. Helmut Lotsch of Springer-Verlag deserves recognition for his leadership in scientific publication. Working with Herr Reinhold Michels from the production department in Heidelberg proved a pure delight.

William O. Baker and Arno Penzias fostered and maintained the research climate at Bell Laboratories in which an endeavor such as the writing of this book could flourish. Max Mathews provided support when it was needed. John R. Pierce inspired me to write by his writings.

Contents

Part II   Some Simple Applications

Part III   Congruences and the Like

Part IV Cryptography and Divisors

Part V   Residues and Diffraction

Part VI   Chinese and Other Fast Algorithms

Part VII   Pseudoprimes, Möbius Transform, and Partitions

Part VIII   Cyclotomy and Polynomials

Part IX   Galois Fields and More Applications

Part X   Self-Similarity, Fractals and Art

1. Introduction

"Die ganzen Zahlen hat der liebe Gott gemacht, alles andere ist Menschenwerk."
— *Leopold Kronecker*

Hermann Minkowski, being more modest than Kronecker, once said "The primary source (Urquell) of all of mathematics are the integers." Today, integer arithmetic is important in a wide spectrum of human activities and natural phenomena amenable to mathematic analysis.

Integers have repeatedly played a crucial role in the evolution of the natural sciences. Thus, in the 18th century, Lavoisier discovered that chemical compounds are composed of fixed proportions of their constituents which, when expressed in proper weights, correspond to the ratios of *small integers*. This was one of the strongest hints to the existence of atoms, but chemists for a long time ignored the evidence and continued to treat atoms merely as a conceptual convenience devoid of any physical meaning. (Ironically, it was from the statistical laws of *large* numbers in Einstein's analysis of Brownian motion at the beginning of our own century, that the irrefutable reality of atoms and molecules finally emerged.)

In the analysis of optical spectra, certain integer relationships between the wavelengths of spectral lines emitted by excited atoms gave early clues to the *structure of atoms,* culminating in the creation of matrix mechanics in 1925, an important year in the growth of *integer physics.*

Later, the near-integer ratios of atomic weights suggested to physicists that the *atomic nucleus* must be made up of an *integer* number of similar nucleons. The *deviations* from integer ratios led to the discovery of elemental *isotopes.*

And finally, small divergencies in the atomic weight of pure isotopes from exact integers constituted an early confirmation of Einstein's famous equation $E = mc^2$, long before the "mass defects" implied by these integer discrepancies blew up into widely visible mushroom clouds.

On a more harmonious theme, the role of integer ratios in musical scales has been widely appreciated ever since Pythagoras first pointed out their importance. The occurrence of integers in biology — from plant morphology to the genetic code — is pervasive. It has even been hypothesized that the North American 17-year cicada selected its life cycle because 17 is a prime number, prime cycles offering better protection from predators than nonprime cycles. (The suggestion that the 17-year cicada "knows" that 17 is a *Fermat* prime has yet to be touted though.)

Another reason for the resurrection of the integers is the penetration of our lives achieved by that 20th-century descendant of the abacus, the *digital computer*. (Where did all the slide rules go? Ruled out of most significant places by the ubiquitous pocket calculator, they are sliding fast into restful oblivion. — Collectors of antiques take note!)

An equally important reason for the recent revival of the integer is the congruence of *congruential arithmetic* with numerous modern developments in the natural sciences and digital communications — especially "secure" communication by cryptographic systems. Last not least, the proper protection and security of computer systems and data files rests largely on congruence relationships.

In congruential arithmetic, what counts is not a numerical value per se, but rather its remainder or *residue* after division by a *modulus*. For example, in wave interference (be it of ripples on a lake or of electromagnetic fields on a hologram plate) it is not path differences that determines the resulting interference pattern, but rather the residues after dividing by the wavelength. For perfectly periodic events, there is no difference between a path difference of half a wavelength or one and a half wavelengths: in either case the interference will be destructive.

One of the most dramatic consequences of congruential arithmetic is the existence of the chemical elements as we know them. In 1913, Niels Bohr postulated that certain integrals associated with electrons in "orbit" around the atomic nucleus should have integer values — a requirement that 10 years later became comprehensible as a wave interference phenomenon of the newly discovered *matter* waves: essentially, integer-valued integrals meant that path differences are divisible by the electron's wavelength without leaving a remainder.

But in the man-made world, too, applications of congruential arithmetic abound. In binary representation and error-correcting codes, the important Hamming distance is calculated from the sum (or difference) of corresponding places *modulo* 2. In other words, $1 + 1$ equals 0 — just as two half-wavelength path differences add up to no path difference at all.

Another example of a man-made application of congruential arithmetic is a public-key encryption method in which messages represented by integers are raised to a given power and only the residue, modulo a preselected encryption modulus, is transmitted from sender to receiver. Thus, in a simplified example, if the message was 7, the encrypting exponent 3 and the modulus 10, then the transmitted message would not be $7^3 = 343$, but only the last digit: 3. The two most significant digits are discarded. Of course, in a serious application, the prime factors of the modulus are so large (greater than 10^{200}, say) that knowledge of their product gives no *usable* clue to the factors, which are needed for decryption.

Thus, we are naturally led to the most important concept of number theory: the distinction between *primes* and nonprimes, or *composites,* and to the properties of divisibility, the greatest common divisor and the least common multiple — precursor to the art of factoring.

These topics form the theme of the first four chapters of the book. The reader will find here an emphasis on *applications* (musical scales and human pitch perception, for example). Also, the useful *probabilistic viewpoint* in number theory is introduced here (Chap. 4) and applied to "derivations" of the prime density and coprime probability.

The book continues with a discussion of *continued fractions, Fibonacci series* and some of their *endless* applications (Chap. 5).

In Chap. 6 *congruences* are introduced and put to work on *Diophantine equations* — both linear and nonlinear (Chap. 7). Again, there are interesting applications both within mathematics (*Minkowski's* geometry of numbers) and without.

In Chap. 8, in preparation for one of our main themes — *cryptography* (Chaps. 9, 12, 14) — we become acquainted with the pivotal *theorems of Fermat and Euler*.

Some of the basic concepts of *factoring* — the various *divisor functions* — are taken up in Chaps. 10 and 11, again stressing the probabilistic viewpoint.

Chapter 13 is devoted to the important concepts of *order, primitive roots* and the number theoretic *index* function. Among the applications encountered are period lengths of periodic fractions and, surprisingly, a prescription for better *concert hall acoustics*.

In Chap. 15 we are introduced to *quadratic residues, Gauss sums* and their *Fourier* spectrum properties.

The *Chinese remainder theorem* (Chap. 16) allows us to tackle *simultaneous congruences* and *quadratic congruences* (Chap. 18). The so-called *Sino-representation* of integers leads to important *fast algorithms*, a theme that is continued in Chap. 17 where we examine the basic principle (*Kronecker* factorization, direct products) behind such indispensable algorithms as the *fast Hadamard* (FHT) and *Fourier transforms* (FFT).

Since in encryption there will always be a need for *very large primes* beyond the "primality horizon", that need may have to be satisfied by *pseudoprimes* which we define in Chap. 19. The techniques of testing the so-called *strong* pseudoprimes are related to such "games" as coin tossing over the telephone and *certified digital mail*.

Chapter 21 introduces *generating functions* and *partitions*, an important topic in *additive* number theory, which is a difficult field and poses many unsolved problems.

In preparation for our excursion into the art and science of dividing the circle (Chap. 22), we acquaint ourselves with the *Möbius* function (Chap. 20) and then put *polynomials* (Chaps. 23 and 24) to use to construct *finite number fields* (Chap. 25). Some of their numerous applications based on the spectrum and correlation properties of *"Galois"* sequences are discussed in Chap. 26 (*error-correcting codes, precision measurements and antenna theory)*, Chap. 27 (*random number generators*) and Chap. 28 (special *waveforms* for *sonar, radar* and *computer speech synthesis*). Minimum redundancy arrays, important in radio astronomy, underwater sound detection and real-time diagnostic tomography, are also mentioned here.

The book concludes with a brief excursion into the world of *artistic design* based on the aesthetically desirable mixture of regularity and randomness inherent in the distribution of primes and primitive polynomials.

In the sections which follow some of these topics are introduced informally, together with the leading dramatis personae. The reader interested in widening her or his arithmetic horizon is referred to [1.1-28].

1.1 Fibonacci, Continued Fractions and the Golden Ratio

In 1202, the Italian mathematician Fibonacci (also known as Leonardo da Pisa) asked a simple question. Suppose we have a pair of newly born rabbits who, after maturing, beget another pair of rabbits. These children, after they mature, beget another pair. So we have first one pair of rabbits, then two pairs, and then three pairs. How will this continue, if it does, supposing that each new pair of rabbits, after one season of maturing, will beget another pair each and every breeding season thereafter? To make things simple, Fibonacci also assumed that rabbits never die.

Obviously, the number of rabbit pairs in the nth breeding season will be equal to the number of pairs one season earlier (because none have died) *plus* the number of rabbits two seasons earlier because all of those rabbits are now mature and each pair produces a new pair.

Calling the number of rabbit pairs in the nth season F_n (F as in Fibonacci), we then have the recursion formula

$$F_n = F_{n-1} + F_{n-2}$$

Starting with one immature pair of rabbits ($F_1 = 1, F_2 = 1$), it is easy to calculate the number of rabbit pairs each successive season:

1, 1, 2, 3, 5, 8, 13, 21, 34, 55, ... ,

where each number is the sum of its two predecessors. Nothing could be much simpler than this — and few things probably seem as useless. At least this is what one might be led to believe by the unreality and triviality of this rabbit story. And yet this apparently harmless example of numerology touches more domains of mathematics and more diverse applications than many other simply defined collections of numbers.

For one, the ratio of successive numbers approaches the so-called Golden ratio $g = 1.618 \ldots$, often considered the ideal proportion in art. (Strangely, the very same ratio also plays a role in modern theories of deterministic chaos, where g is considered the "most irrational" of all ratios.)

The Golden ratio is defined in geometry by sectioning a straight line segment in such a way that the ratio of the total length to the longer segment equals the ratio of the longer to the shorter segment. In other words, calling the total length of the line l, and the longer segment a, g is determined by the equation

$$g := \frac{l}{a} = \frac{a}{l-a} .$$

Substituting g for l/a on the right side of the equation yields

$$g = \frac{1}{g-1} ,$$

which leads to the quadratic equation for g:

$$1 + g - g^2 = 0 .^{[1]}$$

The only positive solution (length ratios are positive) is

$$g = \frac{1 + \sqrt{5}}{2} ,$$

which also happens to be the ratio of a diagonal to a side in the regular pentagon. Since no higher roots than square roots ($\sqrt{5}$) appear in this ratio, it follows that the regular pentagon can be constructed by straightedge and compass alone. So Fibonacci's rabbits have now brought us into contact with one of the classical problems of geometry — the problem of circle division or cyclotomy, a problem that will occupy us again and again in this book because of its importance for fast computational algorithms and numerous other applications.

Incidentally, there is also a fast algorithm making use of g to compute F_n without repeated additions:

$$F_n \approx \frac{g^n}{\sqrt{5}} ,$$

where the wavy equal sign here means "take the nearest integer".

---

[1] The connoisseur may recognize the similarity between this equation, which defines the Golden ratio, and the generating function of the Fibonacci sequence: $1/(1 - x - x^2)$.

The remarkable numbers F_n can also be used to represent any positive integer N in a unique way by the simple rule that N is written as the greatest F_n not exceeding N and a (positive) remainder. Then this remainder is represented by the greatest Fibonacci number not exceeding *it* and another remainder, and so forth. For example, with the aid of the above sequence,

$$83 = 55 + 28$$
$$28 = 21 + 7$$
$$7 = 5 + 2$$

Thus, in the Fibonacci system,

$$83 = 55 + 21 + 5 + 2 .$$

Note that no two of the Fibonacci numbers occurring in the expansion of 83 are adjacent F_n. This is a general rule and distinguishes the Fibonacci system from the binary system where adjacent powers of 2 are often needed. (For example, $83 = 64 + 16 + 2 + 1$, where 2 and 1 are adjacent powers of 2.)

This property of the Fibonacci system has two interesting (and somewhat unexpected) consequences. The first is that in fast search algorithms it is often more efficient to base sequential searches (for the zero of a given function, for example) on Fibonacci numbers rather than binary segmentation. As a second consequence, the Fibonacci system leads to a winning strategy for a game of Fibonacci Nim. [Rules of the game: 1) Take at least one chip, but not more than twice the number of chips just removed by your opponent. 2) On the first move, don't take all the chips. 3) He who takes the last chips wins.] In our numerical example, if you have before you a pile of 83 chips, take away 2 chips (the least term in the decomposition of 83) to start your winning strategy.

Interestingly, the ratios of successive Fibonacci numbers, are given by a very simple *continued fraction* involving only the integer 1. To wit: 1 is the ratio of the first two Fibonacci numbers. Then $1 + 1/1 = 2$ is the ratio of the next two Fibonacci numbers. Next $1 + 1/(1 + 1/1) = 3/2$ gives the next two numbers. Repeating this process yields successively 5/3, 8/5, 13/8, etc. and converges on the Golden ratio. Numbers, like the Golden ratio, whose continued fraction expansion ends in infinitely many 1's have recently been christened "noble" numbers, because such numbers characterize nonlinear dynamic systems that are most resistant to chaos.

Continued fractions, treated in detail in Chap. 5, are a powerful mathematical tool playing a role of increasing importance in modern physics and other quantitative sciences. Indeed they are much more "natural" and inherently mathematical than, say, decimal fractions.

1.2 Fermat, Primes and Cyclotomy

No one doubts that 9 plus 16 equals 25, or, somewhat more elegantly,

$$3^2 + 4^2 = 5^2 .$$

In fact, even the Pythagoreans knew that

$$x^2 + y^2 = z^2 , \qquad\qquad (1.1)$$

has solutions for many values of x,y,z other than the (trivial) $x = y = z = 0$ and the above solution $x = 3$, $y = 4$, $z = 5$; another example is $x = 5$, $y = 12$, $z = 13$. Such triplets of numbers, of which there are infinitely many, are called *Pythagorean numbers*. The problem of finding *all* such numbers is treated in Diophantus's *Arithmetica,* a Latin translation of which was published in 1621.

About fifteen years later, *Pierre de Fermat* (1601 - 1665), son of French leather merchants, accomplished Toulouse jurist, eminent part-time mathematician and inventor of the *principle of least action* (Fermat's principle) scribbled in his personal copy of Diophantus's book the following now famous note: "I have discovered a truly remarkable proof of this theorem which this margin is too small to contain." By "this theorem" Fermat meant that

$$x^n + y^n = z^n \qquad\qquad (1.2)$$

has no integral solution with $xyz \neq 0$ for $n > 2$. "This theorem" is now known as Fermat's last theorem or simply FLT. Unfortunately, Fermat's note notwithstanding FLT is *not* a theorem, but only a conjecture, because it has never been verifiably proved — in spite of massive mathematical efforts spanning several centuries.

As a result of these efforts, which have spawned much new mathematics, it is now known that up to very large values of the exponent n ($n \approx 10^8$ if xyz and n have no common divisor) there are no solutions of (1.2). But 350 years after Fermat's casual remark, a general proof is still not in sight. In fact, the question of FLT's validity may remain unresolved forever, because if it is false, it is very unlikely that a counterexample will be found — even with the aid of today's fastest computers. (Of course, a single counterexample would suffice to inter FLT for good.)

However, some progress on FLT *has* been made. In 1922, Lewis Joel Mordell conjectured that (1.2) for $n > 3$ has at most a finite number of solutions. In spite of many attempts by the world's leading mathematicians, it was not until 1983 that a young German mathematician, *Gerd Faltings,* proved the recalcitrant conjecture true.

Fermat was luckier with an earlier edict, now called *Fermat's theorem* (without the "last"). It states that a prime number p divides $a^{p-1} - 1$ without remainder if a and p have no common divisor. Symbolically Fermat's theorem looks as follows

$$p \mid a^{p-1} - 1 \ , \quad \text{if} \ (p,a) = 1 \ . \tag{1.3}$$

For example, $p = 7$ divides $2^6 - 1 = 63$.

Although the ancient Chinese thought otherwise, a converse of Fermat's theorem is not true: if $p > 2$ divides $2^{p-1} - 1$, p is not *necessarily* a prime. Since large prime numbers are in great demand these days (for cryptographic purposes, for example) such a converse would have been very useful as a sieve for primes. This is regrettable, but primes are simply not that easy to catch. Nevertheless, Fermat's theorem forms the basis of sophisticated primality tests that are capable of stilling our hunger for primes a hundred or more digits long.

Another great invention of Fermat's are the so-called *Fermat numbers*, defined by

$$F_m := 2^{2^m} + 1 \ , \tag{1.4}$$

which Fermat mistakenly thought to be prime for all $m \geqslant 0$. Indeed $F_0 = 3$, $F_1 = 5$, $F_2 = 17$, $F_3 = 257$, $F_4 = 65537$ are all prime numbers (called *Fermat primes*) but, as Leonhard Euler showed, $F_5 = 4\,294\,967\,297$ is divisible by 641. This is easily verified with a good pocket calculator — which neither Fermat nor Euler possessed. But Euler knew that $5 \cdot 2^7 + 1 = 641$ was a *potential* factor of F_5 and he could do the necessary calculations in his head (which stood him in good stead when he went blind).

To this day, no Fermat prime larger than F_4 has been found. At present the smallest Fermat number whose primality status remains unknown is F_{20}, a 315653-digit number. On the other hand, a numerical monster such as F_{3310} is known to be composite. In fact, modern factoring algorithms have shown it to be divisible by $5 \cdot 2^{3313} + 1$. This is no small achievement, because F_{3310} is unimaginably large, having more than 10^{990} digits! (This is not to be confused with the extremely large number 10^{990} which, however, is miniscule by comparison with F_{3310}.)

On March 30, 1796, the Fermat primes, until then largely a numerical curiosity (the mathematical sleeper of the century, so to speak), were raised from dormancy and took on a new beauty embracing number theory and geometry. On that day the young Carl Friedrich Gauss showed that a circle could be divided into n equal parts by straightedge and compass alone, if n was a Fermat prime. Progress on this problem had eluded mathematicians since classical Greece, in spite of many efforts over the intervening 2000 years.

Since the time of Gauss the art of *cyclotomy,* the learned name for dividing the circle, has been studied and perfected in numerous other number fields with spectacular successes. Among the many ingenious applications are methods for designing highly effective error-correcting codes and fast computational algorithms, making precision measurements of exceedingly weak general relativistic effects, and even designing unique phase gratings for scattering sound in concert halls to achieve better acoustic quality. Sometimes the range of applications is even more dazzling than the mathematical invention itself!

1.3 Euler, Totients and Cryptography

Fibonacci, the lone mathematical luminary of the European Dark Ages, and Fermat, man of the Renaissance, were followed in the 18th century by an almost baroque flowering of mathematical genius, of whom Leonhard Euler was the most colorful and prolific centerpiece. Euler's collected works, currently numbering 70 published volumes, are still incomplete. Besides being foremost in geometry and analysis, and a pioneer in the most varied applications of mathematics to physics and astronomy, he added much to the theory of numbers.

Among his myriad contributions while working at the Academies of St. Petersburg and Berlin, few epitomize his genius as concisely as the formula

$$e^{i\pi} = -1 , \tag{1.5}$$

which he not only discovered as a mathematical identity but which also comprises three of Euler's enduring notational inventions: e for the base of the natural logarithms, i for the square root of -1, and π for the ratio of a circle's circumference to its diameter.

One of Euler's fundamental contributions to number theory was a generalization of Fermat's theorem (1.3) which he extended to composite (nonprime) exponents m:

$$m \,|\, a^{\phi(m)} - 1 , \tag{1.6}$$

if m and a have no common divisor. Here $\phi(m)$ is *Euler's totient function,* which counts how many integers from 1 to $m-1$ have no common divisor with m. If m is prime, then $\phi(m)$ equals $m-1$, and we are back to Fermat's theorem (1.3). But for composite m we obtain totally new and often unexpected results. For example, for $m = 10$, $\phi(m) = 4$ and we have

$$10 \,|\, a^4 - 1 .$$

In other words, the fourth power of *any* number not containing the factors 2 or 5 has 1 as the last digit, to wit: $3^4 = 81$, $7^4 = 2401$, $9^4 = 6561$, $13^4 = 28\ 561$, and so on.

Euler's totient function pervades much of number theory, including such statistical concepts as the probability that two numbers selected at random will have no common divisor. This probability tends to the average value of $\phi(m)/m$, which in turn approaches $6/\pi^2 \approx 0.61$ even for relatively small ranges of m ($m = 1$ to 10, say). The same asymptotic average of $\phi(m)/m$, incidentally, also describes the probability that a random integer will not contain a square as a factor. In fact, even among the 10 smallest natural numbers (1 to 10) the proportion of "squarefree" numbers (1, 2, 3, 5, 6, 7, i.e., 6 out of 10) is already close to the asymptotic value $6/\pi^2$.

Recently, Euler's function has taken on a new and most practical significance: it is the basis of a promising *public-key encryption* system. In public-key encryption the encrypting keys of all participants are, as the name implies, publicly known. Thus, anyone wishing to send a secret message to Mr. Sorge can encrypt it with Sorge's published key. And Sorge can, of course, decrypt the message intended for him, because he has the proper "inverse" key. But no one else can do this, since the published key contains no useful cue to the inverse key. If the key consists of an "exponent" k and a "modulus" m (typically the product of two 50-digit numbers), the inverse exponent k' is in principle computable from the expression

$$k^{\phi(\phi(m))-1}.$$

$$(1.7)$$

But in practice this is impossible, because to calculate $\phi(m)$, the factors of m must be known. And those factors are *not* published — nor are they easily derived from m. Even with today's fastest factoring algorithms it could take eons to find the factors of properly chosen 100-digit numbers. And random guessing is no help either, because $\phi(m)$ fluctuates between $m-1$ (for prime m) and values small compared to m. In other word, the chances of a random guess are about 1 in 10^{100}.

On the other hand, for one who *knows* the factors of m, Euler's function quickly gives the answer with the aid of (1.7). Did Euler ever dream of such an application when he invented the totient function?

1.4 Gauss, Congruences and Diffraction

Carl Friedrich Gauss, the son of a poor laborer in the German ducal city of Brunswick, was one of the greatest mathematicians in history. Yet he almost missed becoming a mathematician because he excelled in languages, being fluent at an early age in classical Greek and Latin, in addition to French and his native German. But his desire to study philology came to an abrupt end the day he discovered how to divide a circle into 17 equal parts. Gauss was not yet 19 years old when he made this epochal discovery. From his extensive reading, Gauss knew that no progress in "cyclotomy" had been made since the time of the ancient Greeks, though no effort had been spared in the course of two millenia. Gauss therefore reasoned that he must be a pretty good mathematician, and resolved to devote his life to mathematics — without, however, forsaking his love of literature and languages. In fact, when he was in his fifties he added another foreign language, Russian, to his repertoire enabling him to read Pushkin in the original as well as some innovative mathematical papers that had never been translated into a Western language.

One of Gauss's many mathematical inventions was the congruence notation. Instead of writing $m|a$ for "m divides a," Gauss introduced a kind of equation, a *congruence:*

$$a \equiv 0 \quad (\text{mod } m) , \tag{1.8}$$

which reads "a is congruent 0 modulo m." This notation can be immediately generalized to

$$a \equiv b \quad (\text{mod } m) , \tag{1.9}$$

which means that $a-b$ is divisible by m or, equivalently, a divided by m leaves the remainder b.

Just as calculus was much advanced by the seductive notation of Leibniz (the differential quotient and the integral sign), so the congruence notation of Gauss had an import beyond expectation. One reason was that the congruence notation brings out *remainders,* and remainders are often more important than the rest of a problem. This is true not only in wave interference, as has already been pointed out, but also in many other situations in science, technology and daily life. For example, the trick of many cryptographic systems is to work with remainders: the secret information is locked up in remainders, so to speak, to be released only with a secret decrypting key.

One of the mathematical entities that occupied Gauss a long time are certain complex sums like

$$S(n) = \sum_{k=0}^{n-1} e^{i2\pi k^2/n} \tag{1.10}$$

now called Gauss sums. It is relatively easy to show that $|S(n)|^2 = n$ for odd n. But it took even the great Gauss, to his own dismay, years to prove his guess of the value of $S(n)$ itself.

The sequence defined by the individual terms in the sum (1.10), which repeats with a period of n if n is prime, has a noteworthy correlation property: the periodic correlation is strictly zero for nonzero shifts. As a consequence, the Fourier spectrum of such a periodic sequence has components whose magnitudes are all alike. Such spectra are called flat or *white*. And of course the terms of the sequence itself also have equal magnitudes, namely 1. These facts have led to numerous interesting applications in physics and communications.

If the above sequence represents wave amplitudes at uniform intervals along a linear spatial coordinate, then at some distance, the wave will break up into many wavelets with (nearly) equal energies! Thus, if the terms of a Gauss sum are put to proper use in what may be called quadratic-residue phase gratings, then coherent light, radar beams or sound waves can be very effectively scattered. For light this leads to the ultimate in frosted glass, with interesting applications in coherent optical processing. For radar and sonar detection this means less visibility. In noise control, such phase gratings permit one to disperse offending sound where it cannot be absorbed (along highways, for example). And in concert hall acoustics, the possibility of diffusing sound without weakening it means better acoustic quality for more musical enjoyment.

1.5 Galois, Fields and Codes

We all hear in school about fractions or *rational numbers,* easy objects of adding, subtracting, multiplying and dividing (except by zero). Like whole numbers, the rational numbers form a countable, albeit infinite, set. Such a corpus of numbers is called a *field.*

Interestingly, there are also *finite fields.* Consider the seven numbers 0, 1, 2, 3, 4, 5, 6. If we disregard differences that are multiples of 7 (i.e., if we identify 3, 10, -4, and so forth), we can add, subtract and multiply to our heart's content and stay within these seven numbers. For example $5 + 4 = 2$, $3 - 6 = 4$, $2 \cdot 4 = 1$. These operations obey the customary commutative, associative and distributive laws. To illustrate, $4 \cdot (5+4) = 4 \cdot 5 + 4 \cdot 4 = 6 + 2 = 1$; and we obtain the same result if we first add then multiply: $4 \cdot (2) = 1$.

But what happens with division; for example, what is 1 divided by 2, 1/2, in this finite number field? It must be one of the numbers 0, 1, 2, 3, 4, 5, 6, but which? In other words, what number times 2 equals 1? By trial and error (or consulting one of the above examples) we find $1/2 = 4$ and, of course, $1/4 = 2$, and consequently $3/2 = 3 \cdot 4 = 5$.

How can we find reciprocals in a systematic way; what is the general law? The answer is

$$1/n = n^5$$

for $n \neq 0$. (As usual, 0 has no inverse.) Thus, $1/3 = 3^5 = 5$, and therefore

$1/5 = 3$. Likewise $1/6 = 6^5 = 6$, i.e., 6 is its own inverse! But this is not unusual, because the rational numbers, too, have a self-inverse other than 1, namely -1. (And incidentally, in our finite number system -1 is equivalent to 6.)

While it may seem odd that the sum of 2 odd numbers is sometimes odd $(3 + 5 = 1$, for example) and sometimes even $(1 + 5 = 6)$ or that $1/4$ of $1/4$ is $1/2$, all of these results are nevertheless perfectly consistent, and it is *consistency* that counts in mathematics.

But what good is consistency? Are finite fields just consistent nonsense? The answer is an emphatic *no*. For one thing, finite fields have held a secure place in algebra ever since the young Frenchman *Evariste Galois* used them to show the conditions under which algebraic equations have solutions in radicals (roots).

Such "radical" solutions had been known for quadratic equations since antiquity and for cubic and quartic equations since the Renaissance. But curiously enough, no one had been able to find a formula in radicals for the general equation of the fifth degree or higher. Then, in 1830, the 19-year-old Galois, generalizing a result obtained by the very able Norwegian mathematician Niels Abel, gave the general and definitive answer to a problem that had baffled many generations of mathematicians. Unfortunately, Galois was killed in a duel (probably a political provocation in amorous disguise) before he turned 21, but his name will be enshrined forever in the annals of algebra and immortalized by the addition to the mathematical vocabulary of such important terms as *Galois group* and *Galois field*, the latter a synonym for finite field.

Of particular importance in modern digital applications are the finite fields having p^m members, where p is some prime number, $p = 2$ being the case most relevant to computers, communication and encryption. The simplest such field with $m > 1$ has $2^2 = 4$ elements (members). But these elements cannot be the *scalar* numbers (0, 1, 2, 3, say) if we want to interpret them as usual. Because then we would have, for example, $2 \cdot 2 = 4$; and 4 is congruent 0 modulo 4. As a consequence the equation $2x = 0$ has *two* solutions in such a "nonfield": $x = 0$ and $x = 2$. On the other hand, the equation $2x = 1$ has *no* solution; in other words, 2 has no inverse. This is no way to construct a field!

How then are we to construct a field with four elements? The problem is reminiscent of the situation in quantum mechanics when physicists tried to convert the Schrödinger wave equation into a relativistically invariant form. In order to preserve both quantum essence and Lorentz invariance, P.A.M. Dirac discovered that he had to replace the Schrödinger scalar ψ-function by a vector function with four components, representing two possible spins and two electric charges (spin "up" or "down" and electron or positron).

Similarly, with our four-element field, the solution is to choose *vectors* or "*n*-tuples" as field elements. For a four-element field we may try *binary* vectors with two components, each component being 0 or 1:

0 0
1 0
0 1
1 1

Addition is componentwise, *without* carries. The 00-element is the zero-element of the field, and the 10 is the one-element. The two remaining elements must be multiplicative inverses of each other:

0 1 · 1 1 = 1 0 .

Any nonzero element raised to the third power must equal the one-element 10. This fixes the multiplication table of the field:

·	1 0	0 1	1 1	
1 0	1 0	0 1	1 1	
0 1	0 1	1 1	1 0	
1 1	1 1	1 0	0 1	.

$$(1.11)$$

Here the entry 01 · 01 = 11 is the only nonredundant one; all others follow or are fixed *a priori*.

Instead of proving the consistency of the resulting algebra, we give a simple example of the distributive law:

0 1 · (1 1 + 1 0) = 1 0 + 0 1 = 1 1 ,

which yields the same result as adding *before* multiplying:

0 1 (1 1 + 1 0) = 0 1 · (0 1) = 1 1 .

The choice of symbols for our finite field is of course arbitrary. To save ink, we might write simply 0 instead of 00 and 1 instead of 10. However, for 01 we need a new symbol, say ω. We must then identify 11 with ω^2 because of our multiplication table (1.11). With this notation, one of the entries of (1.11) tells us that $\omega \cdot \omega^2 = \omega^3 = 1$. Furthermore, $1 + \omega + \omega^2 \triangleq 10 + 01 + 11 = 00 \triangleq 0$.

This is all we need to know about our new symbol ω to calculate in our four-element field $GF(4)$. There is a convenient interpretation of ω in terms of *complex* numbers. Because $\omega^3 = 1$, ω must be a third root of 1, and, because $\omega \neq 1$, it must be either $\exp(2\pi i/3)$ or $\exp(-2\pi i/3)$. This is of course no accident: the three nonzero elements of $GF(4)$ form a cyclic group of order three, of which the three roots of 1 are another representation.

(However, this analogy does not carry over to addition. For example, our $-\omega$ equals ω. To wit: $-\omega = 0 - \omega \overset{\wedge}{=} 00 - 01 = 01 \overset{\wedge}{=} \omega$.)

The ω-representation of $GF(4)$ is used in Sect. 29.3 to construct a pretty necklace with four different kinds of pearls.

Of course, a four-element Galois field is not very exciting, but the eight-element field $GF(2^3)$ has interesting applications. But first we must give a rule for its construction. We start with the zero- and one-elements:

0 0 0
1 0 0

and generate the following two elements by rightward shifts, calling them g^1 and g^2, respectively:

$g^1 = 0\ 1\ 0$

$g^2 = 0\ 0\ 1$.

Then we continue shifting, and for each 1 "disappearing" on the right, we add (modulo 2) a 1 to each of the two left places. Thus, our field elements are

$$
\begin{aligned}
g^{-\infty} &= &0\ 0\ 0 \\
g^{0} &= &1\ 0\ 0 \\
g^{1} &= &0\ 1\ 0 \\
g^{2} &= &0\ 0\ 1 \\
g^{3} &= &1\ 1\ 0 \\
g^{4} &= &0\ 1\ 1 \\
g^{5} &= &1\ 1\ 1 \\
g^{6} &= &1\ 0\ 1
\end{aligned}
\qquad (1.12)
$$

$$
\begin{aligned}
g^{7} &= &1\ 0\ 0\ = &g^{0} \\
g^{8} &= &0\ 1\ 0\ = &g^{1}, \text{ etc.}
\end{aligned}
$$

Another (equivalent) method of generating $GF^*(2^3)$, i.e., the nonzero elements of $GF(2^3)$, is to focus on the *columns* of (1.12). The columns are generated by the recursion

$$
a_{n+3} = a_{n+1} + a_n \qquad (1.13)
$$

with the *initial conditions* 100, 010, 001, respectively. All columns have period $2^3 - 1 = 7$ and are identical within a cyclic shift. Each of the seven nonzero triplets occurs exactly once per period.

An important application of $GF(2^m)$ occurs in binary error-correcting codes. One of the simplest codes of block lengths m is obtained by using m information bits as the initial condition computing the remaining $2^m - m - 1$ "check bits" with an appropriate recursion. For $m = 3$, for example, the three information bits 111 (say) are supplemented by the four check bits, which according to (1.13) are 0010. Thus, the total seven-bit code word for the information bits 111 is

$$1\ 1\ 1\ 0\ 0\ 1\ 0 . \tag{1.14}$$

Except for the 0000000-word, all code words are cyclic shifts of (1.14). Using our finite field algebra, it is easy to show that the "Hamming distance" (defined as the modulo 2 component-wise sum) between any two code words so generated is 4. Thus, we have constructed a code capable of *correcting* a single error and *detecting* a second error. The occurrence and location of a single error is found by the occurrence and position of a single 1 in the sum of the erroneous code word with all possible code words.

Many other powerful codes can be derived from Galois-field algebra. For example, interchanging the roles of information and check bits in the above example results in the famous Hamming error-correcting code of block length 7.

Other interesting applications arise when we choose +1 and −1 as elements of $GF(2^m)$. The above code words then become periodic pulse trains whose power spectrum is "white"; i.e., all finite-frequency components of the Fourier transform have equal magnitudes. Thus, Galois-field theory has given us a method of generating "signals" with *constant power in both Fourier domains* (time and frequency or space and direction). Such signals are of great value in making precision measurements at very low energies — such as in the observation of general relativistic effects in radar echoes from planets and interplanetary spacecraft.

Interestingly, by interchanging the time and frequency domains, the same Galois-field algebra also yields continuous periodic waveforms with near-minimal amplitude ranges and flat power spectra. Such waveforms have found useful application in the synthesis by computer of more human-sounding speech.

And by applying these "Galois" sequences of ±1's to spatial coordinates, diffraction gratings of high scattering power can be realized. Such gratings are ideal diffusors for all kinds of waves: laser light, radar radiations, noise emissions and musical sound waves. Here is another example of how a relatively abstract mathematical theory — initiated in the early 19th century by a precocious teenager — can yield useful results in a variety of applications for which the theory was *not* constructed. As is so often the case in human affairs, it is the *unexpected* that matters most.

2. The Natural Numbers

'Ο θεὸς ἀριθμητίζει
— *Carl Friedrich Gauss*

Here we encounter such basic concepts as *integers, composites,* and *primes,* and we learn the very fundamental fact that the composites can be represented in a *unique* way as a product of primes.

The *least common multiple* and the *greatest common divisor* of two or more integers may be familiar from high school, but they are ideas that pervade all of number theory. Here we demonstrate some of their basic properties and point to some natural phenomena in the real world of gears, planetary motion, and musical pitch.

If integers can be prime, pairs of integers can be "mutually prime" or *coprime* if they have no common factors, in other words, if their greatest common divisor is 1. Coprimality is another important property of two (or more) integers.

One of the very early tools of number theory is *Euclid's algorithm;* it allows us to find, in a systematic manner, the greatest common divisor of two integers without solving the often difficult problem of factoring the two integers. As we shall later see, Euclid's algorithm generalizes to polynomials and allows us to solve important integer equations, the so-called Diophantine equations.

2.1 The Fundamental Theorem

We will speak here of the "whole numbers" or *integers* ... $-3, -2, -1, 0, 1, 2, 3,$ ..., denoted by the letter **Z**, and more often of the so-called "natural" numbers or positive integers: 1, 2, 3, 4, 5 and so forth. Some of these are divisible by others without leaving a remainder. Thus, $6 = 2 \cdot 3$, i.e., 6 is divisible by 2 and by 3 without a remainder. Such numbers are called *composites.*

Other numbers have no divisors other than 1 and themselves, such as 2, 3, 5, 7, 11, 13, 17, etc. These numbers are called *prime* numbers or simply *primes.* All primes are odd, except 2 — the "oddest" prime (a designation alluding to the special role which 2 plays among the primes. The number 1 is considered neither prime nor composite. Otherwise some theorems would require very awkward formulations — such as the following.

The *fundamental theorem of arithmetic* states that each natural number n can be uniquely factored into primes:

$$n = p_1^{e_1} \, p_2^{e_2} \, \cdots \, p_k^{e_k} \, \cdots \, p_r^{e_r} = \prod_i p_i^{e_i} \, . \tag{2.1}$$

Here the order of the factors is considered irrelevant.

Equation (2.1) can be read in two ways:

1) p_i is the ith prime — in which case the exponent e_i has to be zero if p_i is not a factor of n.

2) Only those primes that are factors of n appear in (2.1). We will use either reading of (2.1) and state which if it makes a difference.

There is no corresponding theorem for the *additive* decomposition of natural numbers into primes. This is one of the reasons why additive number theory, for example partitions (Chap. 21), is such a difficult subject. In this book we will be mostly concerned with *multiplicative* number theory, which has many more applications.

2.2 The Least Common Multiple

Two integers n and m have a least common multiple (LCM) $[n,m]$. The LCM is needed to combine two fractions with denominators n and m into a single fraction. In fact, that is where the everyday expression "to find the least common denominator" (of divergent views, for example) comes from. For example, for $n = 6$ and $m = 9$, $[6,9] = 18$.

Example: $\dfrac{1}{6} + \dfrac{2}{9} = \dfrac{3}{18} + \dfrac{4}{18} = \dfrac{7}{18}$.

It is easily seen that with n as in (2.1) and

$$m = \prod_i p_i^{f_i} \, , \tag{2.2}$$

$$[n,m] = \prod_i p_i^{\max(e_i,f_i)} \, , \tag{2.3}$$

because in the LCM each prime factor p_i must occur at *least* as often as it does in either n or m. Thus, for $n = 6 = 2^1 \cdot 3^1$ and $m = 9 = 3^2$, $[6,9] = 2^1 \cdot 3^2 = 18$.

There are numerous applications of the LCM. Consider two gears with n and m teeth meshing, and suppose we mark with a white dot one of the n teeth on the first gear and one of the m spaces between teeth of the second gear. When the gears turn, how often will the two white dots meet? Perhaps never! But if they meet once, they will meet again for the first time after $[n,m]$ teeth have passed the point of contact, i.e., after the first gear has undergone $[n,m]/n$ (an integer!) revolutions and the second gear an integer $[n,m]/m$ revolutions.

2.3 Planetary "Gears"

Our "gears", of course, could be any of a plethora of other objects that can be modeled as meshing gears even if no teeth are visible. Thus, the revolutions of the planet Mercury around itself and the Sun are locked by gravitational forces as if geared: during two revolutions around the Sun, Mercury revolves three times around itself. (As a consequence, one day on Mercury lasts two Mercury years. Strange gears — and even stranger seasons!) Similarly, the Earth's moon revolves exactly once around itself while completing one orbit around the Earth; that is why it always shows us the same side. On the moon, Earth day (or night) lasts forever.

The "teeth" that keep the moon locked to the Earth are, as in the case of Mercury and the sun, gravitational forces. But these "gravitational teeth" are relatively weak and would not "engage" if unfavorable initial conditions were not damped out by friction such as that provided by the ocean tides on Earth. (Eventually, the Earth day may lock in with the Earth year, which will play havoc with night and day as we know it.)

And not long ago, it was discovered that even the distant planets Pluto and Neptune are coupled to each other strongly enough to be locked into an integer "resonance" (in the astronomer's lingo).

Another question answered by the LCM, although no teeth are in evidence, has to do with the coincidence of dates and weekdays. Because the number of days per year (365) is not divisible by the number of days per week (7), coincidences of dates and weekdays do not recur from one year to the next. Furthermore, because every fourth year is a leap year, coincidences are not equally spaced in years. However, even without knowing when leap years occur, we can always guarantee that a coincidence will recur after 28 years, 28 being the LCM of 4 and 7. (In the year 2100 the leap day will be dropped, temporarily violating the 28-year cycle.)

Equation (2.3) easily generalizes to more than two integers: the max function in (2.3) then contains as many entries as there are integers whose LCM we want to determine.

As we indicated above when introducing the meshing gear picture, the two white markers may never meet. More learnedly, we would say that a certain linear Diophantine equation (see Chap. 7) has no solution. This can happen only if n and m have a greatest common divisor greater than 1. This brings us to our next topic.

2.4 The Greatest Common Divisor

Another important relation between integers is their greatest common divisor (GCD). For two integers n and m given by (2.1) and (2.2), the GCD is

$$(n,m) = \prod_i p_i^{\min(e_i, f_i)} , \tag{2.4}$$

because for the GCD to divide both n and m it cannot have the factor p_i more often than it is contained in either n or m, whichever is *less*.

Example: $n = 10 = 2^1 \cdot 5^1$ and $m = 25 = 5^2$. Thus $(10,25) = 5$.

Two numbers n and m that have no common factors are called relatively prime, mutually prime or *coprime*. In this case the GCD equals 1.

Example: $(6,35) = (2 \cdot 3, 5 \cdot 7) = 1$.

For any two numbers n and m, the product of the GCD and the LCM equals the product of n and m:

$$(n,m)[n,m] = nm ,$$

because whenever the formula (2.4) for the GCD picks the exponent e_i for p_i, the formula (2.3) for the LCM picks the exponent f_i, and vice versa.
Thus,

$$(n,m)[n,m] = \prod_i p_i^{e_i + f_i} = nm . \tag{2.5}$$

Example:

$$(4,10) = (2^2, \underline{2} \cdot 5) = 2; \quad [4,10] = [2^2, 2 \cdot \underline{5}] = 20; \quad 2 \cdot 20 = 4 \cdot 10 . \quad Check!$$

The generalization of (2.5) to three integers is

$$(n,m,k)[nm,mk,kn] = nmk , \tag{2.6}$$

which is easily verified. Assume that a given prime p occurring in the prime factorization of the product nmk occurs e_n times in n, e_m times in m and e_k times in k and that, without loss of generality,

$$e_n \leq e_m \leq e_k .$$

Then the exponent of p in (n,m,k) it is e_n, and in $[nm,mk,kn]$ it is $e_m + e_k$.

Thus, the left side of (2.6) has the prime p with the exponent $e_n + e_m + e_k$, as does the right side of (2.6). The same is true for all primes occurring in nmk. The correctness of (2.6) then follows from the fundamental theorem of arithmetic.

The *dual* of (2.6) is

$$[n,m,k](nm,mk,kn) = nmk , \tag{2.7}$$

which is proved by the same reasoning. Generalizations of (2.6) and (2.7) to more than three factors should be obvious.

For more than two integers, some particularly interesting relations between GCD and LCM exist. For example, two such relations are the "distributive law"

$$(k[m,n]) = [(k,m),(k,n)] , \tag{2.8}$$

and its *dual*

$$[k,(m,n)] = ([k,m],[k,n]) , \tag{2.9}$$

both of which are a direct consequence of the properties of the min and max functions in (2.4) and (2.3).

There is even a very pretty *self-dual* relation:

$$([k,m],[k,n],[m,n]) = [(k,m),(k,n),(m,n)] , \tag{2.10}$$

i.e., in the expressions appearing on either side of (2.10), the operations LCM and GCD can be completely interchanged without affecting their validity!

However, from a practical point of view there *is* a difference: The right-hand side of (2.10), i.e., doing GCDs before the LCMs, is usually easier to figure out. Thus, (2.10) can be exploited to computational advantage.

It is interesting to note that relations such as (2.6 - 10) occur in many other mathematical fields, such as mathematical logic or set theory, where our LCM corresponds to the set-theoretic *union* $\cup$ and the GCD corresponds to *intersection* $\cap$.

But what, in number theory, corresponds to the set-theoretic relation

$$\overline{(A \cup B)} = \overline{A} \cap \overline{B} ,$$

where the bar stands for complement?

For additional relations see [2.1].

2.5 Human Pitch Perception

An interesting and most surprising application of the GCD occurs in human perception of pitch: the brain, upon being presented with a set of harmonically related frequencies, will perceive the GCD of these frequencies as the pitch. Thus, the subjective pitch of the two-tone chord (320 Hz and 560 Hz) is (320,560) = 80 Hz, and *not* the difference frequency (240 Hz).

Upon a frequency shift of +5 Hz applied to both frequencies, the GCD drops to 5 Hz; and for an irrational frequency shift, the GCD even drops to 0 Hz. But that is not what the ear perceives as the pitch. Rather it tries to find a *close match* in the range of pitches above 50 Hz. For the frequencies 325 Hz and 565 Hz such a match is given by 81 Hz, which is the GCD of 324 Hz and 567 Hz — close to the two given frequencies.

Note that the concept that pitch is given by the difference frequency or "beat" frequency has been beaten: if both frequencies are shifted by the same amount, their difference remains unchanged. Yet psychoacoustic experiments clearly show that the perceived pitch is increased, from 80 Hz to about 81 Hz in our example, just as the amplified GCD model predicts [2.2].

What this tells us is that the human brain switches on something like a GCD-spectral matching computer program when listening to tone complexes. Fascinating? Indeed. Unbelievable? Well, the brain has been caught doing much trickier things than that.

2.6 Octaves, Temperament, Kilos and Decibels

The Pythagoreans discovered that subdividing the string of a musical instrument into the ratio of small integers resulted in pleasing musical intervals. Thus, dividing the string into 2 equal parts gives a frequency ratio (compared with the full-length string) of 2:1 — the musical octave. Shortening the string by one third gives rise to the frequency ratio 3:2 — the musical fifth. And dividing the string into 4 equal parts results in the frequency ratio 4:3 — the musical fourth.

The Pythagorean musical scale was constructed from these simple ratios. How do they fit together? How many fifths make an integral number of octaves? Or, what is x in

$$\left(\frac{3}{2}\right)^x = 2^y \quad ,$$

or, equivalently,

$$3^x = 2^z ,$$

where $z = y + x$? The fundamental theorem (Sect. 2.1) tells us that there are *no* integer solutions. But there are *approximate* solutions, even in small integers. Thus,

$$3^5 = 243 \approx 256 = 2^8 . \qquad (2.11)$$

Consequently, 5 musical fifths equal *about* 3 octaves. To make the octave come out correctly, we would have to tamper with the ratio $3:2 = 1.5$, increasing it by about 1% to 1.515..., to achieve a well-tempered temperament.

The fact that $(3/2)^5$, with a little tampering, equals 2^3 also has its effect on the musical fourth: from

$$\left(\frac{3}{2}\right)^5 \approx 2^3$$

follows directly

$$\left(\frac{4}{3}\right)^5 \approx 2^2 ;$$

in other words, 5 fourths make about 2 octaves. The tampering required on the fourth to make it fit 2 octaves exactly is, as in the case of the fifth, only one part in a hundred.

We shall leave the musical details to J. S. Bach and his well-tempered clavier and ask ourselves the more general question of how we can find approximate integer solutions to equations like $a^x = b^y$ in a more systematic way. The answer: by expanding logarithms into continued fractions, as will be explained in Sect. 5.1. There we learn that for $a = 3$ and $b = 2$, for example, the next best approximation (after $3^5 \approx 2^8$) is $3^{12} \approx 2^{19}$, requiring an adjustment of the musical fifth by only one part in a thousand so that 12 "tampered" fifths will make 7 octaves, thereby avoiding the *Pythagorean comma*. This is of great interest to musicians, because it allows the construction of a complete key from ascending fifths (the famous Circle of Fifths).

A much closer numerical coincidence, with important consequences in music, computer memory, photography and power measurement, is the approximation

$$5^3 = 125 \approx 128 = 2^7 . \qquad (2.12)$$

Musically, this means that 3 major thirds (frequency ratio $= 5:4$) equal about *one* octave:

$$\left(\frac{5}{4}\right)^3 \approx 2 ,$$

which requires an adjustment of less than 8 parts in a thousand in the major third so that 3 of them match the octave exactly.

Another consequence of (2.12) is that

$$2^{10} = 1024 \approx 10^3 .$$

According to international standards, the factor 10^3 is denoted by the prefix *kilo*, as in kilometer. But computer memories are not measured in kilometers or weighed in kilograms; rather they are *addressed*, and the proper form of address is *binary*. As a consequence, memory sizes are usually powers of 2, and in computerese a 256-kilobit memory chip can actually store 262144 bits of information, because to hard- and software types, kilo means 1024 — not 1000.

The near coincidence of 5^3 and 2^7 also shows up among camera exposure times, where 1/125 of a second is 7 lens-aperture "stops" away from 1 second. But 7 stops correspond to a light energy factor of $2^7 = 128$.

Still another application in which $5^3 \approx 2^7$ is exploited is the field of power or intensity measurement. The preferred logarithmic measure of intensity is the *decibel*,[1] 10 decibels being equal to an intensity ratio of 10:1. Thus, twice as much power (of a loudspeaker output, for example) means an extra 3 decibels — almost exactly. (A better figure would be 3.01 decibels, but who can hear a hundredth of a decibel?)

J. R. Pierce, lately of Stanford University, has recently proposed a new musical scale based on dividing the frequency ratio 3:1 (instead of the 2:1 octave) into 13 (instead of 12) equal parts. This scale matches such simple integer ratios as 5:3 *and* 7:5 (and 9:7) with an uncanny accuracy, resulting from the number-theoretic fluke that certain 13th powers of *both* 5 *and* 7 are very close to integer powers of 3. To wit: $5^{13} = 3.0077^{19}$, and $7^{13} = 3.0037^{23}$. Since the integers appearing in the exponents (13, 19, 23) are also coprime (in fact, all three are prime), it is easy to construct complete musical scales exclusively from the small-integer ratios 5:3 and 7:5. The basic chords of the new scale, 3:5:7 and 5:7:9, are superbly approximated by the equal tempered scale $3^{k/13}$ and were found by M. V. Mathews, A. Reeves, and L. Roberts to provide a strong harmonic foundation for music written in the new scale.

---

[1]   Curiously, one never hears about the full unit, the bel, perhaps because a difference of 10 bel is the difference between the sound of a bubbling brook and an earsplitting screech.

2.7 Coprimes

Two integers are said to be *coprime* if their GCD equals 1. Thus, 5 and 9 are coprime: $(5,9) = 1$, while 6 and 9 are *not* coprime: $(6,9) = 3 \neq 1$.

The *probability* that two "randomly selected" integers will be coprime is $6/\pi^2$ (see Sect. 4.4). This is also the probability that a randomly selected integer is "squarefree" (not divisible by a square).

Of three or more integers it is often said that they are *pairwise coprime* if all possible pairs are coprime. Thus, 2, 5 and 9 are pairwise coprime: $(2,5) = (2,9) = (5,9) = 1$. However, 2, 5 and 8 are *not* pairwise coprime, because $(2,8) = 2$, although the three numbers seen as a *triplet* have no common factor. The probability that three randomly selected integers will be pairwise coprime is 0.28... (see Sect. 4.4).

2.8 Euclid's Algorithm

If the GCD is so important, how does one go about finding it? Answer: by Euclid's algorithm, which is best illustrated by an example. To find the GCD of 35 and 21, first divide the larger number by the smaller:

$$\frac{35}{21} = 1 + \frac{14}{21} ,$$

and repeat the process on the remainder:

$$\frac{21}{14} = 1 + \frac{7}{14} ,$$

until the remainder is 0:

$$\frac{14}{7} = 2 + 0 ,$$

which is guaranteed to happen sooner or later. The GCD is the last divisor, 7 in our case. Thus, $(35,21) = 7$, which is the correct answer.

The philosophy behind Euclid's algorithm is the following. It is easy to show that $(a,b) = (a-kb,b)$, where k is an integer. If $a > b > 0$ and if one picks k as large as possible without making $a-kb$ negative, then $a-kb < b$. Thus, we have reduced the problem of computing the GCD of a and b to that of two smaller numbers, namely $a-kb$ and b. Now b is the larger number of the pair, and it can be reduced by subtracting a proper multiple of $a-kb$. Continuing this simple process generates smaller and smaller number pairs all having the same GCD. Finally we must arrive at two numbers that are multiples of each other and the smaller of the two numbers is the GCD. (If a and b are coprime that "smaller number" is of course 1.) This is how and why the Euclidean algorithm works: it chops large numbers down to manageable size.

3. Primes

As we go to larger and larger integers, primes become rarer and rarer. Is there a largest prime after which all whole numbers are composite? This sounds counter-intuitive and, in fact, it isn't true, as Euclid demonstrated a long time ago. Actually, he did it without demonstrating any primes — he just showed that assuming a finite number of primes leads to a neat contradiction.

Primes are found by sieves, not by formulas, the classical sieve having been designed by Eratosthenes in classic Greece. (Formulas that pretend to give only primes are really shams.) Primality testing has advanced to a stage where the primality or compositeness of 100-digit numbers can now be ascertained by computer in less than a minute, without actually giving any of the factors [3.1]. Factoring, on which the security of certain kinds of cryptographic systems depends (Chaps. 9-14), is still very difficult at this writing.

The largest primes known are of a special form called *Mersenne* primes because they don't hide their compositeness too well, and indeed, some were discovered by high-school students. The largest Mersenne prime known (in mid-1983) has 25,962 digits! Mersenne primes lead to even *perfect numbers* and to prime "repunits," meaning repeated units, i.e., numbers consisting exclusively of 1's in any given base system. (The Mersenne primes are repunits in the binary number system.)

Of special interest are the *Fermat primes* of which, in spite of Fermat's expectations, only 5 are known, the largest one being 65537. Each Fermat prime allows the construction of a regular polygon by using only straightedge and compass — Gauss's great discovery made just before he turned nineteen.

3.1 How Many Primes are There?

Again we turn to Euclid, who proved that there are infinitely many primes by giving one of the most succinct indirect proofs of all of mathematics:

Suppose that the number of primes is finite. Then there is a largest prime p_r. Multiply all primes and add 1:

$$N = p_1 p_2 \cdots p_r + 1 .$$

Now N is larger than p_r and thus cannot be a prime, because p_r was assumed to be the largest prime. Thus N must have a prime divisor. But it cannot be any of the *known* primes because by construction of N, all known primes divide $N - 1$ and therefore leave the remainder 1 when dividing N. In other words, none of the known primes divides N. Thus, there is a prime larger than p_r — a contradiction! We must therefore conclude that there is no largest prime, i.e., that there are infinitely many.

In actual fact, the above construction *often* (but not always) does give a prime. For example:

$$2 + 1 = 3$$

$$2 \cdot 3 + 1 = 7$$

$$2 \cdot 3 \cdot 5 + 1 = 31$$

$$2 \cdot 3 \cdot 5 \cdot 7 + 1 = 211$$

$$2 \cdot 3 \cdot 5 \cdot 7 \cdot 11 + 1 = 2311 ,$$

all of which are prime. *But*

$$2 \cdot 3 \cdot 5 \cdot 7 \cdot 11 \cdot 13 + 1 = 30031 = 59 \cdot 509 .$$

Suppose we set $P_1 = 2$ and call P_{n+1} the largest prime factor of $P_1 P_2 \dots P_n + 1$. Is the sequence P_n monotonically increasing? No! Both P_9 and P_{10} have 16 decimal digits but P_{10} equals only about 0.3 P_9.

3.2 The Sieve of Eratosthenes

Like gold nuggets, primes are mostly found by sieves — the first one having been designed in ancient Greece by Eratosthenes of Kyrene around 200 B.C. Eratosthenes's sieve idea is charmingly simple.

To find the primes below 100, say, write down the integers from 1 to 100 in order. Then, after 2, cross out every second one (4,6,8...), in other words all the even numbers, because they are divisible by 2 and therefore not prime (except 2 itself). Then, after 3, cross out every third number that is still standing (9,15,21...), because these numbers are divisible by 3 and therefore also not prime. Repeat the crossing out process for every fifth number after 5 and every seventh number after 7. The remaining numbers (except 1, which is not considered a prime) are the 25 primes below 100:

2, 3, 5, 7, 11, 13, 17, 19, 23, 29, 31, 37, 41, 43, 47,

53, 59, 61, 67, 71, 73, 79, 83, 89, 97 .

Roughly speaking, to find the primes below a given integer N, we only have to use sieving primes smaller than $\sqrt{N}$. (This rule would tell us that by sieving with 2, 3, 5 and 7 we will find all primes below $11^2 = 121$, while actually we have found 4 more primes.)

In applying Eratosthenes's sieve method there is an additional trick that simplifies matters considerably: we write the integers in six columns starting with 1 (Fig. 3.1). Then only the first and the fifth columns (no pun) contain primes, because all numbers in the second, fourth, and sixth columns are divisible by 2, and those in the third column are divisible by 3.

	2	3	4	5	6
7	8	9	10	11	12
13	14	15	16	17	18
19	20	21	22	23	24
25	26	27	28	29	30
31	32	33	34	35	36
37	38	39	40	41	42
43	44	45	46	47	48
49	50	51	52	53	54
55	56	57	58	59	60
61	62	63	64	65	66
67	68	69	70	71	72
73	74	75	76	77	78
79	80	81	82	83	84
85	86	87	88	89	90
91	92	93	94	95	96
97	98	99	100		

Fig. 3.1. The sieve of Eratosthenes (modulo 6)

To eliminate the numbers divisible by 5 and 7 as well, a few 45° diagonals have to be drawn, as shown in Fig. 3.1.

Things become a little more complicated if we include the next two primes, 11 and 13, in our sieve, because the numbers divisible by 11 and 13 follow "knight's-move" patterns as known from chess. But then we have already eliminated all composite numbers below $17 \cdot 19 = 323$. In other words, we have caught the 66 primes up to 317 in our 6-prime sieve.

Sieving may connote a child playing in a sandbox or a gold digger looking for a prime metal, but sieving in number theory is a very respectable occupation and sometimes the only method of finding an elusive prime or unmasking a composite as such. Of course, the sieving algorithms employed today are becoming increasingly sophisticated. We will hear more about the search for primes, especially the urgently needed very large ones, in subsequent chapters.

3.3 A Chinese Theorem in Error

The ancient Chinese had a test for primality. The test said that n is prime iff[1] n divides $2^n - 2$:

$$n \mid (2^n - 2) . \tag{3.1}$$

As we shall prove later, (3.1) is indeed true if n is an odd prime (by Fermat's theorem). Of course, for $n = 2$, (3.1) is trivially true.

Example: for $n = 5$, $2^n - 2 = 30$, which is indeed divisible by 5.

Conversely, for odd $n < 341$, if n is *not* prime n does *not* divide $2^n - 2$.

Example: for $n = 15$, $2^n - 2 = 32{,}766$, which is *not* divisible by 15.

Fortunately (for their self-esteem!), the ancient Chinese never tried $n = 341$, which is composite: $341 = 11 \cdot 31$ and yet 341 divides $2^{341} - 2$ without remainder. This might be a bit hard to check by abacus, but the test is within reach of many a programmable pocket calculator. Of course, the calculation does not give the quotient $(2^{341} - 2)/341$, a number 101 digits long, but rather the remainder, which is 0, thereby falsely asserting that $341 = 11 \cdot 31$ is prime.

The rules for calculating high powers efficiently will be given later, together with a "fast" calculator program. Explicit instructions for calculator programs described in the text are given in the Appendix.

3.4 A Formula for Primes

In 1947, *Mills* [3.2] showed that there is a constant A, such that $\lfloor A^{3^n} \rfloor$ is[2] prime for every n. Here we have a formula that, although it does not generate each and every prime, could be used to generate arbitrarily large primes — for which the sieve methods are less suited.

For anyone who has an appreciation of what a precious thing a prime is, this seems impossible. And indeed, there is trickery at play here, albeit cleverly hidden trickery: determination of the constant A presupposes prior

---

[1] Here and in the rest of this book, "iff" means *if and only if.* Further, x "divides" y means that x divides y *without leaving a remainder.* As a formula this is written with a vertical bar: $x \mid y$.

[2] The so-called "Gauss bracket" or "floor function" $\lfloor x \rfloor$ is defined as the largest integer not exceeding x. Thus, $\lfloor 4.9 \rfloor = 4$; but $\lfloor 5.0 \rfloor = 5$. The Gauss bracket (for $x \geqslant 0$) corresponds to the instruction "take integer part," often designated by INT in computer programs.

knowledge of the primes! This trickery is explained in the excellent little book by *Nagell* [3.3], but it *is* a bit tricky, and we will illustrate the point by another, not quite so surreptitious, trick. Consider the real constant

$$B = 0.2030005000000070000000000000000110 \cdots . \tag{3.2}$$

Upon multiplying by 10 and taking the integer part, one obtains 2 — the first prime. Dropping the integer part, multiplying by 100, and taking the integer part then gives 3 — the second prime. In general, after the nth prime has been extracted from B, multiplying by 10^{2^n} and taking the integer part yields the $(n+1)$th prime. Thus, we have specified a (recursive) algorithm[3] for specifying not only primes but *all* the primes, and in proper order at that!

Of course, here the trick is patently transparent: we have simply "seeded" the primes, one after another, into the constant B, interspersing enough 0's so that they do not "run into" each other. In other words, the constant B does not yield any primes that are not already known. (How many 0's between seeded primes are required to guarantee that adjacent primes do not overlap in B? If 0's are considered expensive because they make B very long, that question is not easy to answer and, in fact, requires a little "higher" number theory.)

Apart from (3.2) and Mill's formula, there have been many other prescriptions for generating primes or even "all" primes. Most of these recipes are just complicated sieves in various disguises, one of the few really elegant ones being Conway's Prime Producing Machine (cf. R. K. Guy, Math. Mag. *56*, 26-33 (1983)). Other attempts, making use of Wilson's theorem (Sect. 8.2), are hilarious at best and distinguished by total impracticality. All this (non)sense is reviewed by U. Dudley in a delightful article ("Formulas for Primes", Math. Mag. *56*, 17-22 (1983)).

3.5 Mersenne Primes

A *Mersenne number* is a number $M_p = 2^p - 1$, where p is *prime*. If M_p itself is prime, then it is called a *Mersenne prime*. Note that numbers of the form $2^n - 1$, where n is composite, can never be prime, because for $n = pq$,

$$2^n - 1 = (2^p - 1)(2^{p(q-1)} + 2^{p(q-2)} + \cdots + 1) . \tag{3.3}$$

However, not all primes p yield Mersenne primes, the first exception being $p = 11$, because $2^{11} - 1 = 2047 = 23 \cdot 89$. Still, there is a fairly simple primality test for numbers of the form $2^p - 1$, the so-called Lucas Test: $2^p - 1$ is prime *iff* (note the double f, meaning *if and only if*) M_p divides

---

[3]   Can the reader specify a nonrecursive algorithm, i.e., one that gives the nth prime directly, without calculating all prior ones?

$S_p (p > 2)$, where S_n is defined by the recursion

$$S_n = S_{n-1}^2 - 2 ,$$ (3.4)

starting with $S_2 = 4$.

Thus, for example, S_{11} is given by the 10th number in the sequence

4, 14, 194, 37634,... ,

which is not divisible by $M_{11} = 2047$. Thus, M_{11} is composite.

While this test does not reveal any factors, there is another test that *can* give a factor for M_p with $p = 4k + 3$: for $q = 2p + 1$ prime, $q \mid M_p$ *iff* $p \equiv 3$ mod 4.

Example: $p = 11 = 4 \cdot 2 + 3$; $M_{11} = 2047$ is not divisible by $(p - 1)/2 = 5$ and is therefore divisible by $2p + 1 = 23$. Check: $2047 = 23 \cdot 89$. Check! Similarly, 47 is discovered as a factor of $M_{23} = 8388607$, etc.

2^0	2^1	2^2	2^3	2^4	2^5	2^6	2^7
2^8	2^9	2^{10}	2^{11}	2^{12}	2^{13}	2^{14}	2^{15}
2^{16}	2^{17}	2^{18}	2^{19}	2^{20}	2^{21}	2^{22}	2^{23}
2^{24}	2^{25}	2^{26}	2^{27}	2^{28}	2^{29}	2^{30}	2^{31}
2^{32}	2^{33}	2^{34}	2^{35}	2^{36}	2^{37}	2^{38}	2^{39}
2^{40}	2^{41}	2^{42}	2^{43}	2^{44}	2^{45}	2^{46}	2^{47}
2^{48}	2^{49}	2^{50}	2^{51}	2^{52}	2^{53}	2^{54}	2^{55}
2^{56}	2^{57}	2^{58}	2^{59}	2^{60}	2^{61}	2^{62}	2^{63}

Fig. 3.2. Mersenne primes on a checkerboard

Figure 3.2 shows the first 9 Mersenne primes arranged on a checkerboard. On January 17, 1968, the largest known prime was the Mersenne prime $2^{11213} - 1$, an event that was celebrated with a postmark (Fig. 3.3) from Urbana, Illinois (at no profit to the U.S. Post Office, considering the zero value of the stamp).

In the meantime, much larger Mersenne primes have been found. The record on November 18, 1978, stood at $2^{21701} - 1$, a prime with 6533 decimal

Fig. 3.3. The largest known Mersenne prime on January 17, 1968

digits found by two California high-school students, Laura Nickel and Curt Noll, using 440 hours on a large computer. The next Mersenne prime is $2^{23209} - 1$. By early 1982 the largest known prime was $2^{44497} - 1$, having 13395 digits [3.4].

Recently, one more Mersenne prime was discovered by D. Slowinski, the 28th known specimen: $2^{86243} - 1$. Assuming that there are no other Mersenne primes between it and $M(27) = 2^{44497} - 1$, then $2^{86243} - 1$ is, in fact, $M(28)$.

Are there more Mersenne primes beyond $2^{86243} - 1$? The answer is almost certainly *yes*. Curiously, we can even say roughly how large the next Mersenne prime is: 10^{38000} — give or take a dozen thousand orders of magnitude. How can we make such a seemingly outrageous statement?

Fermat and Euler proved that all factors of M_p must be of the form $2kp + 1$ and simultaneously of the form $8m \pm 1$. Thus, *potential* factors of M_p are spaced on average $4p$ apart. Assuming that, subject to this constraint, the number of factors of a Mersenne number is governed by a Poisson process, Gillies [3.5] conjectured that of all the primes in the "octave" interval $(x, 2x)$, on average approximately 2 give Mersenne primes. More precisely, the density of primes near p giving rise to Mersenne primes $M_p = 2^p - 1$ would be asymptotic to

$$\frac{2}{p \ln 2} .$$

In a recent paper S. S. Wagstaff, Jr. (Divisors of Mersenne primes. Math. Comp. *40*, 385-397 (1983)), following an argument by H. W. Lenstra, Jr., suggested that the expected number of primes p in an octave interval is $e^\gamma = 1.78 \dots$. Thus, the correct asymptotic density would be

$$\frac{e^\gamma}{p \ln 2} .$$

Comparing this with the general prime density for primes near p, $1/\ln p$, we see that of

$$\frac{p}{e^\gamma \log_2 p}$$

primes, one prime on average leads to a Mersenne prime. For $p \approx 100000$, this means that roughly every 3000th prime gives a Mersenne prime. (The appearance of the factor e^γ is a consequence of Merten's theorem, see Sect. 11.1, and its relevance to prime sieving.)

The distribution of primes p that generate Mersenne primes is expressed even more simply if we consider the density of $\log_2 p$: it is constant and should equal e^γ. Since $\log_2 p$ very nearly equals $\log_2(\log_2 M_p)$, these statements are equivalent to the following: if $\log_2(\log_2 M(n))$ is plotted as a function of n, we can approximate the empirical "data" by a straight line with a slope of about $1/e^\gamma = 0.56$. In fact, for the 27 smallest Mersenne primes $(2 \leqslant p \leqslant 44497)$ the average slope is 0.57, remarkably close to $1/e^\gamma$. The correlation coefficient between $\log_2(\log_2 M(n))$ and n in this range exceeds 0.95.

Figure 3.4 shows $\log_2(\log_2 M(n))$ versus n for the 28 presently known Mersenne primes (assuming $n = 28$ for $p = 86243$). The great regularity is nothing short of astounding.

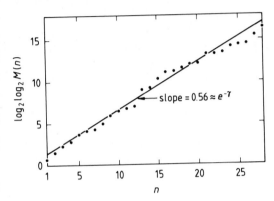

Fig. 3.4. $\log_2(\log_2 M(n))$ versus n

As a Poisson process, the cumulative distribution P of the *intervals* between successive values of $\log_2(\log_2(M(n))$ should go according to the exponential law:

$$P = 1 - e^{-d/\bar{d}}$$

with $\bar{d} = 1/e^\gamma$. This function is plotted in Fig. 3.5 together with the empirical evidence (in interval ranges of 0.2). Here again, the correspondence with the theoretical result expected from a Poisson process is

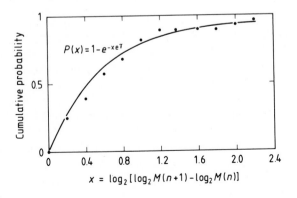

Fig. 3.5. Interval statistics

very good. Specifically, the mean interval (0.57) is close to the standard deviation (0.51) and (beyond the information contained in Fig. 3.5) successive intervals are nearly uncorrelated (correlation coefficient $= -0.17$).

Using 0.56 as the average increment of $\log_2(\log_2(M(n)))$ with n, we expect the next Mersenne prime above $2^{86243} - 1$ in the "neighborhood" of $2^{130000} \approx 10^{38000}$. Of course, to find a prime in this vast "haystack" that gives a Mersenne prime is no small order.

More accurately, we can say that the *probability* of finding the next Mersenne prime either below or above 10^{34000} is about 0.5, and the probability that it exceeds 10^{65000} is less than 10%. But where is it, *exactly* — not with an uncertainty of thousands of orders of magnitude? Even the fastest number crunchers available today, using the most efficient search algorithms, will have a hey (hay?) day.[4]

Unfortunately, the Mersenne primes are very thinly seeded. Thus, if one is looking for a 50-digit prime among the Mersenne primes, he is out of luck: $2^{127} - 1$ has 39 digits and the *next* Mersenne prime, $2^{521} - 1$, has 157 digits — an awesome gap!

Does a Mersenne prime M_p always yield another Mersenne prime by the formula

$$2^{M_p} - 1 ?$$

This had been widely conjectured, but a counterexample is now known: the prime $p = 13$ gives a Mersenne prime $M_{13} = 8191$, but $2^{8191} - 1$ is composite. Too bad! The nearest Mersenne primes, $2^{4423} - 1$ and $2^{9689} - 1$, have 1332 and 2917 decimal digits, respectively — leaving another great void.

3.6 Repunits

Expressed as binary numbers, all numbers of the form $2^n - 1$ consist exclusively of 1's, namely exactly n 1's. For example $M_4 = 2^4 - 1 = 15$, or 1111 in binary. Similarly, for any a, $(a^n-1)/(a-1)$ expressed in base a consists of precisely n 1's and no other digits. Such numbers are called "repunits," and apart from those to the base 2, those to base 10 have been the most widely studied. For a repunit to be prime, n must be prime, but as with the Mersenne numbers, that is not sufficient.

Examples: $(10^5-1)/9 = 11111 = 41 \cdot 271$ and $(10^7-1)/9 = 1111111 = 239 \cdot 4649$.

---

[4] A 29th Mersenne prime was found in the fall of 1983 with $p = 132049$. It has 39751 decimal places. In September 1985, the 30th was found with $p = 216091$.

More examples for which $(10^p-1)/9$ is composite can be created by finding primes $q > 3$ such that $\text{ord}_q\, 10 = p$ (see Chap. 13 for the definition of ord). In fact, q then divides $(10^p-1)/9$.

Which prime exponents p give repunit primes and whether there are infinitely many are two of the many unsolved problems of number theory. With some luck $(10^{317} - 1)/9$ was proved prime, but not until almost 50 years after $(10^{23} - 1)/9$ was found to be prime. Primality testing of large numbers is not easy and *factoring* is even more difficult! In fact, the factoring of $(10^{71} - 1)/9$ (into two primes with 30 and 41 digits, respectively) had to wait for 1984 machines and algorithms.

3.7 Perfect Numbers

Each Mersenne prime has a companion *perfect number* $P = M_p\, 2^{p-1}$. Already *Euclid* showed that *all even* perfect numbers are of this form. A perfect number is a number for which the sum of all divisors (not including P itself) equals P. Thus, for example, $M_2 = 2^2 - 1 = 3$ leads to the perfect number $P = 6$; and indeed, the sum of the divisors of 6: $1 + 2 + 3$ equals 6 itself. Euclid showed that all *even* perfect numbers are of this form.

The next Mersenne prime, M_3, equals $2^3 - 1 = 7$ and the corresponding perfect number is 28. Check: $1 + 2 + 4 + 7 + 14 = 28$. Check!

It is easy to see why this is so. Since M_p is by definition prime, the only divisors of the perfect number $P = M_p\, 2^{p-1}$ are

$$1, 2, \ldots , 2^{p-1}, M_p, 2M_p, \ldots , 2^{p-1}\, M_p ,$$

and their sum equals

$$\sum = 1 + 2 + \cdots + 2^{p-1} + M_p(1 + 2 + \cdots + 2^{p-1}) , \quad \text{or} \tag{3.5}$$

$$\sum = (1 + M_p)(2^p - 1) = 2^p M_p = 2P . \tag{3.6}$$

(The factor 2 appears here because we included, in the sum, P itself as a divisor of P.) The remarkable fact that all *even* perfect numbers are of the form $M_p 2^{p-1}$, where M_p is a Mersenne prime, was first *proved* by Leonhard Euler (1707-1783), the great Swiss mathematician (and not just that!) from Basel who worked for most of his life in St. Petersburg, the then new capital of all the Russias.

Because 28 Mersenne primes are known, at the time of this writing, there are exactly 28 known perfect numbers, all of them even, and the largest one having 51924 decimal places. (See footnote 4.) No odd perfect numbers are known and, tantalizingly, it is not known whether there are any such. As of

1971, no odd perfect number had been found among all the numbers up to 10^{36}. P. Hagis, Jr., showed recently that an odd perfect number not divisible by 3 has at least eleven prime factors (Math. Comp. *40*, 399-404 (1983)).

Apart from perfect numbers, there are pseudoperfect numbers (Sect. 5.9) and *amicable* numbers. Amicable numbers come in pairs. The sum of divisors of one amicable number equals its mate and vice versa. The smallest amicable pair is 220 and 284. Another pair is 17296 and 18416. In a sense, perfect numbers are "self-amicable."

In a further generalization, certain number sequences are called *sociable*. In these, each number equals the sum of the divisors of the preceding number, and the first number equals the sum of divisors of the last number. One such five-member sociable group is 14288, 15472, 14536, 14264, 12496. There is a sociable chain of length 28 whose smallest member is 14316.

A frequently used concept in number theory is the sum of some function f taken over all divisors of a number n, *including* n itself. This is usually shown by the following notation:

$$\sum_{d \mid n} f(d) \ .$$

Using this notation, our statement about perfect numbers P reads

$$\sum_{d \mid p} d = 2P \ , \tag{3.7}$$

the factor 2 appearing here because by definition P itself is a divisor of itself and therefore is included in the sum.

There was a time when the present author was much impressed by the fact that the sum of the *reciprocal* divisors of P is *always* 2:

$$\sum_{d \mid p} \frac{1}{d} = 2! \tag{3.8}$$

(Here, for once, the exclamation mark does not do any harm because 2! — read "two factorial" — still equals 2.) However, (3.8) is a "trivial" consequence of (3.7), because in a sum over all divisors d of a given number n, the divisor d may be replaced by n/d. This reverses the *order* of the terms in such a sum, but does not affect its value:

$$\sum_{d \mid n} f(d) = \sum_{d \mid n} f(n/d) \ . \tag{3.9}$$

Indeed, for $n = 6$ and $f(d) = d$,

$$1 + 2 + 3 + 6 = 6 + 3 + 2 + 1 \ . \tag{3.10}$$

Applying (3.8) to (3.6), we have

$$2P = \sum_{d|P} d = \sum_{d|P} \frac{P}{d} , \tag{3.11}$$

confirming (3.8). Check: $1 + \frac{1}{2} + \frac{1}{3} + \frac{1}{6} = 2$. Check!

It is remarkable that the sum of the reciprocal divisors of a perfect number always equals 2, no matter how large it is. This implies that perfect numbers cannot have too many small divisors, as we already know.

3.8 Fermat Primes

Besides the Mersenne primes $2^p - 1$, which lead to perfect numbers, and of which only 28 are presently known (see footnote 4), there is another kind of prime family with even fewer known members: the Fermat primes. Only 5 such primes are currently known.

$$F_n = 2^{2^n} + 1 \quad \text{for} \quad n = 0, 1, 2, 3, 4 . \tag{3.12}$$

They are $F_0 = 3$, $F_1 = 5$, $F_2 = 17$, $F_3 = 257$ and $F_4 = 65537$.

Incidentally, for $2^m + 1$ to be prime, m must be a power of 2. In fact, for any $a^m + 1$ to be prime, a must be even and $m = 2^n$.

All numbers of the form $2^{2^n} + 1$, whether prime or composite, are called Fermat numbers. They obey the simple (and obvious) recursion

$$F_{n+1} = (F_n - 1)^2 + 1 , \quad \text{or} \tag{3.13}$$

$$F_{n+1} - 2 = F_n(F_n - 2) , \tag{3.14}$$

which leads to the interesting product

$$F_{n+1} - 2 = F_0 F_1 ... F_{n-1} . \tag{3.15}$$

In other words, $F_n - 2$ is divisible by all lower Fermat numbers:

$$F_{n-k} \mid (F_n - 2) , \quad 1 < k \leqslant n . \tag{3.16}$$

With (3.16) it is easy to prove that all Fermat numbers are coprime to each other, and the reader may wish to show this.

Fermat thought that all Fermat numbers F_n were prime, but Euler showed that $F_5 = 4294967297 = 641 \cdot 6700417$, which can easily be confirmed with a good pocket calculator.

The fact that F_6 and F_7 are also composite is a little harder to show because F_6 has 20 decimal digits and F_7 has 39. Nevertheless, *complete* factorizations of F_6, F_7 and, since 1981, F_8 are now known. Further, it is now known that F_{11}, F_{12} and F_{13} are composite, and *some* of their factors are known. Another Fermat number known to be composite is F_{73}, which has more than 10^{21} digits! For special primality tests for Fermat numbers, see *Hardy* and *Wright* [3.6].

At present the smallest, and so far most enduring mystery is presented by F_{20}: its primality status remains unknown. However, the latest progress in primality testing, reported by Walter Sullivan in The New York Times in February 1982 and in [3.1], might yet reveal other Fermat primes, although the next candidate, F_{20}, has 315653 digits. One helpful clue which has been utilized in the past is that if F_n is composite, then it is divisible by $k \cdot 2^{n+2} + 1$ for some k. In fact, Euler knew this, and that is how he discovered the factor $641 = 5 \cdot 2^7 + 1$ in F_5.

In this manner the compositeness of some *very* large Fermat numbers has been established. For example, $5 \cdot 2^{3313} + 1$ is a factor of F_{3310}. By the way, F_{3310} has more than 10^{990} *digits* — not to be confused with the comparatively miniscule number 10^{990}.

3.9 Gauss and the Impossible Heptagon

In March 1796, the Fermat primes suddenly took on a new and overwhelming significance. A precocious teenager from the German ducal town of Brunswick had just discovered that the circle could be divided into 17 equal parts by purely "geometric means," i.e., by straightedge and compass — something that had eluded professional mathematicians and amateurs alike for over two millennia. In fact, nobody had even suspected that such a feat could be possible. After the cases of 2, 3, 4, 5 and 6 had been solved by the ancient Greeks, "everybody" had been working on the "next" case: the regular heptagon (7-gon). But the Brunswick youth proved that that was impossible and that the only regular n-gons that could be constructed were those derivable from the Fermat primes.

The young person, of course, was none other than Carl Friedrich Gauss [3.7], who was himself so impressed by his feat of unlocking a door that had been closed for 2000 years that he decided to become a mathematician rather than a philologist, to which fate his excellence in the classical languages seemed to have "condemned" him.[5]

---

[5] His love of books and languages never left Gauss for the rest of his life. At the age of 62 he learned yet another foreign language — Russian — and began to read Pushkin in the original. Gauss selected the University of Göttingen rather than his "state" university, Helmstedt, for his studies, mainly because of Göttingen's open library policy. Even in his first semester at Göttingen, Gauss spent much time in the university library, which was well stocked and where he had access to the writings of Newton and Euler and many others of his

We shall return to the important subject of dividing the circle, or *cyclotomy* by its learned name, in several other contexts later in this book. Let it only be said here that for the circle to be divisible into n parts, n must be the product of *different* Fermat primes or 1 and a nonnegative power of 2. Thus, regular polygons of n = 2,3,4,5,6,8,10,12,15,16,17,20,24,30, ... sides can be geometrically constructed, while n = 7,9,11,13,14,18,19,21,22,23,25, ... are *impossible* to construct in this manner. Here the "impossible" part of Gauss's assertion is as significant as his positive statement.

---

predecessors. Much of what Gauss read there he had already derived himself, but he still felt that reading was essential — in stark contrast to other scientific geniuses, notably Einstein, who was convinced he could create most of the correct physics from within himself and who is supposed to have said, in jest, that if nature was not the way he felt it ought to be, he pitied the Creator for not seeing the point ("Da könnt' mir halt der liebe Gott leid tun, die Theorie stimmt doch." [3.8].)

4. The Prime Distribution

How are the primes distributed among the integers? Here "distribution" is a misleading term because a given positive integer either is a prime or is not a prime — there is nothing chancy about primality. Yet superficially, the occurrence of primes appears to be rather haphazard, and indeed, many properties can be derived by playing "dumb" and assuming nothing more than that "every other integer is divisible by 2, every third is divisible by 3," etc., and letting complete randomness reign beyond the most obvious. The result of this loose thinking suggests that the average interval between two successive primes near n is about $\ln n$. This is not easy to prove rigorously, especially if one forgoes such foreign tools as complex analysis. Yet fairly simple probabilistic arguments come very close to the truth. In fact, probabilistic thinking as introduced here can reveal a lot about primality and divisibility [4.1], and we shall make ample use of the probabilistic approach throughout this book to gain an intuitive understanding of numerous number-theoretic relationships. For a formal treatment of probability in number theory see [4.2].

4.1 A Probabilistic Argument

Two facts about the distribution of the primes among the integers can be noticed right away:

1) They become rarer and rarer the larger they get.

2) Apart from this regularity in their mean density, their distribution seems rather irregular.

In fact, their occurrence seems so unpredictable that perhaps probability theory can tell us something about them — at least that is what the author

---
[1] In number theory *elementary* methods are often the most difficult, see [4.4].

thought in his second (or third) semester at the Georg-August University in Göttingen. He had just taken a course in *Wahrscheinlichkeitsrechnung* at "Courant's" famous Mathematics Institute, and one afternoon in 1948, in the excruciatingly slow "express" train from Göttingen to his parents' home in the Ruhr, he started putting some random ideas to paper. His train of thought ran roughly as follows.

The probability that a given "arbitrarily" selected integer is divisible by p_i is $1/p_i$. In fact, starting with 1, precisely every p_ith number is divisible by p_i (every third is divisible by 3, every fifth by 5 and so forth). Thus, the "probability" that a given selected number is *not* divisible by p_i is $1 - 1/p_i$.

Assuming that divisibility by different primes is an *independent*[2] property, the probability that x is not divisible by any prime below it is given by the product

$$W(x) \approx \left[1 - \frac{1}{2}\right]\left[1 - \frac{1}{3}\right]\left[1 - \frac{1}{5}\right] \cdots \approx \prod_{p_i < x} \left[1 - \frac{1}{p_i}\right]. \qquad (4.1)$$

If x is not divisible by any prime below it, it is, of course, not divisible by *any* smaller number, i.e., x is prime.

More strictly, we could limit the product to primes $p_i < \sqrt{x}$ (see Sect. 3.2 on the sieve of Eratosthenes). In fact, in that 1948 train the author did limit the product to primes smaller than the square root of x. But since the end result is not much affected, we will not bother about this "refinement."

If one feels uncomfortable with a product, it can be quickly converted into a sum by taking (naturally) logarithms:

$$\ln W(x) \approx \sum_{p_i < x} \ln \left[1 - \frac{1}{p_i}\right]. \qquad (4.2)$$

If one does not like the natural logarithm on the right-hand side, expanding it and breaking off after the first term does not make much difference, especially for the larger primes:

$$\ln W(x) \approx - \sum_{p_i < x} \frac{1}{p_i}. \qquad (4.3)$$

There is something about the sum that is still bothersome: it is not over consecutive integers, but only over the primes. How can one convert it into a

---

[2]   Simultaneous independence for all primes is never exactly true, but there is near independence that suffices for our argument.

sum over *all* integers below x? Again, one can use a probability argument: a given term $1/n$ in the sum occurs with probability $W(n)$. Thus, let us write (and this is the main trick here):

$$\ln W(x) \approx - \sum_{n=2}^{x} \frac{W(n)}{n} .$$
(4.4)

By now sums may have become boring, and one wishes the sum were an integral. Thus, we write with our now customary nonchalance:

$$\ln W(x) \approx - \int_{2}^{x} \frac{W(n)}{n} \, dn .$$
(4.5)

The next thing that may strike one as offensive is the minus sign on the right-hand side. Introducing the *average distance* $A(x) = 1/W(x)$ between primes, we get a positive expression:

$$\ln A(x) \approx \int_{2}^{x} \frac{dn}{nA(n)} .$$
(4.6)

Now, suddenly, the integral has served its purpose and can go; most people would rather solve differential equations than integral ones. Differentiating will of course be the appropriate integral vanishing trick:

$$\frac{A'(x)}{A(x)} \approx \frac{1}{x \, A(x)} , \quad \text{or}$$
(4.7)

$$A'(x) \approx \frac{1}{x} .$$
(4.8)

And the unexpected has happened: we have an answer (fortuitously correct)! The average distance between primes ought to be

$$A(x) \approx \ln(x) ,$$
(4.9)

and the mean density becomes

$$W(x) \approx \frac{1}{\ln x} .$$
(4.10)

Example: $x = 20$, $\ln 20 \approx 3.00$, and indeed, the average spacing of the 3 primes closest to 20, namely 17, 19 and 23, is exactly 3.

Around $x = 150$, the average spacing should be about 5, and in the neighborhood of $x = 10^{50}$, every 115th number, on average, is a prime.

4.2 The Prime-Counting Function $\pi(x)$

If we accept the estimate (4.10) of the average prime density, the number of primes smaller than or equal to x, usually designated by $\pi(x)$, is approximated by the "integral logarithm":

$$\pi(x) \approx \int_{2}^{x} \frac{dx'}{\ln x'} =: Li(x) , \qquad (4.11)$$

where the sign =: indicates that the notation $Li(x)$ is *defined* by the integral on the left.

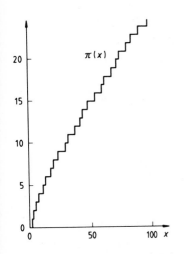

Fig. 4.1. The prime-counting function $\pi(x)$ for $2 \leqslant x \leqslant 100$

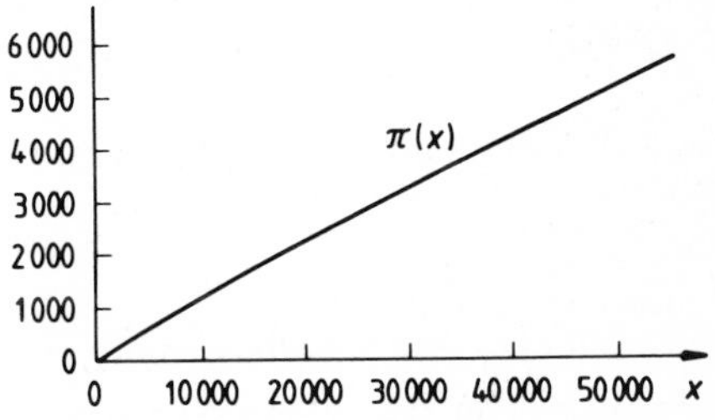

Fig. 4.2. Same as Fig. 4.1 but $x \leqslant 55{,}000$

The prime-counting function $\pi(x)$ is plotted in Fig. 4.1 for $x \leqslant 100$. Every time x equals a prime, $\pi(x)$ jumps up by 1. But apart from the "jumpiness" of $\pi(x)$, a smoother, slightly concave trend is also observable. This smoothness becomes more obvious when we plot $\pi(x)$ for x up to 55,000 as in Fig. 4.2. On this scale, the jumpiness has disappeared completely.

The inadequacy of Gauss's original estimate $\pi(x) \approx x/\ln x$ is illustrated by Fig. 4.3. By contrast, the integral logarithm, which we "derived" above (and which was also conjectured by Gauss) gives seemingly perfect agreement with $\pi(x)$ in the entire range plotted in Fig. 4.4.

However, even $Li(x)$, labelled "Gauss" in Fig. 4.5, shows noticeable deviations when we expand the ordinate by a factor 10^4 as was done in that figure (see *Zagier* [4.10]). In fact, for $x = 10^7$, the excess of $Li(x)$ over $\pi(x)$ is about 300 and remains positive for all $x < 10^9$. Nevertheless,

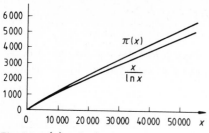

Fig. 4.3. $\pi(x)$ and $x/\ln x$

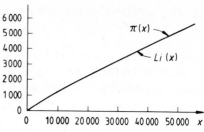

Fig. 4.4. $\pi(x)$ and the integral logarithm $Li(x)$

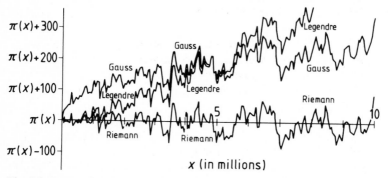

Fig. 4.5. Comparison of formulas by Gauss: $Li(x)$, Legendre: $x/(\ln x - 1.08366)$, and Riemann: $Li(x) - 1/2\,Li(x^{1/2}) - 1/3\,Li(x^{1/3}) - \ldots$ (Courtesy of D. Zagier)

$\pi(x) - Li(x)$ has infinitely many zeroes, at least one of which occurs below $x = 10^{10^{10^{34}}}$; in fact, it may be near $x = 10^{370}$. (A number such as $10^{10^{10^{34}}}$, introduced by S. Skewes in 1933, was once considered a large number. But *much much larger* numbers have now become important in connection with Gödel's famous "incompleteness" theorem [4.3].)

Legendre, independently of Gauss, gave the following formula in 1778:

$$\pi(x) \approx \frac{x}{\ln x - 1.08366} \, , \qquad (4.12)$$

a closer approximation than (4.11) up to about $x = 4 \cdot 10^6$, as can be seen in Fig. 4.5. However, above $x = 5 \cdot 10^6$ Legendre's formula begins to go to pieces. (Expanding $Li(x)$ gives 1 as the constant in (4.12), but Legendre missed that.)

Either formula (4.11) or Legendre's (4.12) says that there are about $7.9 \cdot 10^{47}$ 50-digit primes — plenty to go around for the "trap-door" encryption schemes to be discussed later in Chap. 9.

In our "derivation" of $\pi(x)$, we considered primes up to x and pointed out that consideration of primes up to $\sqrt{x}$ would have sufficed. This idea was further pursued by Bernhard Riemann, who showed that

$$\pi(x) \approx R(x) := Li(x) - \frac{1}{2} Li(\sqrt{x}) - \frac{1}{3} Li(\sqrt[3]{x}) - \cdots . \qquad (4.13)$$

Figure 4.5 demonstrates how good an approximation $R(x)$ is; the curve labelled "Riemann" does not seem to have any deviant trend up to $x = 10^7$.

The closeness of $R(x)$ to $\pi(x)$ is further emphasized by Table 4.1, which shows that even for $x = 10^9$, the error of $R(x)$ is only 79 (out of $5 \cdot 10^7$).

Table 4.1. Comparison of prime-counting function $\pi(x)$ and Riemann's approximation $R(x)$

x	$\pi(x)$	$R(x)$
100,000,000	5,761,455	5,761,552
200,000,000	11,078,937	11,079,090
300,000,000	16,252,325	16,252,355
400,000,000	21,336,326	21,336,185
500,000,000	26,355,867	26,355,517
600,000,000	31,324,703	31,324,622
700,000,000	36,252,931	36,252,719
800,000,000	41,146,179	41,146,248
900,000,000	46,009,215	46,009,949
1,000,000,000	50,847,534	50,847,455

It is interesting to note that it was not until 1896, almost a hundred years after Gauss's and Legendre's conjectures, that Hadamard and de la Vallée Poussin proved the "Prime Number Theorem" in the form

$$\lim_{x \to \infty} \frac{\pi(x) \ln(x)}{x} = 1 \qquad (4.14)$$

using "analytic" methods, i.e., mathematical tools from outside the domain of integers. The first "elementary" proof not using such tools did not come until 1948 and is due to *Erdös* [4.4] and Selberg. This illustrates the vast gap between obtaining an easy estimate, as we have done in the preceding pages, and a hard proof.

Perhaps one of the most surprising facts about $\pi(x)$ is that there "exists" an *exact* formula, given by a limiting process of analytic functions $R_k(x)$:

$$\pi(x) = \lim_{k \to \infty} R_k(x) , \qquad \text{where} \qquad (4.15)$$

$$R_k(x) := R(x) - \sum_{l=-k}^{k} R(x^{\rho_l}) . \qquad (4.16)$$

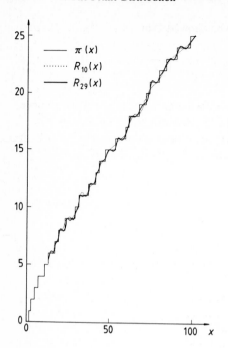

Fig. 4.6.Riemann's approximation to $\pi(x)$ (After H. Riesel, G. Göhl: Math. Comp. **24**, 969–983 (1970)

Here ρ_l is the lth zero of the Riemann zetafunction [4.5]:

$$\zeta(s) := \sum_{n=1}^{\infty} \frac{1}{n^s} \, . \tag{4.17}$$

Figure 4.6 shows $\pi(x)$ and the two approximations $R_{10}(x)$ and $R_{29}(x)$, the latter already showing a noticeable attempt to follow the jumps of $\pi(x)$.

The zeroes for $l = 1,2, ..., 5$ of $\zeta(s)$ are shown in Table 4.2. The real parts are all equal to 1/2. In fact, more than 100 years ago Riemann

Table 4.2. The first five zeroes of the zetafunction with real part equal to 1/2

$$\rho_1 = \tfrac{1}{2} + 14.134725 \; i,$$

$$\rho_2 = \tfrac{1}{2} + 21.022040 \; i,$$

$$\rho_3 = \tfrac{1}{2} + 25.010856 \; i,$$

$$\rho_4 = \tfrac{1}{2} + 30.424878 \; i,$$

$$\rho_5 = \tfrac{1}{2} + 32.935057 \; i,$$

enunciated his famous hypothesis that *all* complex zeroes of $\zeta(s)$ have real part 1/2. Riemann thought at first that he had a proof, but the Riemann Hypothesis (and the so-called Extended Riemann Hypothesis, abbreviated ERH) has remained unproved to this day, although hundreds of millions of zeroes have been calculated, all with real part 1/2. In fact, the ERH is *so* widely believed today that a sizable edifice is based on it, and will collapse when the first $\mathrm{Re}(\rho_l) \neq 1/2$ makes its appearance.

(In late 1984, a *possible* proof was presented by Matsumoto in Paris. Mind boggling! If it can only be confirmed...)

4.3 David Hilbert and Large Nuclei

In conclusion, we mention that David Hilbert once conjectured that the zeroes of the Riemann zetafunction were distributed like the eigenvalues of a certain kind of random Hermitian matrix. This same kind of matrix, incidentally, later gained prominence in the physics of large atomic nuclei, where its eigenvalues correspond to the energy levels of the nucleons (protons and neutrons) [4.6]. In physics the resulting distribution of energy level differences is called the *Wigner distribution* after Eugene Wigner, who derived it. It is shown in Fig. 4.7 as a solid line. The dots are the results of computer calculations by Andrew Odlyzko of Bell Laboratories (private communications) of the zeroes of the Riemann zetafunction around $x = 10^8$. Since the density of zeroes increases logarithmically with their distance from the real line, the spacing of zeroes normalized by their average spacing is

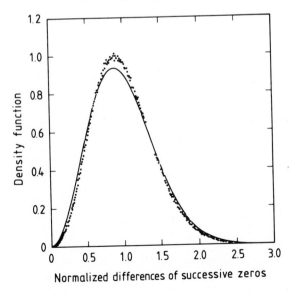

Fig. 4.7. Interval distribution between successive zeroes ($x = 1/2$) of zetafunction. (———) conjecture by Hilbert (Courtesy of A. Odlyzko, Bell Laboratories)

shown. The close agreement between the solid line (Hilbert) and the dots (Odlyzko) shows how close Hilbert's conjecture, made almost a century ago, is.

4.4 Coprime Probabilities

What is the probability that two arbitrarily and independently selected numbers from a large range do not have a common divisor, i.e., that they are coprime? The probability that one of them is divisible by the prime p_i is, as we have seen, $1/p_i$, and the probability that both of them are divisible by the same prime, assuming independence, is $1/p_i^2$. Thus, the probability that they are *not* both divisible by p_i equals $1 - 1/p_i^2$. If we assume divisibility by different primes to be independent, then the probability of coprimality becomes

$$W_2 \approx \prod_{p_i} (1 - 1/p_i^2) , \qquad \text{or} \tag{4.18}$$

$$\frac{1}{W_2} \approx \prod_{p_i} \frac{1}{1 - 1/p_i^2} = \prod_{p_i} \left[1 + \frac{1}{p_i^2} + \frac{1}{p_i^4} + \cdots \right] , \tag{4.19}$$

where we have expanded the denominator into an infinite geometric series.

Now if, for simplicity, we extend the product over *all* primes, then — as Euler first noted — the result is quite simple: one obtains exactly every square integer once (this follows from the *unique* decomposition of the integers into prime factors). Thus,

$$\frac{1}{W_2} \approx \sum_{n=1}^{\infty} \frac{1}{n^2} = \zeta(2) = \frac{\pi^2}{6} , \tag{4.20}$$

and the probability of coprimality W_2 should tend toward $6/\pi^2 \approx 0.608$ for large numbers.

The probability that a randomly selected integer n is "squarefree" (not divisible by a square) also tends to $6/\pi^2$. The reasoning leading to this result is similar to that applied above to the coprimality of two integers: for an integer to be squarefree it must not be divisible by the same prime p_i more than once. Either it is not divisible by p_i or, if it is, it is not divisible again. Thus,

$$\text{Prob}\{p_i^2 \nmid n\} \approx \left[1 - \frac{1}{p_i} \right] + \frac{1}{p_i} \left[1 - \frac{1}{p_i} \right] = 1 - \frac{1}{p_i^2} .$$

Taking the product over all p_i (assuming again independence of the divisibility by different primes) gives the above expression for $W_2 \approx 6/\pi^2$.

How fast is this asymptotic value reached? The sum over the reciprocal squares in (4.20) converges quite rapidly and the value of $6/\pi^2$ might already hold for the coprimality and squarefreeness of small numbers. In fact, 61 of the first 100 integers above 1 are squarefree and of the 100 number pairs made up from the integers 2 to 11, exactly 60 are coprime. This is the closest possible result, because the answer has to be an even number and 62 is farther away from $600/\pi^2$ than 60.

Figure 4.8 shows a computer-generated plot of coprimality in the range from 2 to 256: a white dot is plotted if its two coordinates are coprime. As expected, the density of white dots is quite uniform. All kinds of interesting micropatterns can be observed, and a number of long-range .structures at angles whose tangents are simple ratios: 0, 1/2, 1, 2, etc., are also visible. Does such a plot pose new questions or suggest new relationships for number theory?

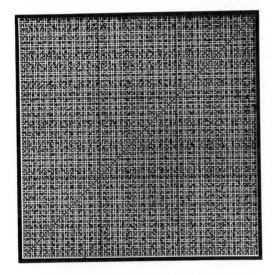

Fig. 4.8. The coprimality function, a simple number-theoretic function, in the range $2 \leqslant x \leqslant 256$ and $2 \leqslant y \leqslant 256$. A white dot is shown if $(x, y) = 1$. Whenever $(x, y) > 1$, there is no dot (*black*)

When the author first had this plot prepared (by Suzanne Hanauer at Bell Laboratories), he thought that a two-dimensional Fourier transform should make an interesting picture because the Fourier transformation brings out periodicities. And, of course, divisibility *is* a periodic property.

Figure 4.9 shows the result, which with its prominent starlike pattern would make a nice design for a Christmas card (and has, in fact, been so used). What is plotted here (as increasing brightness) is the *magnitude* of the two-dimensional discrete Fourier transform of the number-theoretic function $f(n,m)$, for $n,m = 1,2, ..., 256$, with $f=1$ if the GCD $(n,m) = 1$ and $f=0$ otherwise.

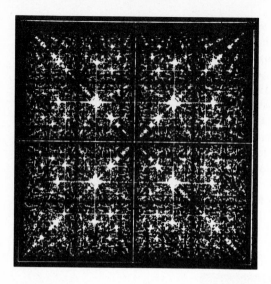

Fig. 4.9. The magnitude of the Fourier transform (simulated by increasing brightness) of the number theoretic function shown in Fig. 4.8. The presence of a white dot, $(x, y) = 1$, is interpreted as $+1$, and the absence of a white dot as -1

Since the original function is symmetric around the 45° diagonal, so is the Fourier transform. Since only magnitude is plotted, there is another symmetry axis: the −45° diagonal. In addition, there are *near* symmetries about the horizontal and vertical axes which are not so easy to explain. We leave it as an exercise to the reader to explain both this near symmetry and each of the stars in Fig. 4.9. (See also [4.7].)

The coprimality probability for more than two randomly selected integers is obtained in the manner that led to (4.20). The general result that k integers are coprimes is

$$W_k := \text{Prob}\{(n_1, n_2 ..., n_k) = 1\} \approx [\zeta(k)]^{-1}, \tag{4.21}$$

where $\zeta(k)$ is Riemann's zetafunction as defined in (4.17). For $k = 3$ one obtains $W_3 \approx 0.832$, and for $k = 4$, $W_4 \approx 90/\pi^4 = 0.9239...$. (The actual proportions in the range from 2 to 101 are 0.85 and 0.93, respectively.)

The probabilities that a randomly selected integer is not divisible by a cube, a fourth power, and in general by a kth power, also tend toward (4.21). Thus, roughly 84% of all integers are "cubefree."

A somewhat more difficult problem is posed by the probability of *pairwise* coprimality of three (or more) randomly selected integers. The probability that none of k integers has the prime factor p_i is

$$\left(1 - \frac{1}{p_i}\right)^k$$

and that exactly one has p_i as a factor is

$$\frac{k}{p_i}\left(1 - \frac{1}{p_i}\right)^{k-1}.$$

The sum of these two probabilities is the probability that at *most* one of the integers has p_i as a factor. The product over all primes p_i then approximates the probability that the k integers are pairwise coprime, i.e., that

$$(n_j, n_m) = 1 \quad \text{for all} \quad j \neq m.$$

The reader may want to show that for $k = 3$, this probability can be written

$$\frac{36}{\pi^4} \prod_{p_i}\left(1 - \frac{1}{(p_i+1)^2}\right) = 0.28 \dots.$$

Thus, only about 28% of three randomly selected integers are *pairwise* coprime. (Compare this with the above result $W_3 \approx 0.832$.)

Jobst von Behr of Hamburg, who read the first edition of this book, generalized this problem by considering the probability $P_k(d)$ that the greatest common divisor (GCD) of k integers equal $d > 1$. By Monte Carlo computation on his home computer he obtained numerical results that looked suspiciously like

$$P_k(d) = d^{-k}\, \zeta^{-1}(k).$$

Can the reader of this edition prove this seductively simple scaling law? Summing our all d gives of course 1, as it should for a proper probability.

For the probability that the GCD of k random integers is *even* the above formula gives 2^{-k}; is this in conformity with elementary probability?

4.5 Twin Primes

Primes not infrequently come in pairs called twin primes, like 11 and 13 or 29 and 31. How often does this happen? An *estimate* [4.2] shows their density to be proportional to $1/(\ln x)^2$, suggesting that they may occur independently. But one must be careful here, because prime triplets of the form $x, x + 2, x + 4$ can never happen (other than 3,5,7), since one member of such a triplet is always divisible by 3. (Reader: try to show this — it is easy.)

On the other hand, triplets of the form $x, x + 2, x + 6$ or $x, x + 4, x + 6$ are not forbidden and do happen, e.g., 11, 13, 17 or 13, 17, 19. Is their asymptotic density $1/(\ln x)^3$? In number theory, what is not explicitly forbidden often occurs, and occurs randomly — resembling the democratic process.

Incidentally, the density estimate for twin primes, $c/(\ln x)^2$, implies that there are infinitely many twins, but this has never been proven!

This brings us to a famous theorem by Dirichlet (1837), Gauss's successor in Göttingen: there are infinitely many primes in every *linear progression*

$$a \cdot n + b, \quad n = 1,2,3, \dots , \tag{4.22}$$

provided the constants a and b are coprime: $(a,b) = 1$. Thus, for example, with $a = 10$ and $b = 1$, 3, 7 or 9, we see that there are infinitely many primes whose last digit is 1, 3, 7 or 9. (In fact, as we shall see later, these four kinds of primes occur in equal proportion.)

The longest sequence known in early 1982 for which $a \cdot n + b$ gives primes for *consecutive n* is the progression

$$223092870 \cdot n + 2236133941 ,$$

which is prime for sixteen consecutive values: $n = 0,1,2, \dots, 15$. Since then a 19-member progression has been discovered.

The record for a *quadratic* progression stands at 80 consecutive primes, namely

$$n^2 + n + 41 \quad \text{for } n = -40,-39, \dots, 0, \dots, 39 . \tag{4.23}$$

This is remarkable because it would ordinarily take a polynomial in n of degree 80 to get 80 primes for consecutive values of n.

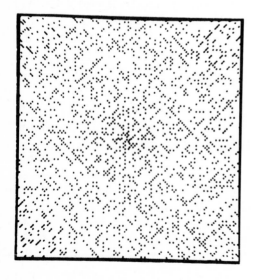

Fig. 4.10. Primes (*dots*) plotted on a spiral. Many primes fall on straight lines

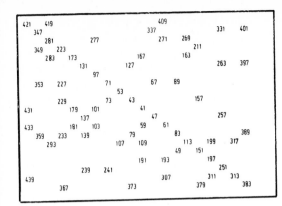

Fig. 4.11. The primes between 41 and 439 plotted on a square spiral beginning with 41 in the center. Note the "solid-prime" diagonal

Many primes are of the form $4n^2 + an + b$, which makes them lie on straight lines if n is plotted along a square spiral. This fact is illustrated by so plotting the primes (see Figs. 4.10, 11).

However, there is no polynomial, no matter how high its degree, which yields primes for all values of n. If there was a polynomial in n of finite degree r generating primes for all n, then $r + 1$ primes would determine the $r + 1$ coefficients of the polynomial and infinitely many other primes could then be calculated from these $r + 1$ primes. The location of primes among the integers is simply too unpredictable to be "caught" by something as regular and finite as a polynomial.

4.6 Primeless Expanses

On the other hand, there is always a prime between n^3 and $(n + 1)^3 - 1$ for large enough n. This fact was exploited by W. H. Mills (1947) to construct a constant A such that $\lfloor A^{3^n} \rfloor$ is prime for all n [see Ref. 4.8, p. 160]. But of course, the Mills expression is not a polynomial, and as we remarked before (Sect. 3.4), the primes thus generated have been "smuggled" into A first.

Somewhat paradoxically, there are also *arbitrarily* large intervals without a single prime! For example, the one *million consecutive* integers

$$(10^6 + 1)! + n, \quad n = 2,3,4, ..., 1,000,001 \tag{4.24}$$

are all composite! In fact, there is even a set of one million somewhat smaller consecutive integers that are all composite, namely those in which the additive term n in the above expression is replaced by $-n$.

In *relative terms,* the primeless expanse of one million integers is, of course, rather small. A (weak) upper bound on the relative size of primeless ranges is an "octave" of integers; i.e., there is always a prime p in the range n and $2n$ (inclusive):

$$n < p \leqslant 2n \, , \tag{4.25}$$

or, equivalently, each prime is less than twice its predecessor [4.9]:

$$p_{k+1} < 2p_k \, . \tag{4.26}$$

Check: $3 < 2 \cdot 2, 5 < 2 \cdot 3, 7 < 2 \cdot 5, 11 < 2 \cdot 7$, etc.

In fact, the number of primes in the interval from n to $2n$ is of the same order as those below n. This follows directly from the asymptotic expression for $\pi(x)$:

$$\pi(x) \approx \frac{x}{\ln x} \, , \tag{4.27}$$

so that

$$\pi(2x) - \pi(x) \approx \frac{2x}{\ln x + \ln 2} - \frac{x}{\ln x}$$

$$\approx \frac{x}{\ln x} - \frac{2x \ln 2}{\ln^2 x} \, . \tag{4.28}$$

4.7 Square-Free and Coprime Integers

The probability that a given integer is squarefree approaches $6/\pi^2$ (see Sect. 4.4) and $6/\pi^2$ is also the asymptotic probability that two randomly chosen integers are coprime. Are these two properties independent? No! Among 500 random integers von Behr found 205 that were both square-free *and* coprime (to another random integer), instead of only $500 \cdot 36/\pi^4 \approx 185$ if these two properties were independent. Thus, there seems to be a *positive* correlation between squarefreeness and coprimality. The reader may wish to show that the joint probability equals $36/\pi^4$ times a peculiar product, which is larger than 1:

$$\prod_i \left[1 + \frac{1}{p_i^3 + p_i^2 - p_i - 1} \approx 1.16 \right] \, .$$

5. Fractions: Continued, Egyptian and Farey

Continued fractions are one of the most delightful and useful subjects of arithmetic, yet they have been continually neglected by our educational factions. Here we discuss their applications as approximating fractions for rational or *irrational numbers* and *functions,* their relations with *measure theory* (and deterministic chaos!), their use in *electrical networks* and in solving the *"squared square;"* and the *Fibonacci* and *Lucas numbers* and some of their endless applications.

We also mention the (almost) useless *Egyptian fractions* (good for designing puzzles, though, including *unsolved* puzzles in number theory) and we resurrect the long-buried *Farey fractions,* which are of considerable contemporary interest, expecially for *error-free computing.*

Among the more interesting recent applications of Farey series is the reconstruction of periodic (or nearly periodic) functions from "sparse" sample values. Applied to two-dimensional functions, this means that if a motion picture or a television film has sufficient structure in space and time, it can be reconstructed from a fraction of the customary picture elements ("pixels"). ("Sufficient structure" in spacetime implies that the reconstructions might not work for a blizzard or a similar "snow job.")

5.1 A Neglected Subject

Continued fractions (CFs) play a large role in our journey through number theory [5.1]. A simple continued fraction

$$b_0 + \cfrac{1}{b_1 + \cfrac{1}{b_2 + \cfrac{1}{b_3 + \dots}}} \qquad (5.1)$$

a typographical nightmare if there ever was one, is usually written as follows: $[b_0; b_1, b_2, b_3, \dots]$. Here the b_m are integers. A finite simple CF then looks like this:

$$[b_0; b_1,..., b_n] .$$

(5.2)

If a finite or infinite CF is broken off after $k < n$, then $[b_0; b_1,..., b_k] = A_k/B_k$ is called the approximating fraction or convergent of order k. Here A_k and B_k are coprime integers; they obey the recursion

$$A_k = b_k A_{k-1} + A_{k-2} ,$$

(5.3)

with $A_0 = b_0$, $A_{-1} = 1$ and $A_{-2} = 0$. The B_k are derivable from the same recursion:

$$B_k = b_k B_{k-1} + B_{k-2} ,$$

(5.4)

with $B_0 = 1$ and $B_{-1} = 0$.

As the order of the approximating fractions increases, so does the degree of approximation to the true value of the fraction, which is approached alternately from above and below.

CFs are unique if we outlaw a 1 as a final entry in the bracket. Thus 1/2 should be written as [0;2] and not [0;1,1]. In general, if a 1 occurs in the last place, it can be eliminated by adding it to the preceding entry.

Continued fractions are often much more efficient in approximating rational or irrational numbers than ordinary fractions, including decimals. Thus,

$$r = \frac{964}{437} = [2; 4, 1, 5, 1, 12] ,$$

(5.5)

and its approximating fraction of order 2, $[2; 4, 1] = 11/5$, approaches the final value within 3 parts in 10^3.

One interesting application of CFs is to answer such problems as "when is the power of the ratio of small integers nearly equal to a power of 2?", a question of interest in designing cameras, in talking about computer memory and in the tuning of musical instruments (Sect. 2.6). For example, what integer number of musical *major thirds* equals an integral number of *octaves*, i.e., when is

$$\left(\frac{5}{4}\right)^n \approx 2^m , \quad \text{or}$$

$$5^n \approx 2^{m+2n} ?$$

By taking logarithms to the base 2, we have

$$\log_2 5 \approx \frac{m}{n} + 2 .$$

The fundamental theorem tells us that there is no exact solution; in other words, $\log_2 5$ is irrational. With the CF expansion for $\log_2 5$ we find

$$\log_2 5 = 2.3219 \ldots = [2; 3,9,\ldots] , \quad \text{or}$$

$$\log_2 5 \approx 2 + \frac{1}{3} ,$$

yielding $m = 1$ and $n = 3$. Check:

$$\left(\frac{5}{4}\right)^3 = 1.953 \ldots \approx 2 .$$

In other words, the well-tempered third-octave $2^{1/3}$ matches the major third within 0.8% or 14 musical cents. (The musical cent is defined as 1/1200 of an octave. It corresponds to less than 0.6 Hz at 1 kHz, roughly twice the just noticeable pitch difference.)

The next best CF approximation gives

$$\log_2 5 \approx 2 + \frac{9}{28} ,$$

or $m = 9$, $n = 28$, a rather unwieldy result .

Because $\log_{10} 2$, another frequently occurring irrational number, is simply related to $\log_2 5$:

$$\log_{10} 2 = 1/(1 + \log_2 5) ,$$

it has a similar CF expansion:

$$\log_{10} 2 = [0;3,3,9, \ldots] .$$

It, too, is well approximated by breaking off before the 9. This yields

$$\log_{10} 2 \approx [0;3,3] = \frac{3}{10} ,$$

a well-known result (related to the fact that $2^{10} \approx 10^3$).

Some irrational numbers are particularly well approximated. For example, the widely known first-degree approximation to π, namely, $[3; 7] = 22/7$, comes within 4 parts in 10^3. The second-order approximation $[3; 7, 16] = 355/113$, known to the early Chinese, approaches π within 10^{-7}.

Euler [5.2] discovered that the CF expansion of $e = 2.718281828\ldots$, unlike that of π, has a noteworthy regularity:

$$e = [2; 1, 2, 1, 1, 4, 1, 1, 6, 1, 1, 8, 1,\ldots] , \tag{5.6}$$

but converges initially very slowly because of the many 1's. In fact, the CF for the *Golden section or Golden ratio* g = [1; 1, 1, 1, 1 ...], which contains infinitely many 1's, is the most slowly converging CF. It is therefore sometimes said, somewhat irrationally, that g is the "most irrational" number. In fact, for a given order of rational approximation the approximation to g is worse than for any other number. Because of this property, up-to-date physicists who study what they call "deterministic chaos" in nonlinear systems often pick the Golden ratio g as a parameter (e.g., a frequency ratio) to make the behavior "as aperiodic as possible." A strange application of number theory indeed!

CFs are also useful for approximating *functions*. Thus, in a generalization of our original bracket notation, permitting noninteger entries,

$$\tan z = \left[\frac{z}{1-} \frac{z^2}{3-} \frac{z^2}{5-} \cdots \right] := \cfrac{z}{1 - \cfrac{z^2}{3 - \cfrac{z^2}{5}} \cdots} \tag{5.7}$$

yields the second-order approximation [5.3]

$$\tan z = z \, \frac{15 - z^2}{15 - 6z^2} \, . \tag{5.8}$$

For $z = \pi/4$, this is about 0.9998 (instead of 1). By contrast, the three-term power series for $\tan z$, $\tan z = z + z^3/3 + z^5/5$, makes an error that is 32 times larger. The reason for the superiority of the CF over the power-series expansion is quite obvious. As we can see from (5.8), the CF expansion makes use of polynomials not only in the numerator but also in the *denominator*. (Not making use of this degree of freedom is as if a physicist or engineer tried to approximate the behavior of a *resonant system* by zeroes of analytic functions only, rather than by zeroes and *poles*: it is possible, but highly inefficient.)

Equally remarkable is the approximation of the error integral by a CF. The third-order approximation

$$\int_0^z e^{-x^2} dx \approx \frac{49140 + 3570z^3 + 739z^5}{49140 + 19950z^2 + 2475z^4} \tag{5.9}$$

makes an error of only 1.2% for $z = 2$, as opposed to a power series including terms up to z^9 which overshoots the true value by 110%.

Incidentally, the fact that e has so regular a CF representation as (5.6), while π does not, does not mean that there is no regular relationship between CFs and some relative of π. In fact, the (generalized) CF expansion of

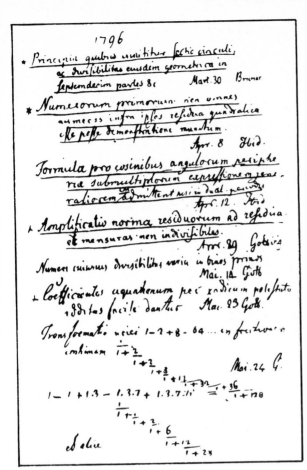

Fig. 5.1. First page of Gauss's notebook, begun in his native city of Brunswick when he was only 18. The first entry concerns the epochal "geometrical" construction of the regular 17-gon which convinced him that he should become a mathematician. The last entry on this page, written like the three preceding ones in Göttingen, shows his early interest in continued fractions

arctan z for $z = 1$ leads to the following neat CF representation:

$$\frac{\pi}{4} = \left[\frac{1}{1+} \quad \frac{1}{2+} \quad \frac{9}{2+} \quad \frac{25}{2+} \quad \frac{49}{2+} \cdots \right].$$

Gauss, the prodigious human calculator, used CFs profusely; even on the first page of his new notebook begun on the occasion of the discovery of the regular 17-gon, CFs make their appearance (Fig. 5.1).

Why are CFs treated so negligently in our high (and low) schools? Good question, as we shall see when we study their numerous uses.

5.2 Relations with Measure Theory

Consider the CF

$$\alpha = [0; a_1, a_2, ...] .\tag{5.10}$$

In 1828, Gauss established that for almost all α in the open interval $(0,1)$ the probability

$$W([0; a_n, a_{n+1}, ...] < x)\tag{5.11}$$

tends to $\log_2(1 + x)$ as n goes to infinity. Gauss also showed that the probability

$$W(a_n = k) \rightarrow \log_2 \left[1 + \frac{1}{k(k+2)} \right] ,\tag{5.12}$$

i.e., the probabilities for $a_n = 1,2,3,...$ decline as 0.42, 0.17, 0.09,..., in contrast to the equal probabilities of the 10 digits for "most" decimal digits.

Khinchin [5.4] showed in 1935 that for almost all real numbers the geometric mean

$$(a_1 a_2 ... a_n)^{\frac{1}{n}} \rightarrow \prod_{k=1}^{\infty} \left[1 + \frac{1}{k(k+2)} \right]^{\log_2 k} = 2.68545 ... ,\tag{5.13}$$

and that the denominators of the approximating fractions

$$(B_n)^{\frac{1}{n}} \rightarrow e^{\pi^2/12 \ln 2} = 3.27582 ... .\tag{5.14}$$

These strange constants are reminiscent of the magic numbers that describe period doubling for strange attractors in deterministic chaos. And perhaps there is more than a superficial connection here.

5.3 Periodic Continued Fractions

As with periodic decimals, we shall designate (infinite) periodic CFs like $[1;2,2,2, ...]$ by a bar over the period:

$$[1;\overline{2}] .\tag{5.15}$$

Incidentally, $[1;\overline{2}]$ has the value $\sqrt{2}$. In general, periodic CFs have values in which square (but no higher) roots appear.

An integer that is a nonperfect square, whose square root has a periodic, and therefore *infinite,* CF, has an irrational square root. However, there are simpler proofs that $\sqrt{2}$, say, is irrational without involving CFs. Here is a simple indirect proof: suppose $\sqrt{2}$ is rational:

$$\sqrt{2} = \frac{m}{n} , \qquad\qquad (5.16)$$

where m and n are coprime:

$$(m,n) = 1 , \qquad\qquad (5.17)$$

i.e., the fraction for $\sqrt{2}$ has been "reduced" (meaning the numerator m and the denominator n have no common divisor). Squaring (5.16) yields

$$2n^2 = m^2 . \qquad\qquad (5.18)$$

Thus, m must be even:

$$m = 2k , \qquad\qquad (5.19)$$

or, with (5.18),

$$2n^2 = 4k^2 , \qquad\qquad (5.20)$$

which implies that n is also even. Consequently,

$$(m,n) > 1 , \qquad\qquad (5.21)$$

contradicting (5.17). Thus, there are *no* integers n,m such that $\sqrt{2} = m/n$; in other words: $\sqrt{2}$ is irrational. Q.E.D.

An even shorter proof of the irrationality of $\sqrt{2}$ goes as follows. Suppose $\sqrt{2}$ is rational. Then there is a *least* positive integer n such that $n\sqrt{2}$ is an integer. Set $k = (\sqrt{2}-1)n$. This is a positive integer *smaller* than n, but

$$k\sqrt{2} = (\sqrt{2}-1)n\sqrt{2} = 2n - \sqrt{2}n$$

is the difference of two different integers and so is a positive integer. Contradiction: n was supposed to be the smallest positive integer such that multiplying it by $\sqrt{2}$ gives an integer! Using this kind of proof, for which s can one show that $\sqrt{s}$ is irrational? What modification(s) does the proof require?

Another exhibition example for CF expansion which we have already encountered is

Fig. 5.2. Golden ratios in a painting by Seurat

$$[1;\overline{1}] = \frac{1}{2}(1 + \sqrt{5}) \, , \tag{5.22}$$

the famous Golden ratio g: if a distance is divided so that the ratio of its total length to the longer portion equals g, then the ratio of the longer portion to the shorter one also equals g. By comparison with (5.13) we see that the expansion coefficients in the continued fraction of g, being all 1, are 2.68... times smaller than the geometric mean over (almost) all numbers.

Golden *rectangles* have played a prominent role in the pictorial arts, and Fig. 5.2 illustrates the numerous appearances of g in a painting by Seurat. Figure 5.3 shows an infinite sequence of "golden rectangles" in which the sides have ratio g. To construct this design, lop off a square from each golden rectangle to obtain the next smaller golden rectangle.

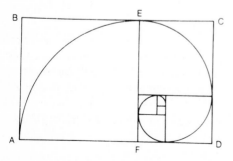

Fig. 5.3. The beginning of an infinite sequence of Golden rectangles. Some sea shells are said to use this construction

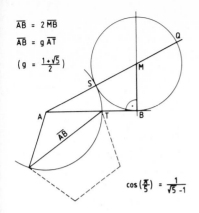

$$\overline{AB} = 2\overline{MB}$$
$$\overline{AB} = g\,\overline{AT}$$
$$(g = \tfrac{1+\sqrt{5}}{2})$$

$$\cos\left(\tfrac{\pi}{5}\right) = \tfrac{1}{\sqrt{5}-1}$$

Fig. 5.4. The Golden ratio and the construction of the regular pentagon

The Golden ratio, involving as it does the number 5 — a Fermat prime — is also related, not surprisingly, to the regular pentagon, as illustrated in Fig. 5.4. It is easily verified (from Pythagoras) that $\overline{AB}/\overline{AT} = g$. Thus, the Golden ratio emerges as the ratio of the diagonal of the regular pentagon to its side.

Finally, the Golden ratio g emerged as the noblest of "noble numbers", the latter being defined by those (irrational) numbers whose continued fraction expansion ends in infinitely many 1's. In fact, the CF of g, see (5.22), has *only* 1's, whence also its nickname "the most irrational number" (because no irrational has a CF approximation that converges more slowly than that for g).

The designation *noble numbers* stems from the fact that in many nonlinear dynamical systems "winding numbers" (the frequency ratios of orbits in phase space) that equal noble numbers are the most resistant against the onset of chaotic motion, which is ubiquitous in nature. (Think of turbulence — or the weather, for that matter.)

Cassini's divisions in the rings of Saturn are a manifestation of what happens when, instead of noble numbers, base numbers reign: rocks and ice particles constituting the rings, whose orbital periods are in simple rational relation with the periods of other satellites of Saturn, are simple swept clean out of their paths by the resonance effects between commensurate orbital periods. In fact, the very stability of the entire solar system depends on the nobility of orbital period ratios.

A double pendulum in a gravitational field is a particularly transparent nonlinear system. As the nonlinearity is increased (by slowly "turning on" the gravitation), the last orbit to go chaotic is the one with a winding number equal to $1/(1+g) = [0; 2, 1, 1, 1, ...]$, a *very* noble number!

For physical systems a winding number $w < 1$ is often equivalent to the winding number $1 - w$. Suppose $w = [0; a_1, a_2, ...]$, what is the CF for $1 - w$? the reader will find it easy to show that

$$1 - [0; a_1, a_2, ...] = [0; 1, a_1 - 1, a_2, ...] . \tag{5.23}$$

Thus, if w is noble, so is $1 - w$. (Note that if $a_1 - 1 = 0$, we need to invoke the rule

$$[... a_m, 0, a_{m+2}, ...] = [..., a_m + a_{m+2}, ...] ,$$

which assures that the CF for $1 - (1-w)$ equals that for w.

Among the most exciting nonlinear systems where CF expansions have led to deep insights are the *fractional quantization* in a two-dimensional electron gas, see *F.D.M. Haddani,* Phys. Rev. Lett. **51**, 605-607 (1983); and "Frustrated instabilities" in active optical resonators (lasers), see *K. Ikeda* and *M. Mitsumo,* Phys. Rev. Lett. **53**, 1340-1343 (1984).

5.4 Electrical Networks and Squared Squares

One of the numerous practical fields where CFs have become entrenched — and for excellent reasons — are electrical networks.

What is the input impedance Z of the "ladder network" shown in Fig. 5.5 when the R_k are "series" impedances and the G_k are "shunt" admittances? A moment's thought will provide the answer in the form of a CF:

$$Z = [R_0; G_1, R_1, G_2, R_2,...] .$$

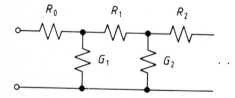

. . . Fig. 5.5. Electrical ladder network. Input impedance is given by continued fraction $[R_0; G_1, R_1, G_2, ...]$

Here, in the most general case, the R_k and G_k are complex-valued rational functions of frequency.

If all R_k and G_k are 1-ohm resistors, the final value of Z for an infinite network, also called the characteristic impedance Z_0, will equal

$$g = \frac{1}{2}(1 + \sqrt{5}) \quad \text{ohm} .$$

The application of CFs to electrical networks has, in turn, led to the solution of a centuries-old teaser, the so-called Puzzle of the Squared Square, i.e., the problem of how to divide a square into unequal squares with integral

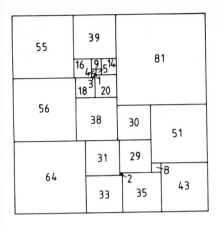

Fig. 5.6. The first squared square, a solution based on the theory of electrical networks and continued fractions (courtesy E. R. Wendorff)

sides. This problem had withstood so many attacks that a solution was widely believed impossible [5.5]. Thus, the first solution, based on network theory, created quite a stir when it appeared (Fig. 5.6).

5.5 Fibonacci Numbers and the Golden Ratio

Another close relative of CFs are the Fibonacci numbers [5.6], defined by the recursion

$$F_n = F_{n-1} + F_{n-2} , \quad \text{with } F_0 = 0 \text{ and } F_1 = 1 , \tag{5.24}$$

which is identical with the CF recursion for the case $b_k = 1$. The first Fibonacci numbers are 0, 1, 1, 2, 3, 5, 8, 13, 21, 34, 55, 89, ... where each number is the sum of its two predecessors.

The ratio of two successive F_n approaches the Golden ratio $g = (1 + \sqrt{5})/2$, which is easily verified in terms of CFs. From the recursion (5.24) it follows that

$$F_{n+1}/F_n = \underbrace{[1; 1, ..., 1]}_{n-1 \; 1's} \quad (n > 1) , \tag{5.25}$$

where the right-hand side of (5.25) is the approximating fraction to the Golden ratio. Equation (5.25) also implies that successive F_n are coprime:

$$(F_n, F_{n+1}) = 1 , \quad n > 0 \tag{5.26}$$

Also, the product of F_n $(n > 1)$ and its predecessor differs by ± 1 from the product of their two neighbors:

$$F_{n-1} F_n - F_{n-2} F_{n+1} = (-1)^n .$$
(5.27)

Examples: $21 \cdot 34 = 13 \cdot 55 - 1$; $34 \cdot 55 = 21 \cdot 89 + 1$.

The reader may wish to prove his prowess by proving these simple statements. Equations such as (5.27) often provide quick answers to a certain class of problems such as the "banking" puzzle described in Sect. 5.11.

A simple alternative recursion for F_n is

$$F_n = 1 + \sum_{k=1}^{n-2} F_k .$$
(5.28)

Because of the internal structure of the F_n, which relates each F_n to its *two* predecessors, the odd-index F_n can be obtained from the even-index F_n alone:

$$F_{2n+1} = 1 + \sum_{k=1}^{n} F_{2k}.$$
(5.29)

It is sometimes said that there is no *direct* (nonrecursive) formula for the F_n, meaning that all predecessors F_k, with $k < n$, have to be computed first. This statement is true, however, only if we restrict ourselves to the integers. If we extend our number field to include square roots, we get the surprising direct formula, discovered by A. de Moivre in 1718 and proved ten years later by *Nicolas Bernoulli:*

$$F_n = \frac{1}{\sqrt{5}} \left[\left[\frac{1}{2}(1 + \sqrt{5}) \right]^n - \left[\frac{1}{2}(1 - \sqrt{5}) \right]^n \right] ,$$
(5.30)

and if we admit even complex transcendental expressions, we obtain a very compact formula:

$$F_n = i^{n-1} \frac{\sin(nz)}{\sin z} , \quad z = \frac{\pi}{2} + i \ln \left[\frac{1+\sqrt{5}}{2} \right] .$$
(5.31)

In (5.30), the first term grows geometrically, while the second term alternates in sign and decreases geometrically in magnitude because

$$-1 < \frac{1}{2}(1 - \sqrt{5}) < 0 .$$

In fact, the second term is so small, even for small n, that it can be replaced by rounding the first term to the nearest integer:

$$F_n = \left\lfloor \frac{1}{\sqrt{5}} \left[\frac{1}{2}(1 + \sqrt{5}) \right]^n + \frac{1}{2} \right\rfloor .\qquad (5.32)$$

A little pocket calculator program based on (5.32) is given in the Appendix.

The result (5.30) is most easily obtained by solving the homogeneous *difference equation* (5.24) by the *Ansatz*

$$F_n = x^n .\qquad (5.33)$$

This converts the difference equation into an algebraic equation:

$$x^2 = x + 1 .\qquad (5.34)$$

(This is akin to solving *differential* equations by an *exponential Ansatz*.)

The two solutions of (5.34) are

$$x_1 = \frac{1}{2}(1 + \sqrt{5}) \quad \text{and} \quad x_2 = \frac{1}{2}(1 - \sqrt{5}) .\qquad (5.35)$$

The general solution for F_n is then a linear combination:

$$F_n = a\, x_1^n + b\, x_2^n ,\qquad (5.36)$$

where with the initial conditions $F_0 = 0$ and $F_1 = 1$,

$$a = -b = \frac{1}{\sqrt{5}} .\qquad (5.37)$$

Equations (5.35-37) taken together yield the desired nonrecursive formula (5.30).

Equation (5.30) can be further compacted by observing that $x_1 = 1/x_2 = g$, so that

$$\sqrt{5}\, F_n = g^n - (-g)^{-n} .\qquad (5.38)$$

The right side of (5.38) can be converted into a trigonometric function by setting

$$t = i \ln g ,\qquad (5.39)$$

yielding

$$\sqrt{5} \; F_n = 2i^{n-1} \sin\left(\frac{\pi}{2} + t\right)n \; , \tag{5.40}$$

which is identical with (5.31), because

$$\sin\left(\frac{\pi}{2} + i \, \ln g\right) = \frac{1}{2}\sqrt{5} \; , \tag{5.41}$$

a noteworthy formula in itself.

There are also numerous relations between the binomial coefficients and the Fibonacci numbers. The reader might try to prove the most elegant of these:

$$F_{n+1} = \sum_{k=0}^{\lfloor n/2 \rfloor} \binom{n-k}{k} \; . \tag{5.42}$$

In other words, summing diagonally upward in Pascal's triangle yields the Fibonacci numbers. (Horizontal summing, of course, gives the powers of 2.)

There is also a suggestive matrix expression for the Fibonacci numbers:

$$\begin{pmatrix} 1 & 1 \\ 1 & 0 \end{pmatrix}^n = \begin{pmatrix} F_{n+1} & F_n \\ F_n & F_{n-1} \end{pmatrix} \; , \tag{5.43}$$

which is obviously true for $n = 1$ and is easily proved by induction. Since the determinant on the left equals -1, it follows immediately that

$$F_{n+1} \, F_{n-1} - F_n^2 = (-1)^n \; , \tag{5.44}$$

which generalizes to

$$F_{n+k} \, F_{m-k} - F_n \, F_m = (-1)^n \, F_{m-n-k} \, F_k \; , \tag{5.45}$$

where any negative-index F_n are defined by the "backward" recursion

$$F_{n-1} = F_{n+1} - F_n \; , \tag{5.46}$$

giving

$$F_{-n} = -(-1)^n \, F_n \; . \tag{5.47}$$

By adding Fibonacci numbers, the positive integers can be represented *uniquely*, provided each F_n ($n > 1$) is used at most once and no two adjacent F_n are ever used. Thus, in the so-called *Fibonacci number system*,

$$3 = 3$$
$$4 = 3 + 1$$
$$5 = 5$$
$$6 = 5 + 1$$
$$7 = 5 + 2$$
$$8 = 8$$
$$9 = 8 + 1$$
$$10 = 8 + 2$$
$$11 = 8 + 3$$
$$12 = 8 + 3 + 1$$
$$1000 = 987 + 13 \text{ etc.}$$

A simple algorithm for generating the Fibonacci representation of m is to find the largest F_n not exceeding m and repeat the process on the difference $m - F_n$ until this difference is zero.

The Fibonacci number system answers such questions as to where to find 0's or 1's or double 1's in the following family of binary sequences:

0

1

1 0

1 0 1

1 0 1 1 0   etc.,

where the next sequence is obtained from the one above by appending the one above *it*. (See Sect. 30.1.)

Another application of the Fibonacci number system is to nim-like games: from a pile of n chips the first player removes any number $m_1 \neq n$ of chips; then the second player takes $0 < m_2 \leq 2m_1$ chips. From then on the players alternate, never taking less than 1 or more than twice the preceding "grab." The last grabber wins.

What is the best first grab? We have to express n in the Fibonacci system:

$$n = F_{k_1} + F_{k_2} + ... + F_{k_r} .$$

The best initial move is then to take

$$F_{k_j} + \ldots + F_{k_r}$$

chips for some j with $1 \leqslant j \leqslant r$, provided $j = 1$ or

$$F_{k_{j-1}} > 2(F_{k_j} + \ldots + F_{k_r}) \ .$$

Thus, for $n = 1000$, the first player should take 13 chips — the *only* lucky number in this case: only for $m_1 = 13$ can he force a victory by leaving his opponent a Fibonacci number of chips, making it impossible for the second player to force a win.

5.6 Fibonacci, Rabbits and Computers

Fibonacci numbers abound in nature. They govern the number of leaves, petals and seed grains of many plants (see Fig. 5.7 [5.7,8]), and among the bees the number of ancestors of a drone n generations back equals F_{n+1} (Fig. 5.8).

Rabbits, not to be outdone, also multiply in Fibonacci rhythm if the rules are right: offspring beget offspring every "season" except the first after birth — and they never die (Fig. 5.9). As already mentioned, this was the original Fibonacci problem [5.9] considered in 1202 by Fibonacci himself.

Leonardo da Pisa, as Fibonacci was also known, was a lone star of the first magnitude in the dark mathematical sky of the Middle Ages. He travelled widely in Arabia and, through his book *Liber Abaci,* brought the Hindu-

Buttercups :	5 petals
Lilies and irises:	3 petals
Some delphiniums:	8 petals
Corn marigolds :	13 petals
Some asters :	21 petals
Daisies :	34,55 and 89 petals

Fig. 5.7.   Flowers have petals equal to Fibonacci numbers

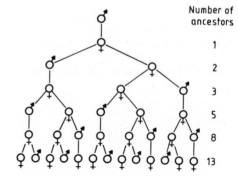

Fig. 5.8. Bees have Fibonacci-number ancestors

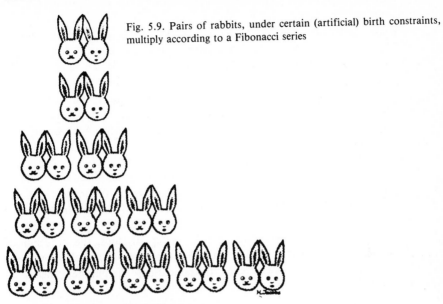

Fig. 5.9. Pairs of rabbits, under certain (artificial) birth constraints, multiply according to a Fibonacci series

Fig. 5.10. Leonardo da Pisa, widely known as Fibonacci ("blockhead"), the great mathematical genius of the Middle Ages – a mathematical dark age outside the Middle East (and the Middle Kingdom!)

Arabic number system and other superior methods of the East to Europe. He is portrayed in Fig. 5.10.

Fibonacci numbers also crop up in computer science and artificial languages. Suppose there is a "language" with variables $A, B, C, \ldots$ and functions of one or two variables $A(B)$ or $A(B, C)$. If we leave out the parentheses, how many ways can a string of n letters be parsed, i.e., grammatically decomposed without repeated multiplication? For a string of three letters, there are obviously two possibilities: $A \cdot B(C)$ and $A(B, C)$. In general, the answer is F_n ways, or so says Andrew Koenig of Bell Laboratories [5.10].

Another area in which Fibonacci numbers have found useful application is that of efficient sequential search algorithms for unimodal functions. Here the kth interval for searching is divided in the ratio of Fibonacci numbers F_{n-k}/F_n, so that after the $(n-1)$st step, the fraction of the original interval (or remaining uncertainty) is $1/F_n \simeq \sqrt{5}/[(1 + \sqrt{5})/2]^n$ as opposed to $(1/\sqrt{2})^n$ for "dichotomic" sequential search. After 20 steps, the precision of the Fibonacci-guided search is 6.6 times higher than the dichotomic one [5.11]. For an extensive treatment of applications of number theory in numerical analysis, see [5.12].

5.7 Fibonacci and Divisibility

It can be proved by induction that

$$F_{n+m} = F_m F_{n+1} + F_{m-1} F_n ,$$

(5.48)

which for $m = 2$ is our fundamental recursion (5.24).

By choosing m as a multiple of n, one can further infer that F_{nk} is a multiple of F_n (and of F_k).

Example: $F_{30} = 832040$, which is divisible by $F_{15} = 610$, $F_{10} = 55$, $F_6 = 8$, $F_5 = 5$, etc.

In other words, every third F_n is even, every fourth F_n is divisible by $F_4 = 3$, every fifth F_n by $F_5 = 5$, etc. As a consequence, all F_n for composite n (except $n = 4$) are composite. However, not all F_p are prime. For example, $F_{53} = 953 \cdot 55945741$.

In 1876 Lucas showed even more, namely that, magically, the two operations "take GCD" and "compute Fibonacci" commute:

$$(F_m, F_n) = F_{(m,n)} ,$$

(5.49)

a "magic" that can be proved with the help of Euclid's algorithm.

Example: $(F_{45}, F_{30}) = (1134903170, 832040) = 610 = F_{15}$.

One of the most interesting divisibility properties of the Fibonacci numbers is that for *each* prime p, there is an F_n such that p divides F_n. More specifically, $p \neq 5$ divides either F_{p-1} [for $p \equiv \pm 1 \pmod 5$] or F_{p+1} [for $p \equiv \pm 2 \pmod 5$]. And of course, for $p = 5$ we have $p = F_p$. In fact, *every integer* divides some F_n (and therefore infinitely many).

Also, for odd prime p,

$$F_p \equiv 5^{\frac{p-1}{2}} \pmod p \tag{5.50}$$

holds.

5.8 Generalized Fibonacci and Lucas Numbers

By starting with initial conditions different from $F_1 = F_2 = 1$, but keeping the recursion (5.24), one obtains the generalized Fibonacci sequences, which share many properties with the Fibonacci sequences proper. The recursion for the generalized Fibonacci sequence G_n, in terms of its initial values G_1 and G_2 and the Fibonacci numbers, is

$$G_{n+2} = G_2 F_{n+1} + G_1 F_n . \tag{5.51}$$

Of course, for $G_1 = G_2 = 1$ the original Fibonacci sequence is obtained. For some initial conditions, there is only a shift in the index, as for example with $G_1 = 1$ and $G_2 = 2$.

However, for $G_1 = 2$ and $G_2 = 1$, one obtains a different sequence:

$$2, 1, 3, 4, 7, 11, 18, 29, \dots , \tag{5.52}$$

the so-called Lucas sequence [5.6]. Of course, obeying the same recursion as the Fibonacci numbers, the ratio of successive Lucas numbers also approaches the Golden ratio. However, they have "somewhat" different divisibility properties.

For example, statistically only *two out of three primes* divide some Lucas number. This result is deeper than those on the divisibility of Fibonacci numbers that we mentioned; it was observed in 1982 by Jeffrey Lagarias [5.13].

A closed form for the Lucas numbers is

$$L_n = g^n + \left(-\frac{1}{g} \right)^n , \quad \text{with} \quad g = \frac{1}{2}(1 + \sqrt{5}) , \tag{5.53}$$

where the second term is again alternating and geometrically decaying. This suggests the simpler formula obtained by rounding to the nearest integer:

$$L_n = \left\lfloor g^n + \frac{1}{2} \right\rfloor , \quad n \geqslant 2 . \tag{5.54}$$

Lucas numbers can be used to advantage in the calculation of large even Fibonacci numbers by using the simple relation

$$F_{2n} = F_n \, L_n \tag{5.55}$$

to extend the accuracy range of limited-precision (noninteger arithmetic) calculators. Similarly, we have for the even Lucas numbers

$$L_{2n} = L_n^2 - 2(-1)^n . \tag{5.56}$$

It is for this reason that the Lucas numbers, computed according to (5.54), are included in the calculator program in the Appendix. For odd-index F_n, one can use

$$F_{2n+1} = F_n^2 + F_{n+1}^2 \tag{5.57}$$

to reach higher indices.

The "decimated" Lucas sequence

$$\tilde{L}_n = L_{2^n} , \qquad \text{i.e.,}$$

$$3, 7, 47, 2207, 4870847, \dots ,$$

for which the simple recursion (5.56) $\tilde{L}_{n+1} = \tilde{L}_n^2 - 2$ holds, plays an important role in the primality testing of Mersenne numbers M_p with $p = 4k + 3$ (see Chap. 3 for the more general test).

It is not known whether the Fibonacci or Lucas sequences contain infinitely many primes. However, straining credulity, R. L. Graham [5.14] has shown that the generalized Fibonacci sequence with

$$G_1 = 1786 \; 772701 \; 928802 \; 632268 \; 715130 \; 455793$$
$$G_2 = 1059 \; 683225 \; 053915 \; 111058 \; 165141 \; 686995$$

contains no primes at all!

Another generalization of Fibonacci numbers allows more than two terms in the recursion (5.24). In this manner kth order Fibonacci numbers $F_n^{(k)}$ are defined that are the sum of the k preceding numbers with the initial conditions $F_o^{(k)} = 1$ and $F_n^{(k)} = 0$ for $n < 0$. For $k = 3$, the 3rd order Fibonacci numbers sequence starts as follows (beginning with $n = -2$): 0, 0, 1, 1, 2, 4, 7, 13, 24, 44, ...

Generalized Fibonacci numbers have recently made their appearance in an intriguing railroad switch yard problem solved by Ma Chung-Fan of the Institute of Mathematics in Beijing (H. O. Pollak, personal communication). In a freight classification yard, a train arrives with its cars in more or less random order, and before the train leaves the yard the cars must be recoupled in the order of destination. Thus, the cars with the nearest destination should be at the front of the train so they can simply be pulled off when that destination is reached, those with the second stop as destination should be next, etc. Recoupling is accomplished with the aid of k spur tracks, where usually $4 \leqslant k \leqslant 8$. The initial sequence of cars is decomposed into $\leqslant k$ subsequences by backing successive cars onto the various spurs, and the subsequences can then be recombined in an arbitrary order as segments in a new sequence. For any initial sequence, the desired rearrangement should be accomplished with a minimum number of times that a collection of cars is pulled from one of the spurs. These are called "pulls." For example, if 10 cars with possible destinations 1 through 7 are given in the order 6324135726, we wish to get them into the order 1223345667. On two tracks, this can be done by first backing

341526 onto the first track,

6237 onto the second track.

Pull both into the order 3415266237 (that's two pulls); then back them onto the two tracks in order

1223 onto the first track,

345667 onto the second track.

Then pull the first track's content onto the second, and pull out the whole train in the right order. Thus, it takes 4 pulls on 2 tracks to get the train together.

Define the *index* $m(\sigma)$ of the sequence $\sigma = 6324135726$ as follows: Start at the leftmost (in this case the only) 1, put down all 1's, all 2's to the right of the last 1, 3's to the right of the last 2 if you have covered *all* the 2's, etc. In this case, the first subset defined in this way is 12 (positions 5 and 9). The next subset takes the other 2 and the second 3 (positions 3 and 6); it can't get to the first 3. The next subset takes the first 3, the 4, the 5, and the second 6; the last subset is 67. Thus 6324135726 has been decomposed into 4 non-decreasing, non-overlapping, non-descending sequences 12, 23, 3456, 67. The

"4" is the index $m(\sigma)$ of the given sequence σ; the general definition is analogous; it is the number of times the ordering comes to the left end of the sequence.

Now Ma showed that the minimum number of pulls in which a sequence σ can be ordered on k tracks is the integer j such that

$$F_{j-1}^{(k)} < m(\sigma) \leqslant F_j^{(k)} .$$

Fibonacci would be delighted!

5.9 Egyptian Fractions, Inheritance and Some Unsolved Problems

A rich sheik, shortly before his death (in one of his limousines; he probably wasn't buckled up) bought 11 identical cars, half of which he willed to his eldest daughter, one quarter to his middle daughter, and one sixth to his youngest daughter. But the problem arose how to divide the 11 cars in strict accordance with the will of the (literally) departed, without smashing any more cars. A new-car dealer offered help by lending the heirs a brand-new identical vehicle so that each daughter could now receive a whole car: the eldest 6, the middle 3 and the youngest 2. And lo and behold, after the girls (and their retinues) had driven off, one car remained for the dealer to reclaim!

The problem really solved here was to express $n/(n+1)$ as a sum of 3 *Egyptian fractions,* also called *unit fractions:*

$$\frac{n}{n+1} = \frac{1}{a} + \frac{1}{b} + \frac{1}{c} . \tag{5.58}$$

In the above story, $n = 11$ and $a = 2, b = 4, c = 6$.

Interestingly, for $n = 11$, there is another solution (and potential story) with $a = 2, b = 3, c = 12$ because *two* subsets of the divisors of 12 (1,2,3,4,6,12) add to 11. Check: $2 + 3 + 6 = 1 + 4 + 6 = 11$. Check! The inheritance problem is related to *pseudoperfect* numbers, defined as numbers equal to a sum of a subset of their divisors [5.15].

For 3 heirs and 1 borrowed car there are only 5 more possible puzzles, the number of cars being $n = 7, 17, 19, 23$ and 41 [5.16].

As opposed to continued fractions, unit fractions are of relatively little use (other than in tall tales of inheritance, perhaps). In fact, they probably set back the development of Egyptian mathematics incalculably. However, they do provide fertile ground for numerous unsolved problems in Diophantine analysis [5.15].

Of special interest are sums of unit fractions that add up to 1. Thus, for example, it is not known what is the smallest possible value of x_n, called $m(n)$, in

$$\sum_{k=1}^{n} \frac{1}{x_k} = 1, \quad x_1 < x_2 < \cdots < x_n .$$ (5.59)

It is easy to check that $m(3) = 6$, $m(4) = 12$ and $m(12) = 120$. But what is the general law? Is $m(n) < cn$ for some constant c? Unknown!

Is $x_{k+1} - x_k \leqslant 2$ ever possible for all k? Erdös in [5.15] thinks not and offers ten (1971?) dollars for the solution.

Graham [5.17] was able to show that for $n > 77$, a partition of n into distinct positive integers x_k can always be found so that $\sum 1/x_k = 1$.

5.10 Farey Fractions

Another kind of fraction, the *Farey Fractions* have recently shown great usefulness in number theory [5.18].

For a fixed $n > 0$, let all the reduced fractions with nonnegative denominator $\leqslant n$ be arranged in increasing order of magnitude. The resulting sequence is called the Farey sequence of order n or belonging to n.

Example: for $n = 5$, in the interval [0,1] we have:

$$\frac{0}{1}, \frac{1}{5}, \frac{1}{4}, \frac{1}{3}, \frac{2}{5}, \frac{1}{2}, \frac{3}{5}, \frac{2}{3}, \frac{3}{4}, \frac{4}{5}, \frac{1}{1} .$$ (5.60)

For other intervals, the Farey fractions are congruent modulo 1 to the Farey fractions in (5.60). In the interval $(c, c+1]$ there are exactly $\sum_{b=1}^{n} \phi(b) \approx 3n^2/\pi^2$ Farey fractions [see Chap. 8 for the definition of $\phi(n)$].

Calling two successive Farey fractions a/b and c/d, then

$$b + d \geqslant n + 1 , \quad \text{and}$$ (5.61)

$$cb - ad = 1 , \quad for \quad \frac{a}{b} < \frac{c}{d} .$$ (5.62)

One of the outstanding properties of the Farey fractions is that given any real number x, there is always a "nearby" Farey fraction a/b belonging to n such that

$$\left| x - \frac{a}{b} \right| \leqslant \frac{1}{b(n+1)} .$$ (5.63)

Thus, if $b > n/2$ the approximating error (5.63) is bounded by $2/n^2$. This compares well with the approximate approximating error $\pi^2/12n^2$ which would result if the Farey fractions were completely uniformly distributed.

What is the spectrum (Fourier transform) of the process defined by (5.63) when x goes uniformly from 0 to 1?

The following recursion provides a convenient method of generating the Farey fractions x_i/y_i of order n: Set $x_0 = 0$, $y_0 = x_1 = 1$ and $y_1 = n$. Then

$$x_{k+2} = \left\lfloor (y_k + n)/y_{k+1} \right\rfloor x_{k+1} - x_k .$$

$$y_{k+2} = \left\lfloor (y_k + n)/y_{k+1} \right\rfloor y_{k+1} - y_k . \tag{5.64}$$

The *mediant* of two fractions $\dfrac{a}{b}$ and $\dfrac{e}{f}$ is defined by

$$mediant \left[\frac{a}{b}, \frac{e}{f} \right] := \frac{a+e}{b+f} , \tag{5.65}$$

which lies in the interval $\left[\dfrac{a}{b}, \dfrac{e}{f} \right]$. Each term in a Farey series $\cdots \dfrac{a}{b}, \dfrac{c}{d}, \dfrac{e}{f}, \cdots$ is the mediant of its two neighbors:

$$\frac{c}{d} = \frac{a+e}{b+f} . \tag{5.66}$$

In fact, the mediant of any two terms is contained in the Farey series, unless the sum of their (reduced) denominators exceeds the order n of the series.

There is also an interesting geometrical interpretation of Farey fractions in terms of point lattices, especially the fundamental point lattice consisting of all integer pairs (x,y). The Farey fractions a/b belonging to n are precisely all those lattice points $(x = a, y = b)$ in the triangle defined by $y = 0$, $y = x$, $y = n$ which can be "seen" from the origin $x = y = 0$, or, equivalently, which can "see" the origin with no other "Farey points" lying on the line of sight (see also Fig. 4.8).

Farey fractions are useful in rational approximations. Continued fractions give the excellent approximation

$$\frac{1}{\pi} \approx \frac{113}{355} .$$

But suppose we want to construct mechanical gears in the approximate ratio $\pi : 1$ using fewer than 100 teeth on the smaller of the two gears. Continued fractions would then give us

$$\frac{1}{\pi} \approx \frac{7}{22} \, ,$$

but we can do better with Farey fractions. In a table published by the London Royal Society [5.19] of the Farey series of order 1025 we find near 113/355 the entries

$$\frac{99}{311}, \frac{92}{289}, \frac{85}{267}, \frac{78}{245}, \frac{71}{223} \text{ and } \frac{64}{201} \, ,$$

any one of which is a better approximation than 7/22.

Or suppose we want one of the gears to have 2^n teeth. We find in the table

$$\frac{1}{\pi} \approx \frac{163}{512} \, ,$$

with an error of $1.5 \cdot 10^{-4}$. This table is of course quite voluminous, having a total of 319,765 entries (and a guide to locate the fraction nearest to any given number in the interval (0,1) quickly). (With $\sum_{b=1}^{n} \phi(b) \approx (3/\pi^2) \, n^2$, (see Chap. 8) we expect about 320,000 Farey fractions of order 1025.)

Another important practical application of Farey fractions implied by (5.62) is the solution of Diophantine equations (see Chap. 7). Suppose we are looking for a solution of

$$243b - 256a = 1 \tag{5.67}$$

in integer a and b. By locating the Farey fraction just below 243/256, namely 785/827, we find $a = 785$ and $b = 827$. Check: $243 \cdot 827 = 200961$ and $256 \cdot 785 = 200960$. Check!

Of course, we can reduce the above solution for a modulo 243 (Chap. 6) giving the smallest positive solution $a = 56$ and $b = 59$. Thus, a table of Farey fractions of a given order n contains *all* integer solutions to equations like (5.67) with coefficients smaller than n.

Another, and quite recent, application of Farey series is the recovery of undersampled periodic (or nearly periodic) waveforms [5.20]. If we think of "nearly periodic waveforms" as a line-scanned television film, for example, then for most pictorial scenes there are similarities between adjacent picture elements ("pixels"), between adjacent scan lines, and between successive image frames. In other words, the images and their temporal sequence carry redundant information (exception: the proverbial "snowstorm").

Because of this redundancy, such images can, in general, be reconstructed even if the image is severely "undersampled," i.e., if only every nth pixel ($n \gg 1$) is preserved and the others are discarded. The main problem in the reconstruction is the close approximation of the ratio of the sampling period to one of the quasi periods in the sampled information by a rational number with a given maximal denominator — precisely the problem for which Farey fractions were invented!

5.11 Fibonacci and the Problem of Bank Deposits

There is an interesting family of problems, appearing in many guises, to which Fibonacci numbers provide a quick solution. Suppose Bill, a wealthy Texan chemist, opens a new bank account with x_1 dollars. The next (business) day he deposits x_2 dollars, both integer dollar amounts. Thereafter his daily deposits are always the sum of the previous two deposits. On the nth day Bill is known to have deposited x_n dollars. What were the original deposits?

A solution of this problem, posed by L. A. Monzert (cf. *Martin Gardner* [5.21]), argues that, for sufficiently large n, successive deposits should be in the Golden ratio. This reasoning permits one to find the $(n-1)$th deposit x_{n-1} and, together with x_n, by *backward* recursion, all prior ones.

However, with the knowledge gained in this chapter, we can find a *direct* answer to this financial problem, one that is valid even for small n. Since the recursion rule for x_n is like that for the Fibonacci numbers, the x_n must be expressible as a linear combination of Fibonacci numbers. In fact, two such terms suffice:

$$x_n = a\, F_{n+k} + b\, F_{n+m} . \tag{5.68}$$

With $F_0 = 0$ and $F_{-1} = F_1 = 1$, the initial conditions are satisfied by

$$x_n = x_1 F_{n-2} + x_2 F_{n-1} . \tag{5.69}$$

Now, because of (5.27), an integer solution to (5.69) is given by

$$x_1 = (-1)^n x_n\, F_{n-3} , \tag{5.70}$$

$$x_2 = -(-1)^n x_n\, F_{n-4} .$$

Solutions (not only to mathematical problems) become that simple if one knows and *uses* the proper relations!

However, we are not quite done yet. According to (5.70) one or the other initial deposit is negative; but we want all deposits to be positive of course. Looking at (5.69) we notice that we can add to x_1 any multiple of F_{n-1} as long as we subtract the same multiple of F_{n-2} from x_2. Thus, the general solution is

$$x_1 = (-1)^n x_n \, F_{n-3} + m F_{n-1} \, , \tag{5.71}$$

$$x_2 = -(-1)^n x_n \, F_{n-4} - m F_{n-2} \, .$$

We can now ask for what values (if any!) of m *both* x_1 and x_2 are positive. Or, perhaps, for what value of m x_2 is positive and as small as possible. The answer, which leads to the longest chain of deposits to reach a given x_n, is

$$m = \lfloor -(-1)^n x_n \, \frac{F_{n-4}}{F_{n-2}} \rfloor \, . \tag{5.72}$$

For $x_{20} = 1\,000\,000$ (dollars), (5.72) yields $m = -381966$ and (5.71) gives $x_1 = 154$ and $x_2 = 144$.

If we had asked that the twenty-*first* deposit be one million dollars, (5.72) would have given the same absolute value of m, and with (5.71), $x_2 = 154$ and $x_1 = -10$. In other words, we would have posed an illicit problem.

Here we have, unwittingly, solved a Diophantine equation, of which more in Chaps. 6 and 7.

5.12 Error-Free Computing

One of the overriding problems in contemporary computing is the accumulation of *rounding errors* to such a degree as to make the final result all but useless. This is particularly true if results depend on the input data in a *discontinuous* manner. Think of matrix inversion.

The inverse of the matrix

$$A = \begin{bmatrix} 1 & 1 \\ 1 & 1+\epsilon \end{bmatrix} \tag{5.73}$$

for $\epsilon \neq 0$ equals

$$A^{-1} = \begin{bmatrix} 1+1/\epsilon & -1/\epsilon \\ -1/\epsilon & 1/\epsilon \end{bmatrix} \, . \tag{5.74}$$

An important generalization of a matrix inverse, applicable also to singular matrices, is the *Moore-Penrose inverse* A^+ [5.22]. For nonsingular matrices, the Moore-Penrose inverse equals the ordinary inverse:

$$A^+ = A^{-1} \tag{5.75}$$

As $\epsilon \to 0$ in (5.73), the matrix A becomes singular and the Moore-Penrose inverse no longer equals A^{-1} but can be shown to be

$$A^+ = \frac{1}{4} A .$$

(5.76)

In other words, as $\epsilon \to 0$, the elements of A^+ become larger and larger only to drop discontinuously to $1/4$ for $\epsilon = 0$.

Examples of this kind of sensitivity to small errors abound in numerical analysis. For many computations the only legal results are integers, for example, the coefficients in chemical reaction equations. If the computation gives noninteger coefficients, their values are rounded to near integers, often suggesting impossible chemical reactions.

In some applications of this kind, double-precision arithmetic is a convenient remedy. (The author once had to invoke double precision in a very early (ca. 1959) digital filter, designed to simulate concert hall reverberation, because the sound would refuse to die away when the music stopped.) In other situations, number-theoretic transforms can be used, of which the Hadamard transform (see Chap. 17) is only one example.

However, quite general methods for error-free computing have become available in the recent past, and it is on these that we shall focus attention in this section.

Specifically, we want to sketch a strategy for computing that will not introduce *any* rounding errors whatsoever, no matter how long or complex the computation. How is this possible? Of the four basic mathematical operations, three (addition, subtraction and multiplication) are harmless: if we start with integers, we stay with integers — no rounding problems there. But *division* is a real bugbear. Computers can never represent the fraction $1/7$, for example, in the binary (or decimal system) without error, no matter how many digits are allowed. If we could only do away with division in our computations! Surprisingly, this is in fact possible, as we shall see.

Of course, computers cannot deal with continuous data — both input and output are, by necessity, rational numbers, and the rational numbers we select here to represent both input data and final results are *Farey fractions* of a given order N (see Sect. 5.10). Once we have chosen a large enough value of N to describe adequately the input data of a problem and all of the answers to that problem, then within this precision, *no* errors will be generated or accumulated.

In this application, we shall generalize our definition of Farey fractions a/b of order N, where a and b are coprime, to include negative and improper fractions:

$$0 \leqslant |a| \leqslant N , \quad 0 < |b| \leqslant N .$$

(5.77)

The error-free strategy, in its simplest form, [5.22] then proceeds as follows.

A prime modulus m is selected such that

$$m \geqslant 2N^2 + 1 , \tag{5.78}$$

and each Farey fraction a/b, with $(b,m) = 1$, is mapped into an *integer* k modulo m:

$$k = \langle a\ b^{-1}\rangle_m , \tag{5.79}$$

where the *integer* b^{-1} is the inverse of b modulo m and the acute brackets signify the smallest nonnegative remainder modulo m (see Sect. 1.5). It is in this manner that we have abolished division! The inverse b^{-1} can be calculated by solving the Diophantine equation

$$bx + my = 1 , \tag{5.80}$$

using the Euclidean algorithm (see Sect. 7.2). The desired inverse b^{-1} is then congruent modulo m to a solution x of (5.80).

After this conversion to integers, all calculations are performed in the integers modulo m. For example, for $N = 3$ and $m = 19$, and with $3^{-1} = 13$, the fraction $2/3$ is mapped into $26 \equiv 7$, and the fraction $-1/3$ is mapped into $-13 \equiv 6$. The operation $(2/3) + (-1/3)$ is then performed as $7 + 6 = 13$, which is mapped back into $1/3$, the correct answer.

It is essential for the practical application of this method that fast algorithms be available for both the forward and backward mappings. Such algorithms, based on the Euclidean algorithm, were recently described by *Gregory* and *Krishnamurthy* [5.22], thereby reclaiming error-free computing from the land of pious promise for the real world.

Sometimes *intermediate* results may be in error, but with no consequence for the final result, as long as it is an order-N Farey fraction. For example, for $m = 19$, $2^{-1} = 10$, so that $1/2$ maps into 10, and $(1/3) - (1/2)$ maps into $13 - 10 = 3$, which is the image of 3 — an erroneous result because $-1/6$ is the correct answer! But 3 is still useful as an intermediate result. For example, multiplying 3 by 2 produces 6, which is the image of $-1/3$, the correct result.

For the large values of N that are needed in practical applications, the prime m has to be correspondingly large. Since calculating modulo very large primes is not very convenient, a multiple-modulus residue (or Chinese remainder) system, see Chap. 16, is often adopted. For example, for $N = 4$, the smallest prime not smaller than $2N^2 + 1 = 33$ is 37. Instead, one can calculate with the residues modulo the *two* primes $m_1 = 5$ and $m_2 = 7$, whose product $m = m_1 \cdot m_2 = 35$ exceeds $2N^2 + 1 = 33$. Such calculations, described in Chap. 16, are much more efficient than the corresponding operation in single-modulus systems, the savings factor being proportional to

$m/\Sigma m_i$. For decomposition of large m into many small prime factors, the savings can be so large as to make many otherwise impossible calculations feasible.

Another preferred number system for carrying out the calculations is based on the integers modulo a prime *power*: $m = p^r$. For example, for $N = 17$, the modulus m must exceed 578 and a convenient choice would be $p = 5$ and $r = 4$, so that m equals $5^4 = 625$. There is only one problem with this approach: all fractions whose denominators contain the factor 5 cannot be represented because 5 has no inverse modulo 625. However, an ingenious application of p-adic algebra and finite-length *Hensel-codes* has solved the problem and looks like the wave of the future in error-free computing. We shall attempt a brief description; for details and practical applications the reader is referred to [5.22].

Essentially, what the Hensel codes do is to remove bothersome factors p in the denominators, so that the "purified" fractions do have unique inverses.

For *integers,* the p-adic Hensel codes are simply obtained by "mirroring" the p-ary expansion. With the 5-ary expansion of 14, for example,

$$14 = 2 \cdot 5^1 + 4 \, ,$$

the Hensel code for 14 becomes

$$H(5, 4, 14) = .4200 \, . \tag{5.81}$$

In general, $H(p, r, \alpha)$ is the Hensel code of α to the (prime) base p, having precisely r digits.

A *fraction* a/b whose denominator b does *not* contain the factor p is converted to an integer modulo p^r, which is then expressed as a Hensel code. For example, with $p^r = 5^4 = 625$, we get

$$\frac{1}{16} \stackrel{\wedge}{=} \langle 16^{-1} \rangle_{625} = 586 = 4 \cdot 5^3 + 3 \cdot 5^2 + 2 \cdot 5 + 1 \, ,$$

or in Hensel code:

$$H(5, 4, \frac{1}{16}) = .1234 \, . \tag{5.82}$$

Similarly, with $\langle 3/16 \rangle_{625} = 508 = 4 \cdot 5^3 + 1 \cdot 5 + 3$, becomes

$$H(5, 4, \frac{3}{16}) = .3104 \, . \tag{5.83}$$

Of course, the Hensel code for 3/16 can be obtained directly by multiplying

(5.82) with the code for 3:

$$H(5,4,3) = .3000 , \qquad\qquad (5.84)$$

where the multiplication proceeds from *left to right*.   (Remember, Hensel codes are based on a *mirrored p*-ary notation.)   Thus, $H(5,4,1/16) \times H(5,4,3)$ equals

$$
\begin{array}{r}
.1234 \\
\times\ .3000 \\
\hline
.3142 \\
\text{carries} \qquad 112 \\
\hline
=\quad .3104
\end{array}
$$

which agrees with $H(5,4,3/16)$, see (5.83). Note that any carries beyond four digits (the digit 2 in the above example) are simply dropped. It is ironic that such a "slipshod" code is the basis of *error-free* computation!

If the numerator contains powers of p, the corresponding Hensel code is simply right-shifted, always maintaining precisely r digits. For example,

$$H(5,4,\frac{5}{16}) = 0.0123 . \qquad\qquad (5.85)$$

Powers of p in the denominator are represented by a *left*-shift. Thus, with $H(5,4,1/3) = .2313$,

$$H(5,4,\frac{1}{15}) = 2.313 . \qquad\qquad (5.86)$$

To expand the range of the Hensel codes to arbitrary powers of p in the denominator or numerator, a floating-point notion, $\hat{H}(p,r,\alpha)$, is introduced. For example,

$$\hat{H}(5,4,\frac{1}{15}) = (.2313, -1) , \qquad \text{and} \qquad\qquad (5.87)$$

$$\hat{H}(5,4,375) = (.3000,3) , \qquad\qquad (5.88)$$

where the first number on the right is the mantissa and the second number the exponent.

When multiplying floating-point Hensel codes, their mantissas are multiplied and their exponents are added. For example,

$$\frac{1}{3} \times \frac{6}{5} \stackrel{\wedge}{=} (.2313,0) \times (.1100,-1) = (.2000,-1) ,$$

and with

$$.2313 \times .1100 = .2000 \; ,$$

we obtain

$$(.2313,0) \times (.1100,-1) \; = \; (.2000,-1) \; ,$$

which corresponds to 2/5, the correct answer. (Remember, all operations proceed from left to right and Hensel code .1000 corresponds to 1 and not 1/10.)

Strange and artificial as they are, Hensel codes perform numerical stunts and never slip a single digit.

6. Linear Congruences

Suppose a certain airline is consistently 25 hours late in departure and arrival (this has happened, but no names will be mentioned) while another one, flying the same route, is only 2 to 3 hours late. If you were in a hurry, which airline would you fly — food, lack of leg room and all else being equal? Obviously, being 25 hours late is as good (or bad) as being only 1 hour late. In other words, in a *daily recurring* event an extra day, or even several, makes no difference. The mathematics that deals with this kind of situation is called *modular arithmetic*, because only remainders *modulo* a given integer matter.

Another application of modular arithmetic occurs in wave interference phenomena such as ripples on a lake or patterns of light and dark on a hologram. In all these cases a path difference of, say, half a wavelength is indistinguishable from a path difference of one and a half or two and a half, etc., wavelengths.

And of course, there are many applications in mathematics proper. For example, few people care what n^{560} is, but the remainder of n^{560} when divided by 561 for all n coprime to 561 is a question of some actuality. (As it happens, all these remainders are 1 — a terrible thing to happen, as we shall see.) But first we have to know the ground rules of modular arithmetic, so that n^{560} for, say, $n = 500$, a 1512-digit monster of a number, cannot frighten us.

6.1 Residues

When c divided by m leaves the remainder b (not necessarily positive) we write (following Gauss)

$$c \equiv b \pmod{m} ; \tag{6.1}$$

read: c is congruent b modulo m.

More generally, we define the above congruence as meaning

$$m \mid (c - b) ; \tag{6.2}$$

read: m divides c minus b (without remainder).

Example: $16 \equiv -2 \pmod 9$ implies $9|(16+2)$.

Together with c, all

$$b = mq + c \; ; \quad q = \text{integer} \tag{6.3}$$

belong to the same *residue class* modulo m [6.1].

A *complete residue system* modulo m consists of m integers, one representative each from each residue class. The most common residue systems are the *least nonnegative* residue system modulo m, consisting of the integers 0, 1, 2, ..., $m-1$, and the *least absolute* residue system, consisting of the integers 0, ± 1, ± 2, ..., $\pm (m-1)/2$ (for odd m).

For many purposes, one calculates with the congruence sign for a given modulus as if it were an equal sign.

Addition:

Example: $13 \equiv 4 \pmod 9$

$$16 \equiv -2 \pmod 9$$

$$\overline{} \tag{6.4}$$

$$29 \equiv 2 \pmod 9 \qquad \text{Check!}$$

Multiplication of the two upper congruences results in

$$208 = -8 \pmod 9 \qquad \text{Check!} \tag{6.5}$$

The congruence

$$c \equiv b \pmod m \tag{6.6}$$

can be "cancelled" by the GCD of c, b and m. With $(c,b,m) = d$, we may write

$$\frac{c}{d} \equiv \frac{b}{d} \left(\bmod \frac{m}{d} \right). \tag{6.7}$$

Example: $28 \equiv 4 \pmod 6$ can be converted to $14 \equiv 2 \pmod 3$.

Another useful rule is the following. If

$$mc \equiv mb \pmod n \tag{6.8}$$

and the GCD $(m,n) = d$, then

$$c \equiv b \left(\mod \frac{n}{d} \right).$$
(6.9)

Example: $28 \equiv 4 \pmod 6$ implies $7 \equiv 1 \pmod 3$.

Among the many useful applications of linear congruences is the ancient error-detecting algorithm sailing under the name of "casting out 9's" [6.2]. If we add two decimal numbers column by column, then if in any one column the sum exceeds 9, we reduce the result modulo 10 and add 1 (or 2 or 3, etc.) to the next column. Thus, in terms of the sum of the decimal digits, we have added 1 (or 2 or 3, etc.) and subtracted 10 (or 20 or 30, etc.). We have therefore changed the sum of the digits by a multiple of 9; in other words, the sum of the digits has not changed modulo 9.

Example:

86	sum of digits = 14,	sum of sum of digits = 5
+ 57	sum of digits = 12,	sum of sum of digits = 3
= 143	sum of digits = 8,	

Check: $14 + 12 = 26 \equiv 8 \pmod 9$. Check! Of course in the check we can apply the same rule and consider sums of sums of digits: Check: $5 + 3 = 8 \equiv 8 \pmod 9$. Check!

The same rule also holds for multiplication.

Example:

15	sum of digits = 6
× 17	sum of digits = 8
= 255	sum of digits = 12, sum of sum of digits = 3 .

Check: $6 \times 8 = 48$, sum of digits = 12, sum of sum of digits = 3. Check!

The reason why sums of digits when multiplied give the same result modulo 9 as the numbers themselves is that, trivially, any power of 10 is congruent 1 modulo 9:

$$10 \equiv 1 \pmod 9 \text{ and therefore } 10^k \equiv 1 \pmod 9 \text{ for any } k \geq 0 .$$

The only problem with this ancient error-detecting code is that it can fail to signal an error. In fact, for random errors, about 10% of the errors go undetected. Fortunately, though, the casting-out-9's is not restricted to the decimal system; it works for any base b, casting out $(b-1)$'s. Specifically, it also works for a much neglected (and very simple) number system: the base-100 or "hectic" system. One of the advantages of the hectic system is that it needs no new digits. For example, the year of Gauss's birth in hectic notation looks like this:

17 77,   sum of digits = 94 ,

and that of his death

18 55,   sum of digits = 73 ,

and the difference of these two hectic numbers (his age when he died) is

00 78 ,   sum of digits = 78 .

Check: $73 - 94 = -21 \equiv 78 \pmod{99}$. Check!

This example is perhaps too simple, but with the hectic error-detecting algorithm the undetected error rate has dropped to about 1%.

A simple rule exists also for divisibility by 11. It follows from $10 \equiv -1 \bmod 11$ and $100 \equiv 1 \bmod 11$ that divisibility by 11 of an integer and its digital sum, taken with *alternating* signs, are equivalent. Thus 517, for example, is immediately seen to be divisible by 11 because $5 - 1 + 7 = 11$. If the result of this operation is itself a large number, the operation can of course be repeated until manageable numbers, like 0 or 11, are reached.

6.2 Some Simple Fields

Complete residue systems modulo a prime form a *field*, i.e., a set of numbers for which addition, subtraction, multiplication and division (except by 0) are defined and for which the usual commutative, associative and distributive laws apply [6.3].

For the least nonnegative residue system modulo 2, consisting of 0 and 1 (perhaps the most important one in this computer age) we have the addition table:

$$
\begin{array}{c|cc}
 & 0 & 1 \\
\hline
0 & 0 & 1 \\
1 & 1 & 0
\end{array}
\tag{6.10}
$$

which is isomorphic both to the logical operation "exclusive or" and to multiplication of signs (if we identify 0 with $+$ and 1 with $-$).

The multiplication table for 0 and 1:

$$
\begin{array}{c|cc}
 & 0 & 1 \\
\hline
0 & 0 & 0 \\
1 & 0 & 1
\end{array}
\tag{6.11}
$$

is isomorphic to the logical "and" and the set-theoretic "intersection."

Multiplication for a complete residue system modulo a composite number has no inverse for some of its members, as can be seen from the multiplication table modulo 4:

$$
\begin{array}{c|cccc}
 & 0 & 1 & 2 & 3 \\
\hline
0 & 0 & 0 & 0 & 0 \\
1 & 0 & 1 & 2 & 3 \\
2 & 0 & 2 & 0 & 2 \\
3 & 0 & 3 & 2 & 1
\end{array}
\tag{6.12}
$$

Thus, there is no number which when multiplied by 2 gives 1, i.e., 2 has no inverse. Also, division by 2 is not unique: 2 divided by 2 could be either 1 or 3.

This grave defect is rectified by *prime* residue systems which consist only of those residue classes that are coprime to the modulus. Thus, the least nonnegative prime residue system modulo 10 consists of the integers 1, 3, 7 and 9, and their multiplication table is well behaved:

```
    1  3  7  9
   ┌───────────
 1 │ 1  3  7  9
   │
 3 │ 3  9  1  7                                    (6.13)
   │
 7 │ 7  1  9  3
   │
 9 │ 9  7  3  1
```

If f is a polynomial with integer coefficients, then $a \equiv b \pmod{m}$ implies

$$f(a) \equiv f(b) \pmod{m} .$$ (6.14)

For $(m,n) = 1$, if x and y run through complete residue systems modulo n and m, respectively, then $mx + ny$ runs through a complete residue system modulo mn.

Example: $n = 5$, $m = 3$:

```
     x = 0    1    2    3    4
   ┌──────────────────────────
y=0│   0    3    6    9   12
   │
  1│   5    8   11   14    2
   │
  2│  10   13    1    4    7
```

We will encounter this kind of decomposition of a residue class modulo a product of coprimes again when we discuss simultaneous congruences and the Chinese Remainder Theorem with its numerous applications in Chap. 16.

6.3 Powers and Congruences

What is $2^{340} \pmod{341}$? Obviously, the Chinese did not know or they would not have formulated their primality test (Sect. 2.3). With the aid of the congruence notation we may rewrite the Chinese test as

$$2^n \equiv 2 \pmod{n} ,$$ (6.15)

if *and only if* n is prime. Here the first "if" is all right (see Fermat's Theorem, Chap. 8) but the "and only if" is wrong, as we shall presently demonstrate with the composite $n = 341 = 11 \cdot 31$. For odd n we may write, because of (6.9),

$$2^{n-1} \equiv 1 \pmod{n} .$$

Of course, it is foolish actually to calculate 2^{340}, a 103-digit number, if we are interested only in the remainder modulo 341. Instead we will decompose 340 into $10 \cdot 34$ and first raise 2 to the 10th power, giving 1024. Dividing 1024 by 341 leaves the remainder 1:

$$2^{10} \equiv 1 \pmod{341} . \tag{6.16}$$

Now raising the result to the 34th power is easy because $1^{34} = 1$. Thus,

$$2^{340} \equiv 1 \pmod{341} , \quad \text{or} \tag{6.17}$$

$$341 \mid (2^{341} - 2) , \tag{6.18}$$

in spite of the fact that 341 is composite. Woe to the Chinese and three cheers for the congruence notation! We could even have done this in our heads, without recourse to pencil and paper, and thereby demolished a false "theorem" which had stood undisputed for so many centuries.

Composite numbers which masquerade as primes vis-a-vis Fermat's theorem are called *pseudoprimes*. 341 is actually the smallest pseudoprime to the base 2. In a certain sense, pseudoprimes have become almost as important as actual primes in modern digital encryption. We will hear more about pseudoprimes and even *absolute* and *strong pseudoprimes* in Chap. 19.

Of course, in calculating $2^{340} \pmod{341}$ we were lucky, because 2^{10} already gave the remainder 1. In general we will not be so lucky, and we need a universal algorithm that will see us through any base b or exponent n.

More formally, to calculate $b^n \pmod{m}$, we first find the binary decomposition of n:

$$n = \sum_{k=0}^{\lfloor \log_2 n \rfloor} n_k 2^k \quad \text{with } n_k = 0 \text{ or } 1 , \tag{6.19}$$

where $\lfloor \ \rfloor$ is the Gauss bracket signifying the integer part.

The binary expansion coefficients n_k can be found by any of a variety of "analog"-to-digital conversion algorithms. In the Appendix, we give a little program suitable for pocket calculators for obtaining the n_k.

Omitting all terms with $n_k = 0$ in the sum, we may write

$$n = \sum_{m=1}^{M} 2^{c_m} .$$
(6.20)

In other words we write, for example, $6 = 2 + 4$ or $13 = 1 + 4 + 8$, etc.

Using this decomposition into a sum of powers of 2, we write b^n as follows:

$$b^n = \prod_{m=1}^{M} b^{2^{c_m}} = \underbrace{(\dots (b^2) \dots)^2}_{c_1 \text{ squarings}} \dots \underbrace{(\dots (b^2) \dots)^2}_{c_M \text{ squarings}}$$
(6.21)

In words: we calculate b^n by squaring b c_1 times in succession. Then we square b c_2 times. (Of course, we make use of the previous result, i.e., we need square only $c_2 - c_1$ more times, etc.) Then we multiply the results of all these squarings together to obtain b^n.

Apart from the gain in computational efficiency (if n is a power of 2, then only about $\log_2 n$ squarings are required, instead of n multiplications), the main raison d'être for the repeated squaring algorithm is that if we want the result modulo m, then after each squaring we can reduce the intermediate result modulo m without running the risk of calculator "overflow" (as long as m^2 is smaller than the largest number the machine can handle).

Here is another rule that makes working with powers and congruences easier:

$$(x + y)^n \equiv x^n + y^n \pmod{n} , \quad n \text{ prime (or absolute pseudoprime)} \quad (6.22)$$

i.e., of the $n + 1$ terms obtained upon expanding $(x + y)^n$ binomially only two remain, because all others, being multiplied by $\binom{n}{k}$, with $k \neq 0$ or n, are divisible by n and therefore do not contribute to the end result modulo n.

Incidentally, the condition $k \neq 0$ or n can be expressed with the following more widely applicable and succinct notation:

$$k \not\equiv 0 \pmod{n} .$$

Read: k is *not* congruent to zero modulo n.

7. Diophantine Equations

Diophantine equations, i.e., equations with integer coefficients for which integer solutions are sought, are among the oldest subjects in mathematics. Early historical occurrences often appeared in the guise of *puzzles,* and perhaps for that reason, Diophantine equations have been largely neglected in our mathematical schooling. Ironically, though, Diophantine equations play an ever-increasing role in modern applications, not to mention the fact that some Diophantine problems, especially the unsolvable ones, have stimulated an enormous amount of mathematical thinking, advancing the subject of number theory in a way that few other stimuli have.

Here we shall deal with some of the basic facts and rules and get to know *triangular* and *Pythagorean numbers, Fermat's Last Theorem,* an unsolved conjecture by *Goldbach,* and another conjecture by *Euler* — *one* that was refuted, although it looked quite convincing while it lasted.

7.1 Relation with Congruences

The congruence

$$ax \equiv c \pmod{m} \tag{7.1}$$

has a solution *iff* $(a,m) \mid c$. In fact, there are then (a,m) solutions that are incongruent modulo m [7.1].

Example:

$$3x \equiv 9 \pmod 6. \tag{7.2}$$

With $(3,6) = 3$ and $3 \mid 9$, there are exactly three incongruent solutions: $x = 1, 3$ and 5. Adding or subtracting multiples of 6 to these three solutions gives additional solutions *congruent* to the three already found $(7, 9, 11, 13,$ etc.$)$.

For $(a,m) = 1$, the solution is *unique* modulo m.

The solution of the congruence $ax \equiv c \pmod n$ is identical with solutions in integers x and y of the *Diophantine equation* [7.1]:

$$ax = my + c ,\tag{7.3}$$

so named after Diophantus of Alexandria (ca. 150 A.D.) [7.2].

For $(a,m) = d$ and $m = m'd$, $a = a'd$ and $c = c'd$ [note that (a,m) must divide c for a solution to exist], we can write instead of the above equation

$$a'x = m'y + c' ,\tag{7.4}$$

whose solution is unique modulo m', because $(a',m') = 1$. Additional solutions that are incongruent modulo m are obtained by adding km', where $k = 1,2,...,d-1$.

Diophantine equations also have a geometric interpretation, which is illustrated for $3y = 4x - 1$ in Fig. 7.1. The straight line representing this equation goes through only those points of the two-dimensional integer lattice shown in Fig. 7.1 for which x and y are solutions. In the illustration this is the case for the points $\{x,y\} = \{-2,-3\}$, $\{1,1\}$ and $\{4,5\}$. Additional solutions are obviously given by linear extrapolation with multiples of the difference $\{4,5\} - \{1,1\} = \{3,4\}$.

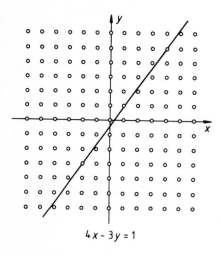

$4x - 3y = 1$

Fig. 7.1. Linear Diophantine equations have simple geometric interpretations: the straight line $3y = 4x-1$ cuts through points $\{-2, -3\}$ and $\{4, 5\}$, two solutions of the equation

7.2 A Gaussian Trick

For $(b,m) = 1$, *Gauss* [7.3] suggested writing the congruence $bx \equiv c \pmod{m}$ as

$$x \equiv \frac{c}{b} \pmod{m}\tag{7.5}$$

and adding or subtracting multiples of m to c and b so that cancellation becomes possible *as if c/b* were, in fact, a fraction.

Example: $27 \, x \equiv 1 \pmod{100}$

Solution:
$$x \equiv \frac{1}{27} \equiv \frac{-99}{27} \equiv \frac{-11}{3} \equiv \frac{-111}{3} \equiv -37 \equiv 63 \pmod{100} \; .$$

Check: $27 \cdot 63 = 1701 \equiv 1 \pmod{100}$. Check!

Another method uses Euclid's algorithm (Sect. 2.7) for solving congruences or Diophantine equations. A congruence is first converted into a Diophantine equation. For example, the congruence

$$15 \, x \equiv 1 \pmod{11} \tag{7.6}$$

has a solution, because $(15,11) = 1$ divides 1. The corresponding Diophantine equation is

$$15 \, x = 11 \, y + 1 \; . \tag{7.7}$$

Solving for y, we obtain

$$y = x + \frac{4x - 1}{11} \; . \tag{7.8a}$$

For (7.8a) to have an integer solution $4x - 1$ must be a multiple of 11:

$$4x - 1 = 11w \; .$$

Now solving for x, we get

$$x = 2w + \frac{3w + 1}{4} \; . \tag{7.8b}$$

By now, the denominator has become so small that a solution is obvious: for x to be an integer, $3w + 1$ must be a multiple of 4, for example $w = 1$. Or, more formally:

$$3w + 1 = 4v \; ,$$

whence

$$w = v + \frac{v - 1}{3} \; . \tag{7.8c}$$

Here an integer solution is even more obvious: $v = 1$.

Solutions for x and y are now obtained by *backward* substitutions: with $v = 1$, (7.8c) gives $w = 1$ (as we noted before); and with $w = 1$, (7.8b) gives $x = 3$ and finally, if we so desire, (7.8a) gives $y = 4$.

Check: $15 \cdot 3 = 11 \cdot 4 + 1 = 45 \equiv 1 \pmod{11}$. Check!

The trick of this method of solution is that, in going from (7.8a) to (7.8b) and (7.8c), we have made the denominators smaller and smaller — just as in the Euclidean algorithm. In fact, the Euclidean algorithm applied to 15/11, the original factors in (7.7), gives precisely the same remainders and denominators as appear in the equations (7.8a-c):

$$15 : 11 = 1 + \frac{4}{11}$$

$$11 : 4 = 2 + \frac{3}{4}$$

$$4 : 3 = 1 + \frac{1}{3} .$$

For numerous calculations with the same modulus, it is most convenient to calculate a *table* of inverses once and for all. Such a table for the modulus 11 would contain, for example, the entry $4^{-1} = 3$. Check: $4 \cdot 3 = 12 \equiv 1 \pmod{11}$. Check!

Thus, the solution of $4x \equiv 7 \pmod{11}$ becomes $x \equiv 4^{-1} \cdot 7 = 21 \equiv 10 \pmod{11}$. Check: $4 \cdot 10 = 40 \equiv 7 \pmod{11}$. Check!

7.3 Nonlinear Diophantine Equations

A simple example of a nonlinear Diophantine equation is

$$x^2 - Ny^2 = \pm 1 , \tag{7.9}$$

where x and y are integers. Two solutions for $+1$ on the right-hand side (the so-called Pell equation) are obviously $x = \pm 1$, $y = 0$. Are there others? For $N = 2$,

$$x^2 - 2y^2 = \pm 1 \tag{7.10}$$

has the solution $x = \pm 3$, $y = \pm 2$. Are there more? Yes, and they are obtained by the continued-fraction (CF) expansion of $\sqrt{N}$. For $N = 2$, we have

$$\sqrt{2} = 1 + \sqrt{2} - 1 , \tag{7.11}$$

$$\frac{1}{\sqrt{2}-1} = 2 + \sqrt{2} - 1 , \tag{7.12}$$

i.e., as we already know, we obtain the periodic CF:

$$\sqrt{2} = [1; \overline{2}] . \tag{7.13}$$

The approximants A_k and B_k are thus obtained recursively from

$$A_k = 2A_{k-1} + A_{k-2} , \tag{7.14}$$

and similarly for the B_k. With the initial conditions $A_0 = A_1 = 1$ and $B_0 = 0$, $B_1 = 1$, we obtain

$$A_k = 1, 1, 3, 7, 17, 41, 99, ...$$
$$B_k = 0, 1, 2, 5, 12, 29, 70, ... \tag{7.15}$$

Here each pair of values (A_k, B_k) corresponds alternately to a solution of $x^2 - 2y^2 = 1$ and $x^2 - 2y^2 = -1$. This is not really too surprising, because we already know that A_k/B_k will tend to $\sqrt{2}$ alternately from above and below. However, the general proof is a bit tedious [7.1].

The first two solutions {1,0} and {1,1} we already know. We will check the third and fourth solutions: $7^2 - 2 \cdot 5^2 = -1$. $17^2 - 2 \cdot 12^2 = 1$. Check!

The CF's of the squareroots of integers are not only periodic; they are also palindromic, i.e., the periods are symmetric about their centers, except for the last number of the period, which equals twice the very first number (left of the semicolon). For example, $\sqrt{29} = [5; 2, 1, 1, 2, 10]$ or $\sqrt{19} = [4; 2, 1, 3, 1, 2, 8]$.

Some periods are very short, such as those of numbers of the form $n^2 + 1$, which have period length 1. For example, $\sqrt{10} = [3; \overline{6}]$ or $\sqrt{101} = [10; \overline{20}]$. But the squareroots of other integers can have rather long periods. For example, $\sqrt{61} = [7; \overline{1,4,3,1,2,2,1,3,4,1,14}]$ has a period length of 11, and $\sqrt{109}$ has a period length of 15. There seems to be no simple formula that predicts long-period lengths.

Solving Pell's equation for such integers is quite tedious, a fact that Fermat exploited when he wrote his friend Frénicle in 1657, rather mischievously, to try $N = 61$ and $N = 109$ "pour ne vous donner pas trop de peine". (The *smallest* solution of the latter problem has 14 and 15 decimal places, respectively. Poor Frénicle!)

7.4 Triangular Numbers

The kth *triangular number* is defined as

$$\Delta_k = 1 + 2 + \dots + k = \frac{1}{2} k(k + 1) . \qquad (7.16)$$

Δ_k is the number of unordered pairs of $k+1$ objects or, more tangibly, the number of handshakings when $k+1$ persons meet (no self-congratulations, please!). The smallest Δ_k are 0, 1, 3, 6, 10, 15, ... Their first differences form a linear progression: 1, 2, 3, 4, 5, ...

On July 10, 1796, Gauss wrote in his still very fresh diary (then in its 103rd day):

$$\text{Eureka! } \quad n = \Delta + \Delta + \Delta , \qquad (7.17)$$

by which he meant that every integer can be represented by the sum of 3 triangular numbers. For example, $7 = 3 + 3 + 1 = 6 + 1 + 0$; $8 = 6 + 1 + 1$; $9 = 3 + 3 + 3 = 6 + 3 + 0$; $10 = 6 + 3 + 1$, etc. What this means is that the Δ_k, although they grow like $k^2/2$, are still distributed densely enough among the integers that three of them suffice to reach any (nonnegative) whole number.

Gauss's discovery implies that every integer of the form $8n + 3$ as a sum of three odd squares, an interesting *nonlinear* Diophantine equation, is always solvable. First we note that the square of an odd number, $2k + 1$, equals 8 times a triangular number plus 1:

$$(2k + 1)^2 = 4k^2 + 4k + 1 = 8\Delta_k + 1 . \qquad (7.18)$$

Hence if

$$n = \Delta_{k_1} + \Delta_{k_2} + \Delta_{k_3} ,$$

then we obtain, in view of Gauss's Eureka discovery, a solution to the following nonlinear Diophantine equation:

$$\sum_{m=1}^{3} (2k_m + 1)^2 = 8n + 3 . \qquad (7.19)$$

Example: $35 = 4 \cdot 8 + 3 = 1 + 9 + 25$.

Finally, there is a connection with the perfect numbers P_p: every even perfect number is also a triangular number:

$$P_p = (2^p - 1)2^{p-1} = (2^p - 1)2^p \frac{1}{2} = \Delta_{2^p - 1} . \qquad (7.20)$$

Triangular numbers can also be illustrated *geometrically* as the number of equidistant points in triangles of different sizes (Fig. 7.2). They were defined this way in antiquity (by the Pythagoreans). These points form a triangular lattice.

In a generalization of this concept, *square numbers* are defined by the number of points in square lattices of increasing size, as illustrated in Fig. 7.3.

Higher n-gonal or *figurate* numbers, such as pentagonal numbers (Fig. 7.4) and hexagonal numbers, are defined similarly. Can the reader derive the general formula for figurate numbers? Calling the kth (beginning with $k = 0$) n-gonal number $g_n(k)$, the answer is

$$g_n(k) = \frac{1}{2}(n-2)k^2 + \frac{1}{2}nk + 1 , \qquad (7.21)$$

which for $n = 4$ gives the square numbers $g_4(k) = (k+1)^2$. Check!

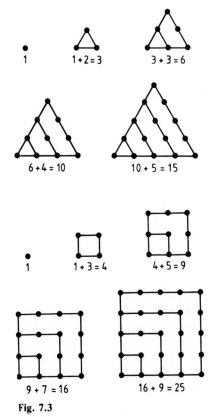

◀ Fig. 7.2. Geometrical interpretation of the triangular numbers (and the reason for their name). Note the simple recursion

Fig. 7.3. The square numbers, their geometrical interpretation and their recursion

Fig. 7.4. The pentagonal numbers, their geometrical interpretation and their recursion. The pentagonal numbers, $n(3n-1)/2$, also play a role in partitioning problems (see Chap. 21)

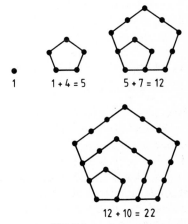

Fig. 7.3

Fig. 7.4

7.5 Pythagorean Numbers

Another nonlinear Diophantine equation is the well-known

$$x^2 + y^2 = z^2 ,$$

(7.22)

expressing Pythagoras's theorem for right triangles. Solutions in integers are called Pythagorean *triplets*, the smallest positive one being 4, 3, 5:

$$4^2 + 3^2 = 5^2 .$$

To avoid redundancy, we shall require x to be even and y to be odd. (If both x and y were even, then z would also be even and the equation could be divided by 4.) All *basic* solutions, i.e., those for which x, y and z do not have a common divisor and x is even, are obtained from the two coprime integers m and n, $m > n > 0$, at least one of which must be even, as follows:

$$x = 2 mn$$

$$y = m^2 - n^2$$

(7.23)

$$z = m^2 + n^2 .$$

It is easy to verify that $x^2 + y^2 = z^2$ and that x is even and $y > 0$ is odd. Of course, z is also odd.

With these conventions, the first case is $m = 2$, $n = 1$, yielding the triplet (4,3;5). The next basic case is $m = 3$, $n = 2$, yielding (12,5;13). The third basic case, $m = 4$, $n = 1$, yields (8,15;17).

Incidentally, at least one pair of basic Pythagorean triplets has the same product xy, meaning that there are (at least) two incongruent right triangles with integer sides and equal areas:

$$m = 5, \quad n = 2 : (20,21;29)$$

$$m = 6, \quad n = 1 : (12,35;37) ,$$

(7.24)

both of which have area 210. Are there other equal-area pairs?

Figure 7.5, prepared by Suzanne Hanauer of Bell Laboratories, shows the x and y values of all (not just the basic) Pythagorean triplets up to $x,y = 52$. The two pronounced straight "lines" are the solutions obtained from the basic triplet (3,4;5). Each plotted point in Fig. 7.5 is one corner of an integer right triangle obtained by connecting it with the origin {0,0} and drawing the normal to the abscissa through it. It is apparent that there is a certain "thinning" out as x and y get larger.

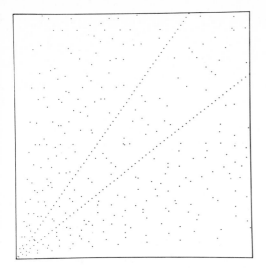

Fig. 7.5. The Pythagorean numbers $x^2 + y^2 = z^2$ in the range $1 \leqslant x \leqslant 52$, $1 \leqslant y \leqslant 52$. The two prominent straight "lines" correspond to the basic triplet $4^2 + 3^2 = 5^2$, the triplet $3^2 + 4^2 = 5^2$ and their multiples

7.6 Exponential Diophantine Equations

Another type of Diophantine equation, in which the unknowns appear in the exponent is exemplified by

$$2^n = 3^m - 1 .$$

(7.25)

The background against which the author encountered this equation is the following. The Fast Fourier Transform (FFT) works most efficiently for data whose length is a power of 2. On the other hand, pseudorandom "maximum-length" sequences, which are ideal for precision measurements (see Chap. 25), have period lengths $p^m - 1$, where p is a prime. In most applications, the preferred sequences are binary, i.e., $p = 2$. Of course, there is no integer solution to the equation $2^n = 2^m - 1$, except the uninteresting one $n = 0$, $m = 1$.

For some physical applications, ternary-valued sequences would also be acceptable. Thus, the question arises whether the equation $2^n = 3^m - 1$ has integer solutions other than the two obvious ones $n = 1$, $m = 1$ and $n = 3$, $m = 2$.[1] Lewi Ben Gerson (1288-1344) proved that these are indeed the only solutions.

---

[1]   In fact, in 1844 *Catalan* [7.4] posed a more general question and conjectured that 2^3 and 3^2 are the only perfect powers that differ by 1.

7.7 Fermat's Last "Theorem"

Perhaps the most famous Diophantine equation is

$$x^n + y^n = z^n , \quad (n > 2) \tag{7.26}$$

for which Fermat asserted that no nontrivial $(xy \neq 0)$ solution in integers exists. Fermat thought he had a proof, but this seems more than doubtful after centuries of vain efforts by some of the greatest mathematicians who came after Fermat.

Some special cases for n are relatively easy to prove, for example the quartic case, $n = 4$. The cubic case, $n = 3$, was solved by Sophie Germain in Paris who mailed her proof to Gauss in Göttingen under the male pseudonym of Monsieur Le Maitre [7.5].

The greatest breakthrough (as some would say today) was made in 1851 by the German mathematician Ernst Eduard Kummer. He showed that Fermat's Last Theorem (FLT) was true for what he called regular primes p_r, defined as those primes which do not divide the numerators of the Bernoulli numbers B_k up to B_{p_r-3} [7.6].

The only irregular primes below 100 are 37, 59, and 67, i.e., three primes out of 25. It is believed that the asymptotic fraction of the irregular primes tends toward $1 - 1/\sqrt{e} = 0.393469 \dots$. Thus, there would be an infinity of irregular primes, roughly 40% of all primes. This is in stark contrast to the Fermat and Mersenne primes, of which only 5 and 29, respectively, are presently known.

In 1908 the Göttingen Academy of Science, established the Wolfskehl Prize to the tune of 100,000 (gold!) marks for proving FLT. In 1958, two World Wars later, this was reduced to 7,600 Deutsche Mark, but FLT has still not been proved or disproved. By 1976, it had been shown to be correct for all prime exponents smaller than 125,000.

If the proof of FLT has proved so difficult, perhaps the theorem is just not true, and one should look for a counterexample. (One counterexample, of course, would suffice to demolish FLT once and for all.) But unfortunately, since the exponent n has to be larger than 125000, and it can also be shown that x must be greater than 10^5, any counterexample would involve numbers *millions* of decimal digits long. Hence the counterexample route seems closed, even if FLT is false. Consequently, we will have to conclude, more than 300 years after its somewhat offhand assertion, that FLT will probably never be disproved. However, in the meantime, work on the FLT has led to some profound mathematical insights and innovations.

FLT, although perhaps the most famous unsolved case, is not an isolated quirk. Another seemingly unprovable (but probably false) conjecture is the following one by Georg Cantor: All numbers generated by the recursion

$$p_{n+1} = 2^{p_n} - 1 , \quad \text{with} \quad p_o = 2 \tag{7.27}$$

are prime. The first Cantor numbers after 2 are $p_1 = 3$, $p_2 = 7$, $p_3 = 127$ and $p_4 = 2^{127} - 1$, all of which are (Mersenne) primes. Unfortunately, little is known about the next Cantor number, p_5, other than that it has more than $5 \cdot 10^{37}$ decimal digits! (This should not be confounded with the large — but "infinitely" smaller — number $5 \cdot 10^{37}$, which has only 38 digits.) Nevertheless, with the latest advances in primality testing (see Chap. 11) perhaps the primality of p_5 can be confirmed — or refuted, thereby demolishing another conjecture. (Note: if p_m is composite, then *all* p_n, $n \geqslant m$ are composite — see Sect. 3.5).

One of the most enduring (if not endearing) conjectures is the famous *Goldbach conjecture*, asserting that every even number >4 is the sum of two primes. Some progress has been made on related weaker assertions, and the Goldbach conjecture itself has been *numerically* confirmed up to very large numbers. But alas, even if it had been shown to hold up to $10^{10^{10}}$, there would be no guarantee that it would not fail for $10^{10^{10}} + 2$.

7.8 The Demise of a Conjecture by Euler

Euler conjectured that (excepting trivial cases)

$$x_1^n + x_2^n + \cdots + x_k^n = z^n \tag{7.28}$$

has nontrivial integer solutions *iff* $k \geqslant n$. For $n = 3$ and $k = 2$, Euler's conjecture corresponds to the proven case $n = 3$ of FLT: the sum of 2 cubes cannot be another cube. For $n = 3$ and $k = 3$, (7.28) asserts that the sum of *three* cubes can be another cube. Euler's conjecture stood for two centuries but fell in 1966 to the joint effort of Lander and Parkin, who found a counterexample for $n = 5$:

$$27^5 + 84^5 + 110^5 + 133^5 = 144^5 , \tag{7.29}$$

which can be verified with a good pocket calculator (and which Euler himself could have done in his head).

Thus Fermat was avenged, whose fifth "prime" Euler had shown to be composite.

7.9 A Nonlinear Diophantine Equation in Physics and the Geometry of Numbers

In 1770, Lagrange proved that "the set of squares is a basis of order 4." This means that *every* positive integer can be represented as the sum of 4 squares. If we allow ourselves just 3 squares, then some integers cannot be so represented.

Which integers n can be expressed as the sum of 3 squares? The author first encountered this problem in the formula for the resonant frequencies of a cube-shaped resonator, which in units of the lowest resonant frequency are:

$$f_{x,y,z}^2 = x^2 + y^2 + z^2 . \tag{7.30}$$

An important question in some areas of physics is whether $f_{x,y,z}^2$ can take on all positive integer values or whether there are gaps. (This problem occurred in the author's Ph.D. thesis on normal-mode statistics.) To answer this question let us consider the complete residue system modulo 8:

$$r = 0, 1, 2, 3, 4, 5, 6, 7 . \tag{7.31}$$

Hence,

$$r^2 \equiv 0, 1 \text{ or } 4 \pmod 8 .$$

Thus, the sum of 3 squares modulo 8 is precisely those numbers that can be generated by adding 3 of the integers 0, 1, 4 (with repetition allowed). This is possible for integers 0 through 6, but *not* 7. Thus, certainly,

$$x^2 + y^2 + z^2 \neq 8m + 7 . \tag{7.32}$$

However, it can be shown that these "forbidden" numbers, when multiplied by a nonnegative power of 4, are also not possible as the sum of 3 squares [7.1]. In fact,

$$x^2 + y^2 + z^2 = n \tag{7.33}$$

has a solution in integers *iff*

$$n \neq 4^k (8m + 7), \quad k \geqslant 0 . \tag{7.34}$$

This means that on average, the fraction

$$\frac{1}{8} + \frac{1}{4 \cdot 8} + \frac{1}{16 \cdot 8} + \cdots = \frac{1}{6} \tag{7.35}$$

of all integers cannot be represented as the sum of 3 squares.

Since any positive integer n can be represented by a sum of 4 squared integers, it is interesting to ask how many ways $r_2(n)$ a given n can be so represented. The answer (due to Jacobi), including permutations, signs and 0's, is

$$r(n) = 8 \sum_{d \mid n, 4 \nmid d} d \, , \tag{7.36}$$

i.e., 8 times the sum of the divisors of n that are not divisible by 4.

Example: $r_2(4) = 8(1+2) = 24$. Check:

$$4 = (\pm 1)^2 + (\pm 1)^2 + (\pm 1)^2 + (\pm 1)^2 \quad \text{(16 cases)} \qquad \text{and}$$

$$4 = (\pm 2)^2 + 0^2 + 0^2 + 0^2 \qquad \text{(+8 cases)} \quad \text{Check!}$$

For a recent proof see *Hirschhorn* [7.7].

On the other hand, Fermat proved that a certain class of integers can be represented by the sum of just 2 squares, and in a *unique* way at that. This class consists of all primes of the form $4k+1$.

This result has an enticing geometrical interpretation. Consider the *integer lattice* in the plane, i.e., all the points (x, y) in the plane with integer coordinates. Draw a circle around the origin $(0,0)$ with radius $p^{1/2}$, where p is a prime with $p \equiv 1 \pmod{4}$. Then there are exactly eight lattice points on the circle.

No solutions exist for the primes $p \equiv -1 \pmod{4}$. For composite n, we have to distinguish between the factor 2 and the two kinds of primes $p_i \equiv 1 \pmod{4}$ and $q_i \equiv -1 \pmod{4}$:

$$n = 2^\alpha \prod_{p_i} p_i^{\beta_i} \prod_{q_i} q_i^{\gamma_i} \, .$$

Solutions for $n = x^2 + y^2$ exist only if all γ_i are even. *Hardy* and *Wright* [7.1, p. 299] give four different proofs, no less, one going back to Fermat and his "method of descent."

If all γ_i are even, the number of primitive solutions, $(x,y) = 1$, is $\prod_i (\beta_i + 1)$, including permutations. ("Trivial" solutions with x or y equal to 0 are counted only once.) Example: $325 = 5^2 \cdot 13$ has 6 solutions. $325 = 1^2 + 18^2 = 6^2 + 17^2 = 10^2 + 15^2 = 15^2 + 10^2 = 17^2 + 6^2 = 18^2 + 1^2$ and no others.

The connection between geometry and number theory was forged into a strong link by Hermann Minkowski. In his *Geometrie der Zahlen* (published in 1896) Minkowski established and proved many beautiful relationships at the interface of geometry and number theory. His most famous result, known as *Minkowski's theorem*, says that *any* convex region symmetrical about $(0,0)$ having an area greater than 4 contains integer lattice points other than $(0,0)$.

This theorem, and its generalization to higher dimensional spaces, is particularly useful in proofs concerning the representation of numbers by quadratic forms, such as the above result on the decomposition of certain primes into sums of squares.

In an address before the Göttingen Mathematical Society commemorating the 100th anniversary of Dirichlet's birth, Minkowski hypothesized that someday soon, number theory would triumph in physics and chemistry and that, for example, the decomposition of primes into the sum of 2 squares would be seen to be related to important properties of matter.[2]

Another intriguing geometrical concept by Minkowski is that of a *Strahlkörper* (literally: ray body) defined as a region in n-dimensional Euclidean space containing the origin and whose surface, as seen from the origin, exhibits only one point in any direction. In other words, if the inner region was made of transparent glass and only the surface was opaque, then the origin would be visible from each surface point of the Strahlkörper (i.e., there are no intervening surface points). Minkowski proved that if the volume of such a Strahlkörper does not exceed $\zeta(n)$, a volume preserving linear transformation exists such that the Strahlkörper has no points in common with the integer lattice (other than the origin). Here $\zeta(n)$ is Riemann's zetafunction which we encountered already in Chap. 4 in connection with the distribution of primes.

The fact that $\zeta(n)$ should determine a Strahlkörper property is not totally surprising. If we look at our "coprimality function," Fig. 4.8, it consists of precisely all those points of the lattice of positive integers from which the origin is visible, i.e., the white dots in Fig. 4.8 define the surface of a (maximal) Strahlkörper. And the asymptotic density of dots is $1/\zeta(2) = 6/\pi^2$ (see Sect. 4.4) or, in n dimension, $1/\zeta(n)$.

7.10 Normal-Mode Degeneracy in Room Acoustics (A Number-Theoretic Application)

The minimum frequency spacing of two nondegenerate normal modes of a cubical room, in the units used above in (7.30), is

$$\Delta f_{min} = \frac{1}{2f_{x,y,z,}} \, . \tag{7.37}$$

---

2 "In letzterer Hinsicht bin ich übrigens für die Zahlentheorie Optimist and hege still die Hoffnung, dass wir vielleicht gar nicht weit von dem Zeitpunkt entfernt sind, wo die unverfälschteste Arithmetik gleichfalls in Physik and Chemie Triumphe feiern wird, und sagen wir z. B., wo wesentliche Eigenschaften der Materie als mit der Zerlegung der Primzahlen in zwei Quadrate im Zusammenhang stehend erkannt werden." [7.8].

Because of the gaps in the numbers representable by the sum of 3 squares, the *average* nondegenerate frequency spacing becomes 7/6 of this value:

$$\overline{\Delta f} = \frac{7}{12f} . \tag{7.38}$$

The asymptotic density of normal modes per unit frequency (using a famous formula on the distribution of eigenvalues, proved in its most general form by Hermann Weyl) is

$$\Delta Z = \frac{\pi}{2} \cdot f^2 . \tag{7.39}$$

Thus, the average degree of degeneracy D (i.e., the number of modes having the *same* resonance frequency) becomes [7.9]:

$$D = \Delta Z \cdot \overline{\Delta f} = \frac{7\pi}{24} \cdot f \simeq 0.92 \cdot f . \tag{7.40}$$

The degree of degeneracy is important because a high degeneracy can be detrimental to good room acoustics: normal modes that coincide in frequency are missing elsewhere and leave gaps in the frequency scale. Consequently, musical notes generated at those frequencies are not well transmitted to the attending audience (or microphones). This problem is most significant for small enclosures such as recording studios, where the mode density, especially at the lower end of the audiofrequency range, is already small and any unnecessary degeneracy impairs the acoustic responsiveness [7.9].

7.11 Waring's Problem

A problem that has stimulated much mathematical thought, by Hilbert among others, is *Waring's problem* [7.1]: given a positive integer $n > 0$, what is the least number of terms $G(k)$ in the sum:

$$n = \sum_j m_j^k \tag{7.41}$$

for all *sufficiently large n*?

Another question is how many terms $g(k)$ are needed so that *all* n can be represented as in (7.41). Of course:

$$g(k) \geqslant G(k) . \tag{7.42}$$

As we saw in Sect. 7.9, $g(2) = 4$. It is also known that $g(3) = 9$. In fact, there are only finitely many n for which 9 third powers are required. Probably the only two cases are

$$23 = 2 \cdot 2^3 + 7 \cdot 1^3 \qquad \text{and}$$

$$239 = 2 \cdot 4^3 + 4 \cdot 3^3 + 3 \cdot 1^3 .$$

Thus, by definition:

$$G(3) \leqslant 8 .$$

Also, $G(3) \geqslant 4$, but the actual value of $G(3)$ is still not known — another example of how a seemingly innocent question can lead to mathematically most intractable problems!

If we ask *how many different* ways $r_k(n)$ an integer n can be represented as the sum of kth powers (including different sign choices and permutations), then for example by (7.36), $r_2(5) = 8$. Of particular interest are the following asymptotic averages [7.1]:

$$\sum_{n=1}^{N} r_2(n) = \pi N + 0(\sqrt{N}) \qquad \text{and} \tag{7.43}$$

$$\sum_{n=1}^{N} r_3(n) = \frac{4\pi}{3} N^{3/2} + 0(N) . \tag{7.44}$$

Both (7.43) and (7.44) are intuitively obvious because they count the number of integer lattice points in a circle and, respectively, a sphere of radius $\sqrt{N}$.

8. The Theorems of Fermat, Wilson and Euler

These three theorems, especially the one by *Euler*, play a central role in many modern applications, such as digital *encryption*. They are deeply related to the theory of groups, and indeed, their most elegant proofs are group theoretic. Here, however, we shall stress the purely arithmetic viewpoint. We also introduce the important *Euler φ function* (or totient function) which reaches into every corner of number theory and which, by way of illustration, tells us how many ways an n-pointed *star* can be drawn by n straight lines without lifting the pen.

8.1 Fermat's Theorem

For prime p and integer b not divisible by p, the following congruence holds:

$$b^{p-1} \equiv 1 \pmod{p}, \quad p \nmid b. \tag{8.1}$$

Let us check this for $p = 5$: $1^4 = 1$, $2^4 = 16 \equiv 1$, $3^4 = 81 \equiv 1$, $4^4 = 256 \equiv 1 \pmod 5$.

One of several proofs of Fermat's theorem relies on the fact that a complete residue system modulo a prime, excluding 0, i.e., the numbers $1, 2, \ldots, p-1$ (if we decide to use least positive residues) form a multiplicative group of order $p - 1$. It is well known from elementary group theory that *any* element of such a group, when raised to the power equal to the order of the group, will yield the unit element.

Another proof starts from the product

$$P = 1b \cdot 2b \cdot 3b \cdots (p-2)b \cdot (p-1)b = b^{p-1}(p-1)! \tag{8.2}$$

Here multiplication modulo p of the numbers $1, 2, 3, \ldots, p-2, p-1$ by any factor b which is coprime to p will only change their sequence, but not the value of the product. Thus, $P \equiv (p-1)! \pmod p$. Since the factor $(p-1)!$ is coprime to p, we can cancel it and conclude that b^{p-1} must be congruent 1 modulo p.

Example: for $p = 5$ and $b = 2$,

$$P = 2 \cdot 4 \cdot 6 \cdot 8 = 2^4 \cdot 4! \, , \qquad\qquad (8.3)$$

but taking each factor in P modulo 5 gives

$$P \equiv 2 \cdot 4 \cdot 1 \cdot 3 = 4! \pmod 5 \, .$$

Hence $2^4 \equiv 1 \pmod 5$.

8.2 Wilson's Theorem

If and only if p is prime, then

$$(p-1)! \equiv -1 \pmod p \, . \qquad\qquad (8.4)$$

For a proof when p is a prime, not relying on group theory, consider the product

$$2 \cdot 3 \cdot 4 \cdots (p-3) \cdot (p-2) \, . \qquad\qquad (8.5)$$

Here each factor has its own inverse (modulo p) somewhere among the other factors. For example, for $p = 7$, in the product $2 \cdot 3 \cdot 4 \cdot 5$ the factors 2 and 4 form a pair of inverses modulo 7, and so do 3 and 5. The product of such a pair is by definition congruent 1 modulo p, and so is the product of all the $(p-3)/2$ pairs in the above product. Thus,

$$2 \cdot 3 \cdot 4 \cdots (p-3) \cdot (p-2) \equiv 1 \pmod{p} \, , \qquad\qquad (8.6)$$

and multiplying by $p-1$ yields:

$$(p-1)! \equiv p-1 \equiv -1 \pmod p \, . \quad \text{Q.E.D.} \qquad\qquad (8.7)$$

Another way to state Wilson's theorem is as follows:

$$p \mid (p-1)! + 1 \, . \qquad\qquad (8.8)$$

Example: for $p = 5$: 5 divides $24 + 1$. Check!

Using this form of Wilson's theorem, we construct the function

$$f(n) = \sin\left[\pi \, \frac{(n-1)! + 1}{n}\right] , \qquad\qquad (8.9)$$

which has a truly remarkable property: $f(n)$ is zero if *and only* if n is prime. Here we seem to have stumbled across the ultimate test for primality!

By contrast, Fermat's theorem, always true for primes, *can* also be true for composite numbers. Not so for Wilson's theorem, which holds if *and only* if p is prime.

Unfortunately, this primality test is of *no* practical advantage, because calculating $f(n)$ takes much longer than even a slow sieve method. For example, to test whether 101 is prime, we would have to compute 101! first — a 160 digit tapeworm!

Another proof of Wilson's theorem uses Fermat's theorem in the following form:

$$x^{p-1} - 1 \equiv 0 \pmod{p}, \tag{8.10}$$

which holds for $x = 1, 2, ..., p-1$. According to the fundamental theorem of algebra these $p-1$ roots must be *all* the roots of the above equation. We can therefore write

$$x^{p-1} - 1 \equiv (x-1) \cdot (x-2) \cdots (x-p+1) \pmod{p}. \tag{8.11}$$

Now letting $x = p$ yields

$$p^{p-1} - 1 \equiv (p-1) \cdot (p-2) \cdots 1 = (p-1)! \pmod{p}. \tag{8.12}$$

Here p^{p-1} is of course congruent 0 modulo p and Wilson's theorem is proved.

8.3 Euler's Theorem

For b and m coprime, i.e., $(b,m) = 1$, Euler asserted that

$$b^{\phi(m)} \equiv 1 \pmod{m}. \tag{8.13}$$

Here $\phi(m)$ is Euler's ϕ function (also called totient function) defined as the number of positive integers r smaller than m that are coprime to m, i.e., for which $1 \leqslant r < m$ and $(r,m) = 1$ holds.

Example: $m = 10$: $r = 1, 3, 7, 9$. Thus $\phi(10) = 4$. Note: $\phi(1) = 1$.

For *prime* moduli, each of the numbers $r = 1, 2, ..., p-1$ is coprime to p and therefore $\phi(p) = p-1$.

For prime powers p^α, one obtains in a similar fashion

$$\phi(p^\alpha) = (p-1)p^{\alpha-1} = p^\alpha \left(1 - \frac{1}{p}\right). \tag{8.14}$$

Example: $\phi(9) = 6$. Check! (There are exactly six positive integers below 9 coprime to 9, namely, 1, 2, 4, 5, 7, and 8.)

Euler's ϕ function is a so-called *multiplicative function*, which in number theory is defined as a function for which

$$\phi(n \cdot m) = \phi(n) \cdot \phi(m) \quad \text{for all } (n,m) = 1 \tag{8.15}$$

holds. Thus, we have for any

$$m = \prod_i p_i^{e_i} \tag{8.16}$$

$$\phi(m) = \prod_i (p_i - 1)p_i^{e_i-1} \quad (e_i > 0) , \tag{8.17}$$

or in a form without the restriction on the exponents:

$$\phi(m) = m \prod_i \left[1 - \frac{1}{p_i} \right] . \tag{8.18}$$

As can be seen from (8.17), the totient function $\phi(m)$ is always even, except $\phi(2) = \phi(1) = 1$.

Example: for $m = 10$, $\phi(10) = 10(1 - \frac{1}{2})(1 - \frac{1}{5}) = 4$; and $1^4 = 1$, $3^4 = 81 \equiv 1$, $7^4 = 49^2 \equiv (-1)^2 = 1$, $9^4 = 81^2 \equiv 1^2 = 1$ (mod 10).

One proof of Euler's theorem parallels that of Fermat's: Consider the prime residue system modulo m:

$$r_1, r_2, ..., r_{\phi(m)} \tag{8.19}$$

and multiply each r_k with b, where $(b,m) = 1$. This multiplication changes the *sequence* of the above residues, but their total product is not affected. For example, for $m = 10$,

$$r = 1, 3, 7, 9 ,$$

which upon multiplication with $b = 7$ becomes

$$rb = 7, 21, 49, 63 \equiv 7, 1, 9, 3 \quad (\text{mod } 10) .$$

This is, of course, a consequence of the fact that prime residue systems form a multiplicative group.

Hence,

$$b^{\phi(m)} r_1 r_2 \cdots r_{\phi(m)} \equiv r_1 r_2 \cdots r_{\phi(m)} \quad (\text{mod } m) . \tag{8.20}$$

Since the residues are by definition prime to m, we can cancel these factors and are left with

$$b^{\phi(m)} \equiv 1 \pmod{m} . \tag{8.21}$$

The reader may wish to show the following curious but important property of the totient function:

$$m \mid \phi(p^m - 1) , \tag{8.22}$$

which holds for any m.

8.4 The Impossible Star of David

How many ways can an n-pointed star be drawn from n straight lines without lifting the pen? An n-pointed star has of course n outer corners. For simplicity we assume that these corners lie equidistantly on a circle. Connecting corners that are adjacent on the circle produces a polygon, not a star. To obtain a star we must connect each corner with one of its $n-2$ non-neighbors. For $n = 6$, we get the six-pointed Star of David: two superimposed triangles.

But suppose we are asked to draw a six-pointed star in six straight consecutive strokes. The Star of David then becomes impossible (the reader is invited to try the following himself). Suppose the points are numbered 1,2,3,4,5,6 and we connect 1 with 3 with 5 with 1. But that is not a six-pointed star; that is a triangle and points 2,4,6 have been left out. Suppose we skip *two* points. Then we connect 1 with 4 with 1, and we have missed four points. Suppose we permit irregular skippings and connect, for example, 1 with 3 with 6 with 2 with 4 — but now we cannot skip anymore because the only missing point is the adjacent point 5.

No matter how hard we try, a six-pointed star cannot be completed in this way. Yet it is easy to draw a five-pointed star by five straight lines without lifting the pen: connect point 1 with 3 with 5 with 2 with 4 with 1. And there are even *two* seven-pointed stars: skipping *one* point at a time, connect point 1 with 3 with 5 with 7 with 2 with 4 with 6 with 1 (Fig. 8.1); or, skipping *two* points at a time, connect point 1 with 4 with 7 with 3 with 6 with 2 with 5 with 1 (Fig. 8.2).

When is a star possible? If $k-1$ is the number of points skipped, then k must be greater than 1 and smaller than $n-1$ (or we would get a polygon) and k must be coprime with n, the number of corners. Because $\phi(5) = 4$, there are four values of k such that $(k,5) = 1$, of which only two differ from 1 or $n-1$: $k = 2$ and $k = 3$.

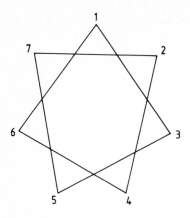

Fig. 8.1 One of two possible regular seven-pointed stars. It is based on the fact that 7 is coprime with 2 (and 5)

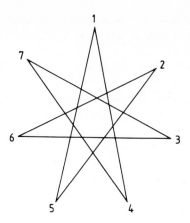

Fig. 8.2. The other possible seven-pointed stars based on the coprimality of 7 with 3 (and 4)

The two stars drawn with $k = 2$ and $k = 3$ are indistinguishable once the drawing is completed. Thus, there is only one possible five-pointed star.

From this discussion it should be clear that for general n, the following holds:

$$\text{Number of possible stars} = \frac{\phi(n) - 2}{2}. \tag{8.23}$$

Let us try this formula for $n = 6$. With $\phi(6) = 2$, we get *no* stars — as we have already discovered by trial and error.

For $n = 7$ we get, with $\phi(7) = 6$, $(6-2)/2 = 2$ stars. Check! (See Figs. 8.1, 2.)

Question: for which n do we get 6 different stars? Answer: for *no n* — the number $2 \cdot 6 + 2 = 14$ is a so-called *nontotient*: $\phi(n)$ can never equal 14. (Can the reader show this?)

The number 6 seems to have an unlucky star: not only is a singly connected six-pointed star impossible, but there is no n for which exactly six different stars are possible. Other nontotients are 26 and 34. What is the general condition for a number to be a nontotient?

8.5 Dirichlet and Linear Progression

Consider the linear progression

$$a_k = mk + c, \quad (c,m) = 1, \quad k = 0,1,2, \dots . \tag{8.24}$$

There are exactly $\phi(m)$ incongruent choices for c and therefore $\phi(m)$ nonoverlapping progressions. In 1837 Dirichlet showed, in one of the most spectacular early applications of analytic number theory, that the primes are equally distributed over these $\phi(m)$ different progressions, i.e., each progression contains asymptotically the fraction $1/\phi(m)$ of all the primes [8.1].

For $m = 3$, for example, $\phi(3) = 2$, and all primes (except 3 itself) belong to one of two residue classes: $3k + 1$ and $3k - 1$. And Dirichlet asserted that these two kinds of primes are equally frequent, each class (on average) laying claim to one-half of all the primes.

For $m = 4$, $\phi(4) = 2$, there are likewise two kinds of primes: $4k \pm 1$, and again they are equally frequent (but see below!).

For $m = 6$, $\phi(6) = 2$, and we conclude that primes of the form $6k \pm 1$ are equally frequent.

For $m = 10$, $\phi(10) = 4$, and all the primes (except 2 and 5) have 1, 3, 7 or 9 as their last digits, and we find that each class asymptotically gets a quarter of the total.

For $m = 100$, $\phi(100) = 40$, the conclusion from Dirichlet's theorem is that the second-to-last digit (the "tens" digit) will be 0, 1, 2, 3, 4, 5, 6, 7, 8 or 9 with equal probability. In fact, in the interval from 7200 to 8800, where we would expect to find a total of $16600/\ln 8000 = 178$ primes, there are actually 176 primes with the following distribution over the tens digits:

0: 15	5: 15
1: 20	6: 18
2: 20	7: 14
3: 17	8: 20
4: 20	9: 17 ,

with a mean value of 17.6 primes and a standard deviation of $\sigma = 2.2$ — considerably less than for a Poisson process, which would have $\sigma = 4.2$.

The author once thought that nothing but equipartition of the primes between different prime residue classes of a given modulus k was even thinkable. After all, what is the difference between primes of the form $4k + 1$ and $4k - 1$? Well, for one, +1 is a quadratic residue of 4 and -1 is not. As a result, there is a subtle "quadratic effect" on the distribution of primes taken modulo 4. As *Chebyshev* [8.2] noted in 1853, there are more primes of the form $4k - 1$ than $4k + 1$, although not as *many* as some people believed. In fact, the difference in number between these two kinds of primes changes sign infinitely often! The number of primes below $x = 2.5 \cdot 10^9$ is about $1.2 \cdot 10^8$, and the absolute excess of primes $4k - 1$ is roughly 2000 — subtle but finite.

There are even subtler cubic residue effects. However, in *relative* terms, all these effects vanish as x approaches infinity, and Dirichlet is ultimately vindicated.

9. Euler Trap Doors and Public-Key Encryption

Here we describe a *public-key digital encryption* scheme that does not require secret-key distribution — one of the Achilles' heels of secret-key cryptographic systems because once the secret encrypting key becomes known, transmissions are no longer secret. In public-key encryption, every potential recipient of secret messages *publishes* his encrypting key (hence the name *public* key). But knowledge of the *en*crypting key is of no practical help in *de*cryption. The public-key is the key to a "trap door" through which messages can vanish, not to be recovered — except through a different route to which only the legitimate receiver holds the required key, the decrypting key. (Trap door means easily *in,* but out only with the proper tools.) It is as if everyone had a (chained!) box in front of his house into which secret messages could be stuffed and to which only the owner had the proper opening key.

The required trap door behavior is realized for digital messages (represented by long strings of digits) by the fact that it is easy to multiply two large numbers, but impossible to factor sufficiently large numbers in a reasonable amount of time.

The public-key cryptographic systems described here are based on modular arithmetic, requiring knowledge of the Euler ϕ function of the modulus n to calculate inverses, i.e., to decrypt. And of course, to determine $\phi(n)$, we have to know the factors of n.

Some exceedingly interesting interrelations between cryptography and *error-correcting codes* (Sect. 26.2) are discussed by *Sloane*, and the reader seeking deeper insights into these seemingly conflicting goals is referred to [9.1].

9.1 A Numerical Trap Door

An important practical application of Euler's theorem today is *public-key encryption* [9.2], and specifically the construction of *trap-door functions* for such secret activities [9.3].

What is a trap-door function? It is, as the name implies, a (mathematical) function that is easy to calculate in one direction but *very* hard to calculate in the opposite direction. For example, it takes a modern computer only microseconds to multiply two 500-digit numbers. By contrast, to take a 1000 digit number having two 500-digit factors and decompose it into its factors can take *ages,* even on the very fastest computer available in 1982 and using the most efficient factoring algorithms known today.

Although no completely satisfactory theoretical definition of trap door functions has been given, or a proof that they exist, reasonable *candidates* for such functions have been proposed, for example the Euler trap door function to be discussed here. These are not *proved* secure, but are considered *empirically* (at present) secure trap doors.

9.2 Digital Encryption

Before encryption, a written message is usually first converted into a positive integer $1 < M < r$, i.e., a string of (decimal or binary) digits. In addition, we will assume that M and r are coprime.

The encryption we want to discuss here consists in raising M to a certain power s and keeping only the remainder modulo r. Thus, the encrypted message E is represented by the number

$$E \equiv M^s \pmod{r}, \quad 1 < E < r . \tag{9.1}$$

Now, if we had chosen a prime for our modulus r, decrypting E, i.e., recovering M from E, would be easy. First we find the inverse, t, of the encrypting exponent s modulo $r-1$ by solving the congruence

$$st \equiv 1 \pmod{r-1} . \tag{9.2}$$

For this congruence to have a solution, s and $r-1$ must be coprime: $(s, r-1) = 1$. One method of obtaining t from s is by an application of Euler's theorem:

$$t \equiv s^{\phi(r-1)-1} \pmod{r-1} , \tag{9.3}$$

because for such a t, according to Euler,

$$st \equiv s^{\phi(r-1)} \equiv 1 \pmod{r-1} , \tag{9.4}$$

provided that $(s, r-1) = 1$.

Now, if we raise the encrypted message E to the power t and reduce the result modulo r, we obtain the original message M back. First,

$$E^t \equiv M^{st} \pmod{r} , \tag{9.5}$$

and since $st \equiv 1 \pmod{r-1}$ we can write

$$st = (r-1)k + 1 \quad \textit{(for some integer k)} \; . \tag{9.6}$$

Thus,

$$E^t \equiv M^{(r-1)k+1} \pmod{r} \; . \tag{9.7}$$

Now Euler intervenes again (or rather Fermat, as long as r is prime), telling us that

$$M^{r-1} \equiv 1 \pmod{r} \tag{9.8}$$

if $(M,r) = 1$, which we initially assumed.

With this intermediate result, we obtain

$$E^t \equiv 1^k M^1 = M \pmod{r} \; , \tag{9.9}$$

i.e., we have recovered the original message M from the encrypted message E, which is what we wanted to show.

Example: $s = 3$, $r = 17$; $\phi(r-1) = 8$.

With this choice of s and r, for which $(s,\phi(r)) = 1$ as required, the decrypting exponent t becomes

$$t \equiv s^{\phi(r-1)-1} = 3^7 \equiv 11 \pmod{16} \; .$$

Check: $st = 3 \cdot 11 = 33 \equiv 1 \pmod{16}$. Check!

Now take a message, say $M = 4$. The encrypted message becomes

$$E \equiv M^s = 4^3 = 64 \equiv 13 \pmod{17} \; .$$

The decrypted message is obtained by raising $E = 13$ to the decrypting exponent $t = 11$:

$$E^t = 13^{11} = (13^5)^2 \cdot 13 = (371{,}293)^2 \cdot 13 \equiv 13^2 \cdot 13$$

$$\equiv 4 = M \pmod{17} \; . \quad \text{Check!}$$

We have now set the stage for our central theme.

9.3 Public-Key Encryption

We shall try to understand one of the greatest advances in the safeguarding and secure transmission of secrets: *public-key encryption*. This story has both "logistic" and mathematical aspects. I shall attempt to illuminate both.

One form of public-key encryption makes use of a very large *composite* number, the encrypting modulus r. Specifically, we shall assume that r has exactly two prime factors, p and q, both of which are also very large

$$r = pq \ . \tag{9.10}$$

More specifically, we may think of p and q as being each about 50 decimals long, so that r has roughly 100 decimal places.

In addition, the method makes use of an encrypting exponent s, which is chosen so that s and $\phi(r)$ are coprime:

$$(s,\phi(r)) = 1 \ . \tag{9.11}$$

Everyone who ever wishes to receive a secret message selects such a triplet of numbers p, q and s and *publishes* s and the product r of p and q. The publishing may take the form of a kind of "telephone" book, where behind the name and address of each prospective secret-message recipient R, two numbers appear an encrypting exponent s and an encrypting modulus r — selected in the manner just described. Thus, anyone — say a person named S — wishing to send a secret message raises its digital form M to the power s and reduces it modulo r.

Now notice something very important here: a key never had to be sent from R to S; R simply published it in a "telephone" book for everyone to see and use if he so desires. Thus, the grave problems of secure key transmission, which fill volumes in the literature of espionage, have been circumvented.

But cannot everyone now decrypt the encrypted message? The answer is *no,* if the encryption has made use of a trap-door (or almost one-way) function.

But what if the encryption was by a true one-way process like, let us say, the randomizing of air molecules in a bicycle tire: if the tire was initially half filled with air — also known among cyclists as (half) flat — and then pumped up full, all the molecules just pumped in could *never* be extracted again while leaving all the original ones in the tire. Here we have an everyday example of a true one-way function. In public-key encryption the function is not totally one-way — or the sender might as well burn the message and scatter it to the winds. No, it is only a trap door, i.e., a door that can still be opened from one side — the side of the legitimate recipient. Here is how.

As in the previous example, the recipient needs a decrypting exponent t, given again by the congruence

$$ts \equiv 1 \pmod{\phi(r)} . \tag{9.12}$$

Now, *if the factors of r are known,* then $\phi(r)$ is also known. For $r = pq$, we have

$$\phi(r) = (p - 1)(q - 1) , \tag{9.13}$$

and t can be calculated from the following congruence:

$$t \equiv s^{\phi(\phi(r))-1} \pmod{\phi(r)} , \tag{9.14}$$

because then

$$ts \equiv s^{\phi(\phi(r))} \pmod{\phi(r)} , \tag{9.15}$$

which according to Euler, is congruent to 1 modulo $\phi(r)$, provided $(s,\phi(r)) = 1$, as initially required of s.

Having demonstrated that t is the proper (and, incidentally, essentially unique) decrypting exponent, decryption now proceeds as before. The received encrypted message is

$$E \equiv M^s \pmod{r} , \tag{9.16}$$

and decrypting with t yields

$$E^t \equiv M^{st} = M^{\phi(r)k+1} \pmod{r} . \tag{9.17}$$

Now, with Euler,

$$M^{\phi(r)} \equiv 1 \pmod{r} . \tag{9.18}$$

Thus,

$$E^t \equiv M \pmod{r} , \tag{9.19}$$

as desired. Note that for the encryption to work, M has to be coprime to r, but the chances of this not being the case are of order 10^{-50}.

The main point here is that the decrypting exponent t cannot be derived from the publicly known s and r, but only from s and the *factors* of r, namely p and q. Thus, as long as the receiver keeps knowledge of p and q to himself and publishes only their product r, he is safe, i.e., the messages he receives cannot be read by unauthorized third parties — *provided that r cannot be factored.*

At present (1983), a 100-digit number that is the product of two 50-digit numbers cannot be factored in any reasonable time — say a couple of minutes. In fact, not so long ago, the most efficient factoring algorithms on a very fast computer were estimated to take 40 trillion years, or 2000 times the present age of the universe. That *sounds* very safe indeed. But algorithms get more efficient by the month and computers become faster and faster every year, and there is no *guarantee* that one day a so-called "polynomial-time" algorithm[1] will not emerge that will allow fast factoring of even 1000-digit numbers [9.4]. Few mathematicians believe that a true polynomial-time algorithm is just around the corner, but there also seems to be no prospect of proving that this will not occur [9.5].

9.4 A Simple Example

Next we will illustrate public-key encryption with a simple example: $s = 7$, $r = 187$. Say the message is $M = 3$; then the encrypted message is

$$E \equiv M^7 = 3^7 = 2187 \equiv 130 \pmod{187} \; .$$

Now for the receiver to decrypt E, he needs to know the decrypting exponent t, which he can obtain from the factors of $r = 187 = 11 \cdot 17$, which only *he* knows (or that only he would know if r were a much larger number).

With $r = 11 \cdot 17$, he has $\phi(r) = 10 \cdot 16 = 160$ and $\phi(\phi(r)) = 64$, and

$$t \equiv s^{\phi(\phi(r))-1} = 7^{63} = (7^9)^7 = (40\,353\,607)^7$$
$$\equiv 7^7 = 823\,543 \equiv 23 \pmod{160} \; .$$

Knowing that $t = 23$, the receiver can now proceed with his decryption:

$$E^t = 130^{23} \equiv 3 = M \pmod{187} \; . \quad \text{Check!}$$

The calculation of 130^{23} (mod 187) was done with the pocket calculator program mentioned earlier; it is described in the Appendix.

9.5 Repeated Encryption

Here are two more examples which illustrate an important point:

$$M_1 = 123, \qquad E_1 = 183, \; E_1' \equiv 123 \pmod{187}. \quad \text{Check!}$$
$$M_2 = E_1 = 183, \; E_2 = 72, \quad E_2' \equiv 183 \pmod{187}. \quad \text{Check!}$$

---

[1] A polynomial-time algorithm is one in which computing time increases like some power or polynomial of the "size" of the problem. Thus, if the polynomial were of the third degree, for example, every doubling of size would increase computation time eightfold.

Thus, for the above choice of parameters the encrypting operation is *not* its own inverse. Symbolically:

$$E\{E\{M\}\} \neq M \; , \quad \text{or} \tag{9.20}$$

$$E^{-1}\{\,\} \neq E\{\,\} \; . \tag{9.21}$$

This is important, because otherwise an eavesdropper who tried, perhaps accidentally, to double-encrypt with the publicly known key would recover the message.

Because of the uniqueness of the decrypting exponent t, a necessary and sufficient condition for $E^{-1} \neq E$ is that the exponents for encryption and decryption be different: $s \neq t$.

If we continue encrypting $M_1 = 123$, the third round will look like this:

$$M_3 = E_2 = 72, \; E_3 = 30, \; E_3^t = 72 \; . \quad \text{Check!}$$

Thus, for this choice of parameters,

$$E\{E\{E\{M\}\}\} \neq M \; , \tag{9.22}$$

or $E^3 \neq 1$, for short. Of course, for some number n of repetitions of the encryption process the original message must reappear. Symbolically: $E^n = 1$. The smallest n for which this is true is called the *order* of E. For practical security considerations the order of E should be fairly high. However, it is easy to see that it cannot be larger than $\phi(r)$, because there are only $\phi(r)$ different possible messages (remember: $(M,r) = 1$).

Whether the maximal order $\phi(r)$ can in fact be obtained is another question for which we require the concept of a "primitive element", to be introduced in Chap. 13. In our example, starting with $M = 123$, the fourth round of encrypting yields

$$M_4 = E_3 = 30, \; E_4 = 123, \; E_4^t = 30 \; . \quad \text{Check!}$$

So after 4 successive encryptions, we get the original message back, i.e., the order of our encrypting exponent is 4. And, indeed $7^4 = 2401 \equiv 1 \pmod{\phi(r) = 160}$.

From a practical point of view, such a small order would be completely unacceptable, but then so would our choice of a three-digit integer as encryption modulus.

The encrypting exponent $s = 3$ and its inverse modulo 160, $s^{-1} = 107$, have order 8. Other triplets of exponent, inverse and order are: (9, 89, 4), (11, 131, 8), (13, 37, 8), (17, 113, 4), (19, 59, 8), (21, 61, 8), (23, 7, 4), (27, 83, 8), (29, 149, 8), (31, 31, 1). As we shall see in Chap. 13, all orders must be some divisor of $\phi(r) = 160$. In the above examples we found the divisors 1, 4 and 8.

9.6 Summary and Encryption Requirements

For public-key encryption according to the method just outlined, proceed as follows:

1) Pick two *large* primes p and q and publish their *product* the modulus $r = pq$. This might make factoring difficult and there will be few "forbidden" messages for which $(M,r) \neq 1$.

2) Select an encrypting exponent s which is coprime to $\phi(r) = (p-1)(q-1)$: $(s,\phi(r)) = 1$. Make sure that the order of s is large, to prevent accidental decryption by repeated encryption.

3) Calculate the decrypting exponent t and *do not* publish it.

Here the following problem occurs: To determine t from s and p and q we have to calculate

$$t \equiv s^{\phi(\phi(r))-1} \pmod{\phi(r)} ,$$

where $\phi(r) = (p-1)(q-1)$. To calculate $\phi(\phi(r))$, we have to factor $p-1$ and $q-1$. Now since p and q are large numbers, so are $p-1$ and $q-1$ and the required factoring might be difficult or impossible.[2]

This problem can be avoided by *constructing* $p-1$ and $q-1$ from *known* primes and then testing the resulting p and q for primality. For example, we might try $p - 1 = 2 \cdot 3 \cdot 11 = 66$, and 67 is indeed prime. The subject of primality testing, which becomes very important here, will be discussed in Chap. 11.

On the other hand, factoring $p-1$ and $q-1$, even though they may be 50-digit numbers, is in general not as difficult as factoring a 100-digit number, whose smallest prime factor has 50 digits. We will also consider the factoring problem in greater detail in Chap. 11 when we discuss prime divisor functions. We will see there that the median value $p_{0.5}$ of the prime factors of n is given approximately by

---

[2] There is an implied logical contradiction here: if we consider subtraction of 1 not to change the attribute "large," then by induction we can show that 0 is a large number. However, it is clear here that we only mean that if 10^{50} is large, then so is $10^{50}-1$. A similar quandary, related to the expression "close enough," is illustrated by the following situation: Suppose all the young gentlemen in a class were to line up on one side of the room, and all the young ladies on the other. At a given signal, the two lines move toward each other, halving the distance between them. At a second signal, they move forward again, halving the remaining distance; and so on at each succeeding signal. Theoretically the boys would never reach the girls; but actually, after a relatively small number of moves, they would be close enough for all practical purposes [9.6].

$$p_{0.5} \approx e^{(\ln n/2.81)^{0.5}}.$$

Thus, for $n \approx 10^{50}$, half the prime factors can be expected to be smaller than roughly 600.

The application of cryptography to computer data security is extensively discussed in [9.7]. Other up-to-date information in this fast-moving field can be found in a new journal, *Cryptologia,* published by Albion College (Albion, MI $2^3 \cdot 3 \cdot 7 \cdot 293$) and in the proceedings of the recurring international symposia on information theory sponsored by the Institute of Electrical and Electronics Engineers (New York, N.Y. 10017).

10. The Divisor Functions

Some numbers have few divisors, such as primes, and some numbers have many divisors, such as powers of 2. The number of divisors, although it fluctuates wildly from one integer to the next, obeys some interesting rules, and averages are quite predictable, as we shall see below.

10.1 The Number of Divisors

By definition, $\phi(n)$ is the number of integers in the range from 1 to n which have 1 as greatest common divisor (GCD) with n (see Chap. 8). To codify this statement we introduce the following notation, making use of the number sign #:

$$\phi(n) := \#\{m : 1 \leqslant m < n, \ (m,n) = 1\} . \tag{10.1}$$

This notation is, of course, equivalent to the more frequently encountered notation using the sum sign:

$$\phi(n) = \sum_{(m,n)=1, m<n} 1 .$$

Let (cf. Chap. 2)

$$n = \prod_{p_i|n} p_i^{e_i} . \tag{10.2}$$

Then all divisors of n are of the form

$$d_k = \prod_{p_i|n} p_i^{f_i} \quad with \ 0 \leqslant f_i \leqslant e_i . \tag{10.3}$$

Here each exponent f_i can take on $e_i + 1$ different values. Thus the number of distinct divisors of n is given by the *divisor function* defined as

$$d(n) := \#\{d_k : d_k|n\} ,$$

and equal to

$$d(n) = \prod_i (e_i + 1) .$$

$$(10.4)$$

Example: $n = 12 = 2^2 \cdot 3^1$, $d(12) = 3 \cdot 2 = 6$.

Check: $d_k = 1, 2, 3, 4, 6, 12$, i.e., there are indeed 6 distinct divisors of 12. Check!

In some applications (10.3) is used more than once. Consider a small (elite) university with $N = 3174$ (or was it only 1734) students, subdivided into groups of equal size, each group being cared for by one of x tutors. The tutors in turn are supervised by y professors, each professor looking after the same number of tutors. It is clear that $y|x|N$. Repeated applications of (10.4) show that there are a total of 54 solutions for x and y. Even requiring that there will be more than one professor and more tutors than professors, leaves 31 possibilities. The university president, unhappy with this extravagant freedom of choice, fixed the number z of students in each tutor's group in such a (unique) way that the solution became unique. Not surprisingly, the president, a former mathematician, called this the perfect solution. (But what *is* his unique "perfect" z ?)

Now take some m, $1 \leqslant m \leqslant n$. Its GCD with n must be one of the divisors of n:

$$(m,n) = d_k$$

$$(10.5)$$

for some k. How many numbers are there that share the same GCD? We shall denote the size of this family by N_k:

$$N_k: = \#\{m : 1 \leqslant m \leqslant n, (m,n) = d_k\} .$$

$$(10.6)$$

Of course, by definition of Euler's function, for $d_k = 1$, $N_k = \phi(n)$. We can rewrite the above definition of N_k as follows:

$$N_k = \#\left\{m : 1 \leqslant m \leqslant n, \left[\frac{m}{d_k}, \frac{n}{d_k}\right] = 1\right\},$$

$$(10.7)$$

and now we see that

$$N_k = \phi\left[\frac{n}{d_k}\right].$$

$$(10.8)$$

Example: for $n = 12$ and $d_k = 4$, $N_k = \phi(3) = 2$, and there are indeed precisely 2 integers in the range 1 to 12 that share the GCD 2 with 12, namely 2 and 10. Check!

Now each m, $1 \leqslant m \leqslant n$, must have one of the $d(n)$ distinct divisors of n as the GCD with n. Hence,

$$\sum_{k=1}^{d(n)} \phi\left[\frac{n}{d_k}\right] = n \ . \tag{10.9}$$

Reverting to our old notation of summing over all divisors, we may write instead

$$\sum_{d|n} \phi\left[\frac{n}{d}\right] = \sum_{d|n} \phi(d) = n \ , \tag{10.10}$$

an interesting and important result.

The sum over all divisors of the argument of a number-theoretic function is called its *summatory* function. Thus, the summatory function of Euler's ϕ function is its argument!

Example: $n = 18$:

Divisors $d = 1$	Integers m for which $(m,n) = d$						Number of such integers
1	1	5	7	11	13	17	$6 = \phi\,(18)$
2	2	4	8	10	14	16	$6 = \phi\,(18/2)$
3	3	15					$2 = \phi\,(18/3)$
6	6	12					$2 = \phi\,(18/6)$
9	9						$1 = \phi\,(18/9)$
18	18						$1 = \phi\,(18/18)$

The divisor function $d\,(n)$ is *multiplicative,* i.e., for coprime n and m:

$$d\,(nm) = d\,(n) \cdot d\,(m) \quad \text{for} \ (n,m) = 1 \ , \tag{10.11}$$

which follows immediately from the formula (10.2) for $d\,(n)$ in terms of prime exponents.

For the special case that n is the product of k distinct primes, none of which is repeated,

$$d\,(n = p_1 p_2 \cdots p_k) = 2^k \ . \tag{10.12}$$

Such n are also called *squarefree,* for obvious reasons. For example, 18 is not squarefree, but 30 is, being the product of 3 distinct primes. Thus, 30 has $2^3 = 8$ divisors.

As we saw in Sect. 4.4, the probability of a large integer being squarefree is about $6/\pi^2 \approx 0.61$. Thus, a (narrow) majority of integers are squarefree. In fact, of the 100 integers from 2 to 101, exactly 61 are squarefree. And even among the first 20 integers above 1, the proportion (0.65) is already very close to the asymptotic value. Thus, in this particular area of number theory, 20 is already a large number. [But the reader should be reminded that in other areas (cf. Chap. 4) even $10^{10,000}$, for example, is not so terribly large.]

Using the notation for the sum over all divisors, we could have introduced $d(n)$ in the following way:

$$d(n) := \sum_{d|n} 1 \ . \tag{10.13}$$

Thus, $d(n)$ is the summary function of n^0.

10.2 The Average of the Divisor Function

Using the Gauss bracket, there is still another way of expressing $d(n)$ which is especially suited for estimating an asymptotic average:

$$d(n) = \sum_{k=1}^{n} \left(\lfloor \frac{n}{k} \rfloor - \lfloor \frac{n-1}{k} \rfloor \right) \ . \tag{10.14}$$

Here, if n is divisible by k, then the difference in the parentheses will be 1; otherwise it will be zero. The important point in the above expression is that it is extended over *all* k, not just the divisors of n, as in the definition of $d(n)$.

Now if we sum, we obtain

$$\sum_{n=1}^{N} d(n) = \sum_{k=1}^{N} \lfloor \frac{N}{k} \rfloor \approx N \sum_{k=1}^{N} \frac{1}{k} \ . \tag{10.15}$$

Of course, the estimate on the right is an upper limit, because by dropping the Gauss bracket we have increased (by less than 1) all summands for which N is not divisible by k. However, for large N, this increase should be relatively small. Hence we expect the average value to go with the sum of the reciprocal integers, i.e., the logarithm

$$\frac{1}{N} \sum_{n=1}^{N} d(n) \approx \ln N \ . \tag{10.16}$$

The exact result is [10.1]

$$\frac{1}{N} \sum_{n=1}^{N} d(n) = \ln N + 2\gamma - 1 + 0(1/\sqrt{N}) . \tag{10.17}$$

Here γ is Euler's constant:

$$\gamma := \lim_{n \to \infty} \left(1 + \frac{1}{2} + \frac{1}{3} + \cdots + \frac{1}{n} - \ln n \right) = 0.57721... , \tag{10.18}$$

which makes the additive constant in (10.17) about 0.15442. The notation $0(1/\sqrt{N})$ means that the absolute error in (10.17) is smaller than $c/\sqrt{N}$, where c is some constant.

10.3 The Geometric Mean of the Divisors

There is also a nice formula for the *product* of all divisors of a given integer n. With

$$n = \prod_{p_i | n} p_i^{e_i} , \tag{10.19}$$

We have

$$\prod_{d | n} d = n^{\frac{1}{2} d(n)} . \tag{10.20}$$

Example: $n = 12$. The product of divisors equals $1 \cdot 2 \cdot 3 \cdot 4 \cdot 6 \cdot 12 = 1728$. The number of divisors $d(12) = 6$ and $1728 = 12^3$. Check!

To obtain the geometric mean of the divisors of n, we have to take the $d(n)$-th root of their product, giving $\sqrt{n}$ according to (10.20).

A curious result? Not really! Divisors come in *pairs:* if d divides n, so does its distinct "mate" n/d (exception: for $n = d^2$, the mate is not distinct). And the geometric mean of each of these pairs equals $\sqrt{n}$, and so does the overall geometric mean.

10.4 The Summatory Function of the Divisor Function

The summatory function

$$\sigma(n) := \sum_{d | n} d , \tag{10.21}$$

like all summatory functions of multiplicative functions, is multiplicative.

Thus, it suffices to consider the problem first only for n that are powers of a single prime and then to multiply the individual results. This yields with (10.3):

$$\sigma(n) = \prod_{p_i \mid n} \frac{p_i^{1+e_i} - 1}{p_i - 1} .$$
(10.22)

The asymptotic behavior of $\sigma(n)$ is given by [10.1]:

$$\frac{1}{N^2} \sum_{n=1}^{N} \sigma(n) = \frac{\pi^2}{12} + 0\left[\frac{\ln N}{N}\right] ,$$
(10.23)

a result that can be understood by dropping the -1 in the denominator in each term of the product (10.23) and converting it into a product over *all* primes with the proper probabilistic factors, and proceeding as in the case of $\phi(n)/n$ (see Sect. 10.6).

The above expression converges quite rapidly. For example, for $N = 5$, the result is 0.84 compared to $\pi^2/12 \approx 0.82$.

10.5 The Generalized Divisor Functions

Generalized divisor functions are defined as follows:

$$\sigma_k(n) := \sum_{d \mid n} d^k .$$
(10.24)

Of course, $\sigma_0(n) = d(n)$ and $\sigma_1(n) = \sigma(n)$. These generalized divisor functions obey a simple symmetry with respect to their index:

$$\sigma_k(n) = \sum_{d \mid n} d^k = \sum_{d \mid n} \left[\frac{d}{n}\right]^{-k} = n^k \sigma_{-k}(n) .$$
(10.25)

Example: $n = 6$, $k = 3$. Divisors $d = 1, 2, 3, 6$.

$$1 + 8 + 27 + 216 = 216\left[1 + \frac{1}{8} + \frac{1}{27} + \frac{1}{216}\right] .$$

The above symmetry relation, for $k = 1$, leads to the following relation between divisor means of n: Arithmetic mean $\bar{d}$ times harmonic mean $\hat{d}$ equals geometric mean $\tilde{d}$ squared equals n:

$$\bar{d} \, \hat{d} = \tilde{d}^2 = n .$$
(10.26)

Example: $n = 6$; $d = 1, 2, 3, 6$. $\bar{d} = 3$, $\hat{d} = 2$, $\tilde{d} = \sqrt{6}$, $3 \cdot 2 = (\sqrt{6})^2 = 6$. Check!

10.6 The Average Value of Euler's Function

Euler's ϕ function is a pretty "wild" function. For example,

$$\phi(29) = 28, \quad \phi(30) = 8, \quad \phi(31) = 30, \quad \phi(32) = 16 .$$

If we are interested in the asymptotic behavior of $\phi(n)$, we had better consider some *average* value. The following probabilistic argument will give such an average automatically, because our probabilities ignore fine-grain fluctuations such as those in the above numerical example.

Consider first

$$\frac{\phi(n)}{n} = \prod_{p_i|n} \left(1 - \frac{1}{p_i} \right) . \tag{10.27}$$

Using our by now customary (but of course unproved) probabilistic argument, we convert the above product over primes that divide n into a product over *all* primes. The probability that "any old" prime will divide n equals $1/p_i$, and the probability that it will *not* equals $1 - 1/p_i$. In that case, the prime p_i "contributes" the factor 1 to the product. Thus, we may write

$$\prod_{p_i|n} \left(1 - \frac{1}{p_i} \right) \approx \prod_{p_i} \left[\left(1 - \frac{1}{p_i} \right) \frac{1}{p_i} + 1 \left(1 - \frac{1}{p_i} \right) \right] \tag{10.28}$$

$$= \prod_{p_i} \left[1 - \frac{1}{p_i^2} \right] ,$$

an infinite product that we have encountered before and which we calculated by converting the reciprocal of each factor into an infinite geometric series and then multiplying everything out. This produces every squared integer exactly once. Thus, we find

$$\text{average of } \frac{\phi(n)}{n} \approx \left(\sum_{n=1}^{\infty} \frac{1}{n^2} \right)^{-1} = \frac{6}{\pi^2} . \tag{10.29}$$

The result corresponds to the asymptotic probability that two arbitrarily selected integers are coprime — as it should, if we remember the definition of $\phi(n)$. In fact, the number of white dots in Fig. 4.8 in a vertical line up to the $45°$ diagonal equals Euler's function for that coordinate.

The formal results for the asymptotic behavior of $\phi(n)$ are as follows [10.1]:

$$\frac{1}{n} \sum_{k=1}^{n} \frac{\phi(k)}{k} = \frac{6}{\pi^2} + 0 \left(\frac{\ln n}{n} \right) , \quad \text{and} \tag{10.30}$$

$$\frac{1}{n^2} \sum_{k=1}^{n} \phi(k) = \frac{3}{\pi^2} + 0\left[\frac{\ln n}{n^2}\right] . \tag{10.31}$$

The result (10.30) corresponds of course to our probabilistic estimate, and the formula (10.31) is likewise unsurprising because the *average* factor inside the sum in (10.31) compared to the sum (10.30) is $n/2$.

Example: $n = 4$

$$\frac{1}{n} \sum_{k=1}^{n} \frac{\phi(k)}{k} = \frac{1}{4}\left[1 + \frac{1}{2} + \frac{2}{3} + \frac{1}{2}\right] = 0.667 ,$$

as compared to the asymptotic value 0.608. And

$$\frac{1}{n^2} \sum_{k=1}^{n} \phi(k) = \frac{1}{16}(1 + 1 + 2 + 2) = 0.375 ,$$

which also compares well with the asymptotic value (0.304).

What are the probabilities that *three* integers will not have a common divisor? And what is the probability that each of the three *pairs* that can be formed with three integers will be made up of coprime integers? The reader can find the (simple) answers or look them up in Sect. 4.4.

11. The Prime Divisor Functions

Here we consider only *prime* divisors of n and ask, for given order of magnitude of n, "how many prime divisors are there typically?" and "how many *different* ones are there?" Some of the answers will be rather counterintuitive. Thus, a 50-digit number (10^{21} times the age of our universe measured in picoseconds) has only about 5 different prime factors on average and — even more surprisingly — 50-digit numbers have typically fewer than 6 prime factors in all, even counting repeated occurrences of the same prime factor as separate factors.

We will also learn something about the distribution of the number of prime factors and its implications for the important factoring problem. Thus, we discover that even for numbers as large as 10^{50}, the two smallest primes, 2 and 3, account for about 25% of all prime factors!

11.1 The Number of Different Prime Divisors

In connection with encrypting messages by means of Euler's theorem, the number of distinct *prime* divisors of a given integer n, $\omega(n)$, is of prime importance. Its definition is similar to that of the divisor function $d(n)$, except that the sum is extended — as the name implies — only over the prime divisors of n:

$$\omega(n) := \sum_{p_i | n} 1 . \tag{11.1}$$

It is easily seen that $\omega(n)$ is additive, i.e., for $(n,m) = 1$,

$$\omega(nm) = \sum_{p_i | nm} 1 = \sum_{p_i | n} 1 + \sum_{p_i | m} 1 = \omega(n) + \omega(m) . \tag{11.2}$$

Of particular interest to our encrypting desires will be the behavior of $\omega(n)$ for large n, i.e., its asymptotic behavior. We shall try to get an idea of this behavior by means of our usual "dirty tricks." First, we will convert the sum of those primes that divide n into a sum over *all* primes up to n, using the "probability" factor $1/p_i$:

$$\omega(n) = \sum_{p_i|n} 1 \approx \sum_{p_i \leqslant n} \frac{1}{p_i} . \tag{11.3}$$

This, in turn, we will convert into a sum over all *integers* up to n, using the probability factor for primality $1/\ln x$:

$$\overline{\omega}(n) \approx \sum_{x \leqslant n} 1/(x \ln x) ,$$

which we will approximate by an integral:

$$\overline{\omega}(n) \approx \int_2^n \frac{dx}{x \ln x} = \ln(\ln n) + 0.367 \dots . \tag{11.4}$$

Of course $\omega(n)$ is a wildly fluctuating function and exact results [11.1] are available only for asymptotic averages, just as in the case of $\phi(n)$ and $d(n)$:

$$\frac{1}{n} \sum_{k=1}^n \omega(k) = \ln(\ln n) + 0(1) , \tag{11.5}$$

where $0(1)$ is a fancy way of writing a bounded quantity.

To get a better grip on this constant, we calculate the sum over the reciprocal primes in (11.3) out to some p_m and convert only the remaining sum to a sum over all integers using the probability factor $\ln x$:

$$\overline{\omega}(n) \approx \sum_{p_i=2}^{p_m} \frac{1}{p_i} + \sum_{x=p_m+1}^n 1/(x \ln x) . \tag{11.6}$$

Approximating the second sum by an integral, we have

$$\overline{\omega}(n) \approx \sum_{p_i=2}^{p_m} \frac{1}{p_i} + \ln \ln n - \ln \ln p_m . \tag{11.7}$$

In other words, our estimate tells us that the difference between $\overline{\omega}(n)$ and $\ln \ln n$, i.e., the constant in (11.5), is given by

$$\overline{\omega}(n) - \ln \ln n \approx \lim_{p_m \to \infty} \sum_{p_i=2}^{p_m} \frac{1}{p_i} - \ln \ln p_m . \tag{11.8}$$

In the last century Kronecker, assuming that the limiting average of $\omega(n)$ existed, obtained

$$\overline{\omega}(n) = \ln(\ln n) + b_1 , \tag{11.9}$$

with

$$b_1 = \gamma + \sum_{p_i=2}^{\infty} \left[\ln \left(1 - \frac{1}{p_i} \right) + \frac{1}{p_i} \right] , \tag{11.10}$$

where γ is again Euler's constant.

To compare Kronecker's constant b_1 with ours, we make use of the following asymptotic result (Merten's theorem [11.1]):

$$\lim_{p_m \to \infty} e^{\gamma} \ln p_m \prod_{p_i=2}^{p_m} \left(1 - \frac{1}{p_i} \right) = 1 , \tag{11.11}$$

which yields for Kronecker's constant

$$b_1 = \lim_{p_m \to \infty} \sum_{p_i=2}^{p_m} \frac{1}{p_i} - \ln(\ln p_m) , \tag{11.12}$$

which is identical with our "crude" estimate (11.8)!

Equation (11.12) is not very suitable to obtain a numerical value for b_1, because it converges rather slowly. (In fact, even for p_m as large as 104759, the relative error is still larger than 10^{-3}.) A faster converging series is obtained by expanding the logarithm in (11.10), which yields

$$\gamma - b_1 = \sum_{p_i=2}^{\infty} \left[\frac{1}{2p_i^2} + \frac{1}{3p_i^3} + \cdots \right] . \tag{11.13}$$

Now, if we remember the Riemann zetafunction (Chap. 4), we have

$$\zeta(k) = \sum_{n=1}^{\infty} \frac{1}{n^k} = \prod_{p_i=2}^{\infty} \left[1 - \frac{1}{p_i^k} \right]^{-1} , \tag{11.14}$$

or

$$\ln \zeta(k) = - \sum_{p_i=2}^{\infty} \ln \left(1 - \frac{1}{p_i^k} \right) . \tag{11.15}$$

Expanding the logarithm, we obtain

$$\ln \zeta(k) = \sum_{p_i=2}^{\infty} \left[\frac{1}{p_i^k} + \frac{1}{2p_i^{2k}} + \cdots \right] . \tag{11.16}$$

Introducing this result into (11.13) yields

$$\gamma - b_1 = \frac{1}{2} \ln \zeta(2) + \frac{1}{3} \ln \zeta(3) + \frac{1}{5} \ln \zeta(5) - \frac{1}{6} \ln \zeta(6) + \dots . \tag{11.17}$$

This sum written in terms of the Möbius function $\mu(m)$ (Chap. 20) is:

$$\gamma - b_1 = - \sum_{m=2}^{\infty} \frac{\mu(m)}{m} \ln \zeta(m) . \tag{11.18}$$

This sum converges very quickly and, for just 7 terms yields a relative accuracy of about 10^{-5}. The result is

$$b_1 = 0.2614 \dots . \tag{11.19}$$

How do Milton Abramowitz and Irene Stegun feel about this? On page 862 of their *Handbook of Mathematical Functions* [11.2] they list the prime factors of the integers from 9000 to 9499 (see Fig. 11.1). I have counted a total of 1260 distinct prime factors for these 500 integers. Thus, $\bar{\omega} = 2.52$, which should be compared to our $\ln(\ln 9250) + 0.26 = 2.47$. Close enough? Certainly, because as we said before, $\omega(n)$ fluctuates and an average, even over 500 consecutive integers, is not completely smooth. (More about the fluctuations of $\omega(n)$ in a moment.)

11.2 The Distribution of $\omega(n)$

The probability that the prime factor p_i does not occur in the prime factor decomposition of $n > p_i$ is given by

$$1 - \frac{1}{p_i} .$$

The probability that it *does* occur (at least once) is therefore

$$\frac{1}{p_i} .$$

The mean occurrence is therefore

$$m_i = \frac{1}{p_i} , \tag{11.20}$$

and its variance, according to the formula for the binomial distribution for two possible outcomes, equals

$$\sigma_i^2 = \frac{1}{p_i} \left(1 - \frac{1}{p_i} \right) = m_i - \frac{1}{p_i^2} . \tag{11.21}$$

COMBINATORIAL ANALYSIS

Table 24.7 **Factorizations**

9000 ≤ n ≤ 9499

N	9	8	7	6	5	4	3	2	1	0
900	$3^{2}\cdot7\cdot11\cdot13$	$2^{4}\cdot563$	9007	$2\cdot3\cdot19\cdot79$	$5\cdot1801$	$2^{2}\cdot2251$	$3\cdot3001$	$2\cdot7\cdot643$	9001	$2^{3}\cdot3^{2}\cdot5^{3}$
901	$29\cdot311$	$2\cdot3^{3}\cdot167$	$71\cdot127$	$2^{3}\cdot7^{2}\cdot23$	$3\cdot5\cdot601$	$2\cdot4507$	9013	$2^{2}\cdot3\cdot751$	9011	$2\cdot5\cdot17\cdot53$
902	9029	$2^{2}\cdot37\cdot61$	$3^{2}\cdot17\cdot59$	$2\cdot4513$	$5^{2}\cdot19^{2}$	$2^{6}\cdot3\cdot47$	$7\cdot1289$	$2\cdot13\cdot347$	$3\cdot31\cdot97$	$2^{2}\cdot5\cdot11\cdot41$
903	$3\cdot23\cdot131$	$2\cdot4519$	$7\cdot1291$	$2^{2}\cdot3^{2}\cdot251$	$5\cdot13\cdot139$	$2\cdot4517$	$3\cdot3011$	$2^{3}\cdot1129$	$11\cdot821$	$2\cdot3\cdot5\cdot7\cdot43$
904	9049	$2^{3}\cdot3\cdot13\cdot29$	$83\cdot109$	$2\cdot4523$	$3^{3}\cdot5\cdot67$	$2^{2}\cdot7\cdot17\cdot19$	9043	$2\cdot3\cdot11\cdot137$	9041	$2^{4}\cdot5\cdot113$
905	9059	$2\cdot7\cdot647$	$3\cdot3019$	$2^{5}\cdot283$	$5\cdot1811$	$2\cdot3^{2}\cdot503$	$11\cdot823$	$2^{2}\cdot31\cdot73$	$3\cdot7\cdot431$	$2\cdot5^{2}\cdot181$
906	$3\cdot3023$	$2^{2}\cdot2267$	9067	$2\cdot3\cdot1511$	$5\cdot7^{2}\cdot37$	$2^{3}\cdot11\cdot103$	$3^{2}\cdot19\cdot53$	$2\cdot23\cdot197$	$13\cdot17\cdot41$	$2^{2}\cdot3\cdot5\cdot151$
907	$7\cdot1297$	$2\cdot3\cdot17\cdot89$	$29\cdot313$	$2^{2}\cdot2269$	$3\cdot5^{2}\cdot11^{2}$	$2\cdot13\cdot349$	$43\cdot211$	$2^{4}\cdot3^{4}\cdot7$	$47\cdot193$	$2\cdot5\cdot907$
908	$61\cdot149$	$2^{7}\cdot71$	$3\cdot13\cdot233$	$2\cdot7\cdot11\cdot59$	$5\cdot23\cdot79$	$2^{2}\cdot3\cdot757$	$31\cdot293$	$2\cdot19\cdot239$	$3^{2}\cdot1009$	$2^{3}\cdot5\cdot227$
909	$3^{3}\cdot337$	$2\cdot4549$	$11\cdot827$	$2^{3}\cdot3\cdot379$	$5\cdot17\cdot107$	$2\cdot4547$	$3\cdot7\cdot433$	$2^{2}\cdot2273$	9091	$2\cdot3^{2}\cdot5\cdot101$
910	9109	$2^{2}\cdot3^{2}\cdot11\cdot23$	$7\cdot1301$	$2\cdot29\cdot157$	$3\cdot5\cdot607$	$2^{4}\cdot569$	9103	$2\cdot3\cdot37\cdot41$	$19\cdot479$	$2^{2}\cdot5^{2}\cdot7\cdot13$
911	$11\cdot829$	$2\cdot47\cdot97$	$3^{2}\cdot1013$	$2^{2}\cdot43\cdot53$	$5\cdot1823$	$2\cdot3\cdot7^{2}\cdot31$	$13\cdot701$	$2^{3}\cdot17\cdot67$	$3\cdot3037$	$2\cdot5\cdot911$
912	$3\cdot17\cdot179$	$2^{3}\cdot7\cdot163$	9127	$2\cdot3^{3}\cdot13^{2}$	$5^{3}\cdot73$	$2^{2}\cdot2281$	$3\cdot3041$	$2\cdot4561$	$7\cdot1303$	$2^{5}\cdot3\cdot5\cdot19$
913	$13\cdot19\cdot37$	$2\cdot3\cdot1523$	9137	$2^{4}\cdot571$	$3^{2}\cdot5\cdot7\cdot29$	$2\cdot4567$	9133	$2^{2}\cdot3\cdot761$	$23\cdot397$	$2\cdot5\cdot11\cdot83$
914	$7\cdot1307$	$2^{2}\cdot2287$	$3\cdot3049$	$2\cdot17\cdot269$	$5\cdot31\cdot59$	$2^{3}\cdot3^{2}\cdot127$	$41\cdot223$	$2\cdot7\cdot653$	$3\cdot11\cdot277$	$2^{2}\cdot5\cdot457$
915	$3\cdot43\cdot71$	$2\cdot19\cdot241$	9157	$2^{2}\cdot3\cdot7\cdot109$	$5\cdot1831$	$2\cdot23\cdot199$	$3^{4}\cdot113$	$2^{6}\cdot11\cdot13$	9151	$2\cdot3\cdot5^{2}\cdot61$
916	$53\cdot173$	$2^{4}\cdot3\cdot191$	$89\cdot103$	$2\cdot4583$	$3\cdot5\cdot13\cdot47$	$2^{2}\cdot29\cdot79$	$7^{2}\cdot11\cdot17$	$2\cdot3^{2}\cdot509$	9161	$2^{3}\cdot5\cdot229$
917	$67\cdot137$	$2\cdot13\cdot353$	$3\cdot7\cdot19\cdot23$	$2^{3}\cdot31\cdot37$	$5^{2}\cdot367$	$2\cdot3\cdot11\cdot139$	9173	$2^{2}\cdot2293$	$3^{2}\cdot1019$	$2\cdot5\cdot7\cdot131$
918	$3^{2}\cdot1021$	$2^{2}\cdot2297$	9187	$2\cdot3\cdot1531$	$5\cdot11\cdot167$	$2^{5}\cdot7\cdot41$	$3\cdot3061$	$2\cdot4591$	9181	$2^{2}\cdot3^{3}\cdot5\cdot17$
919	9199	$2\cdot3^{2}\cdot7\cdot73$	$17\cdot541$	$2^{2}\cdot11^{2}\cdot19$	$3\cdot5\cdot613$	$2\cdot4597$	$29\cdot317$	$2^{3}\cdot3\cdot383$	$7\cdot13\cdot101$	$2\cdot5\cdot919$
920	9209	$2^{3}\cdot1151$	$3^{3}\cdot11\cdot31$	$2\cdot4603$	$5\cdot7\cdot263$	$2^{2}\cdot3\cdot13\cdot59$	9203	$2\cdot43\cdot107$	$3\cdot3067$	$2^{4}\cdot5^{2}\cdot23$
921	$3\cdot7\cdot439$	$2\cdot11\cdot419$	$13\cdot709$	$2^{10}\cdot3^{2}$	$5\cdot19\cdot97$	$2\cdot17\cdot271$	$3\cdot37\cdot83$	$2^{2}\cdot7^{2}\cdot47$	$61\cdot151$	$2\cdot3\cdot5\cdot307$
922	$11\cdot839$	$2^{2}\cdot3\cdot769$	9227	$2\cdot7\cdot659$	$3^{2}\cdot5^{2}\cdot41$	$2^{3}\cdot1153$	$23\cdot401$	$2\cdot3\cdot29\cdot53$	9221	$2^{2}\cdot5\cdot461$
923	9239	$2\cdot31\cdot149$	$3\cdot3079$	$2^{2}\cdot2309$	$5\cdot1847$	$2\cdot3^{5}\cdot19$	$7\cdot1319$	$2^{4}\cdot577$	$3\cdot17\cdot181$	$2\cdot5\cdot13\cdot71$
924	$3\cdot3083$	$2^{5}\cdot17^{2}$	$7\cdot1321$	$2\cdot3\cdot23\cdot67$	$5\cdot43^{2}$	$2^{2}\cdot2311$	$3^{2}\cdot13\cdot79$	$2\cdot4621$	9241	$2^{3}\cdot3\cdot5\cdot7\cdot11$
925	$47\cdot197$	$2\cdot3\cdot1543$	9257	$2^{3}\cdot13\cdot89$	$3\cdot5\cdot617$	$2\cdot7\cdot661$	$19\cdot487$	$2^{2}\cdot3^{2}\cdot257$	$11\cdot29^{2}$	$2\cdot5^{3}\cdot37$
926	$13\cdot23\cdot31$	$2^{2}\cdot7\cdot331$	$3\cdot3089$	$2\cdot41\cdot113$	$5\cdot17\cdot109$	$2^{4}\cdot3\cdot193$	$59\cdot157$	$2\cdot11\cdot421$	$3^{3}\cdot7^{3}$	$2^{2}\cdot5\cdot463$
927	$3^{2}\cdot1031$	$2\cdot4639$	9277	$2^{2}\cdot3\cdot773$	$5^{2}\cdot7\cdot53$	$2\cdot4637$	$3\cdot11\cdot281$	$2^{3}\cdot19\cdot61$	$73\cdot127$	$2\cdot3^{2}\cdot5\cdot103$
928	$7\cdot1327$	$2^{3}\cdot3^{3}\cdot43$	$37\cdot251$	$2\cdot4643$	$3\cdot5\cdot619$	$2^{2}\cdot11\cdot211$	9283	$2\cdot3\cdot7\cdot13\cdot17$	9281	$2^{6}\cdot5\cdot29$
929	$17\cdot547$	$2\cdot4649$	$3^{2}\cdot1033$	$2^{4}\cdot7\cdot83$	$5\cdot11\cdot13^{2}$	$2\cdot3\cdot1549$	9293	$2^{2}\cdot23\cdot101$	$3\cdot19\cdot163$	$2\cdot5\cdot929$
930	$3\cdot29\cdot107$	$2^{2}\cdot13\cdot179$	$41\cdot227$	$2\cdot3^{2}\cdot11\cdot47$	$5\cdot1861$	$2^{3}\cdot1163$	$3\cdot7\cdot443$	$2\cdot4651$	$71\cdot131$	$2^{2}\cdot3\cdot5^{2}\cdot31$
931	9319	$2\cdot3\cdot1553$	$7\cdot11^{3}$	$2^{2}\cdot17\cdot137$	$3^{4}\cdot5\cdot23$	$2\cdot4657$	$67\cdot139$	$2^{5}\cdot3\cdot97$	9311	$2\cdot5\cdot7^{2}\cdot19$
932	$19\cdot491$	$2^{4}\cdot11\cdot53$	$3\cdot3109$	$2\cdot4663$	$5^{2}\cdot373$	$2^{2}\cdot3^{2}\cdot7\cdot37$	9323	$2\cdot59\cdot79$	$3\cdot13\cdot239$	$2^{3}\cdot5\cdot233$
933	$3\cdot11\cdot283$	$2\cdot7\cdot23\cdot29$	9337	$2^{3}\cdot3\cdot389$	$5\cdot1867$	$2\cdot13\cdot359$	$3^{2}\cdot17\cdot61$	$2^{2}\cdot2333$	$7\cdot31\cdot43$	$2\cdot3\cdot5\cdot311$
934	9349	$2^{2}\cdot3\cdot19\cdot41$	$13\cdot719$	$2\cdot4673$	$3\cdot5\cdot7\cdot89$	$2^{7}\cdot73$	9343	$2\cdot3^{3}\cdot173$	9341	$2^{2}\cdot5\cdot467$
935	$7^{2}\cdot191$	$2\cdot4679$	$3\cdot3119$	$2^{2}\cdot2339$	$5\cdot1871$	$2\cdot3\cdot1559$	$47\cdot199$	$2^{3}\cdot7\cdot167$	$3^{2}\cdot1039$	$2\cdot5^{2}\cdot11\cdot17$
936	$3^{3}\cdot347$	$2^{3}\cdot1171$	$17\cdot19\cdot29$	$2\cdot3\cdot7\cdot223$	$5\cdot1873$	$2^{2}\cdot2341$	$3\cdot3121$	$2\cdot31\cdot151$	$11\cdot23\cdot37$	$2^{4}\cdot3^{2}\cdot5\cdot13$
937	$83\cdot113$	$2\cdot3^{2}\cdot521$	9377	$2^{5}\cdot293$	$3\cdot5^{5}$	$2\cdot43\cdot109$	$7\cdot13\cdot103$	$2^{2}\cdot3\cdot11\cdot71$	9371	$2\cdot5\cdot937$
938	$41\cdot229$	$2^{2}\cdot2347$	$3^{2}\cdot7\cdot149$	$2\cdot13\cdot19^{2}$	$5\cdot1877$	$2^{3}\cdot3\cdot17\cdot23$	$11\cdot853$	$2\cdot4691$	$3\cdot53\cdot59$	$2^{2}\cdot5\cdot7\cdot67$
939	$3\cdot13\cdot241$	$2\cdot37\cdot127$	9397	$2^{2}\cdot3^{4}\cdot29$	$5\cdot1879$	$2\cdot7\cdot11\cdot61$	$3\cdot31\cdot101$	$2^{4}\cdot587$	9391	$2\cdot3\cdot5\cdot313$
940	97^{2}	$2^{6}\cdot3\cdot7^{2}$	$23\cdot409$	$2\cdot4703$	$3^{2}\cdot5\cdot11\cdot19$	$2^{2}\cdot2351$	9403	$2\cdot3\cdot1567$	$7\cdot17\cdot79$	$2^{3}\cdot5^{2}\cdot47$
941	9419	$2\cdot17\cdot277$	$3\cdot43\cdot73$	$2^{3}\cdot11\cdot107$	$5\cdot7\cdot269$	$2\cdot3^{2}\cdot523$	9413	$2^{2}\cdot13\cdot181$	$3\cdot3137$	$2\cdot5\cdot941$
942	$3\cdot7\cdot449$	$2^{2}\cdot2357$	$11\cdot857$	$2\cdot3\cdot1571$	$5^{2}\cdot13\cdot29$	$2^{4}\cdot19\cdot31$	$3^{3}\cdot349$	$2\cdot7\cdot673$	9421	$2^{2}\cdot3\cdot5\cdot157$
943	9439	$2\cdot3\cdot11^{2}\cdot13$	9437	$2^{2}\cdot7\cdot337$	$3\cdot5\cdot17\cdot37$	$2\cdot53\cdot89$	9433	$2^{3}\cdot3^{2}\cdot131$	9431	$2\cdot5\cdot23\cdot41$
944	$11\cdot859$	$2^{3}\cdot1181$	$3\cdot47\cdot67$	$2\cdot4723$	$5\cdot1889$	$2^{2}\cdot3\cdot787$	$7\cdot19\cdot71$	$2\cdot4721$	$3^{2}\cdot1049$	$2^{5}\cdot5\cdot59$
945	$3^{2}\cdot1051$	$2\cdot4729$	$7^{2}\cdot193$	$2^{4}\cdot3\cdot197$	$5\cdot31\cdot61$	$2\cdot29\cdot163$	$3\cdot23\cdot137$	$2^{2}\cdot17\cdot139$	$13\cdot727$	$2\cdot3^{3}\cdot5^{2}\cdot7$
946	$17\cdot557$	$2^{2}\cdot3^{2}\cdot263$	9467	$2\cdot4733$	$3\cdot5\cdot631$	$2^{3}\cdot7\cdot13^{2}$	9463	$2\cdot3\cdot19\cdot83$	9461	$2^{2}\cdot5\cdot11\cdot43$
947	9479	$2\cdot7\cdot677$	$3^{6}\cdot13$	$2^{2}\cdot23\cdot103$	$5^{2}\cdot379$	$2\cdot3\cdot1579$	9473	$2^{8}\cdot37$	$3\cdot7\cdot11\cdot41$	$2\cdot5\cdot947$
948	$3\cdot3163$	$2^{4}\cdot593$	$53\cdot179$	$2\cdot3^{2}\cdot17\cdot31$	$5\cdot7\cdot271$	$2^{2}\cdot2371$	$3\cdot29\cdot109$	$2\cdot11\cdot431$	$19\cdot499$	$2^{3}\cdot3\cdot5\cdot79$
949	$7\cdot23\cdot59$	$2\cdot3\cdot1583$	9497	$2^{3}\cdot1187$	$3^{2}\cdot5\cdot211$	$2\cdot47\cdot101$	$11\cdot863$	$2^{2}\cdot3\cdot7\cdot113$	9491	$2\cdot5\cdot13\cdot73$

Fig. 11.1. The prime factors of n in the range $9000 \leqslant n \leqslant 9499$. The number of distinct prime factors in this range is 1260; the corresponding theoretical expectation equals 1237 ± 32. The number of prime factors, including multiple occurrences, is 1650, compared to a theoretical expectation of 1632 ± 31

Assuming divisibility by different primes to be independent, we get for the overall mean

$$\bar{\omega}(n) \approx \sum_{p_i < n} \frac{1}{p_i} \approx \ln(\ln n) + 0.2614 , \tag{11.22}$$

as before [see (11.9) and (11.19)]. The overall variance becomes, with (11.21),

$$\sigma_\omega^2 \approx \bar{\omega}(n) - \sum_{p_i = 2}^{\infty} \frac{1}{p_i^2} , \tag{11.23}$$

where we have extended the sum out to infinity because it converges quite rapidly.

The numerical value of the sum can be obtained most efficiently with the help of Riemann's zetafunction, expanded as in (11.16). This yields

$$\sum_{p_i = 2}^{\infty} \frac{1}{p_i^2} = \ln \zeta(2) - \frac{1}{2} \ln \zeta(4) - \dots$$

$$= \sum_{m=1}^{\infty} \frac{\mu(m)}{m} \ln \zeta(2m) \approx 0.452248 , \tag{11.24}$$

where $\mu(m)$ is again the Möbius function (see Chap. 20).

Thus,

$$\sigma_\omega^2 \approx \bar{\omega}(n) - 0.45 \tag{11.25}$$

and, because $\sigma_\omega^2 \approx \bar{\omega}$, we expect ω to be approximately *Poisson* distributed [11.3]. Of course, each number has at least *one* prime factor (itself, if it is prime), so that the Poisson distribution must be shifted by 1:

$$\text{Prob } \{\omega(n) = k\} \approx \frac{(\bar{\omega}-1)^{k-1}}{(k-1)!} e^{-\bar{\omega}+1} , \quad k = 1, 2, \dots, \bar{\omega} > 1 , \tag{11.26}$$

with $\bar{\omega}$ from (11.22).

The mode (most probable value) of this distribution occurs for

$$\check{k} = \lfloor \bar{\omega} \rfloor + 1 , \tag{11.27}$$

where $\check{k}$ is read "kay check." Although intended for large n, (11.27) seems to work very well even for small n. Equation (11.27) predicts that the most probable number $\check{k}$ of different prime factors of n is as follows:

$\overset{v}{k} = 1$ for $\qquad\qquad n < 9$

$\overset{v}{k} = 2$ for $\qquad\quad 9 \leqslant n < 296$

$\overset{v}{k} = 3$ for $\qquad 296 \leqslant n < 5 \cdot 10^6$

$\overset{v}{k} = 4$ for $\quad 5 \cdot 10^6 \leqslant n < 2 \cdot 10^{18}$

$\overset{v}{k} = 5$ for $\quad 2 \cdot 10^{18} \leqslant n < 4 \cdot 10^{49}$

$\overset{v}{k} = 6$ for $\quad 4 \cdot 10^{49} \leqslant n < 8 \cdot 10^{134}$ etc.

Thus, up to almost 10^{135} the most likely number of different prime factors is 6 or less!

According to (11.26), the probability that n has exactly one prime factor, i.e., that n is either a prime or a prime power, equals about $2/\ln n$. This value is somewhat larger than the one we would expect from the distribution of primes. But then, we should not expect the Poisson distribution for $\omega(n)$ to be exact. For example, σ_ω^2 should equal $\overline{\omega} - 1$ for the shifted Poisson distribution and not $\overline{\omega} - 0.45$ as in (11.23) and (11.24).

11.3 The Number of Prime Divisors

Apart from the "little" $\omega(n)$ we need a "big" $\Omega(n)$, the number of prime divisors of n, counted with multiplicity. For

$$n = \prod_{p_i|n} p_i^{e_i}, \tag{11.28}$$

we have the definition

$$\Omega(n) := \sum_{p_i|n} e_i. \tag{11.29}$$

The divisor function $\Omega(n)$ is *completely* additive, i.e.,

$$\Omega(mn) = \Omega(m) + \Omega(n), \tag{11.30}$$

whether m and n are coprime or not.

To estimate an average value of $\Omega(n)$, we convert the sum appearing in its definition into a sum over all primes up to n:

$$\Omega(n) \approx \sum_{p_i \leqslant n} e_i \frac{1}{p_i^{e_i}} \left(1 - \frac{1}{p_i}\right), \tag{11.31}$$

recognizing that the probability that p_i occurs e_i times equals $(1 - 1/p_i)/p_i^{e_i}$. Averaging over these values of e_i yields

$$\overline{\Omega}(n) \approx \sum_{p_i \leqslant n} \frac{1}{p_i - 1} . \tag{11.32}$$

Note the closeness of our estimates of $\overline{\Omega}(n)$ and $\overline{\omega}(n)$ according to (11.3)! The difference (which some friends did not even think converged) is given by

$$\overline{\Omega}(n) - \overline{\omega}(n) \approx \sum_{p_i \leqslant n} \frac{1}{p_i(p_i - 1)} , \tag{11.33}$$

in agreement with a result by Kronecker. (This sum is upperbounded by the sum over all integers out to infinity, which equals 1.)

Since the sum does not only converge, but converges quite rapidly, we will only bother about its value taken out to infinity. First we write

$$\sum_{p_i=2}^{\infty} \frac{1}{p_i(p_i-1)} = \sum_{p_i=2}^{\infty} \left[\frac{1}{p_i^2} + \frac{1}{p_i^3} + \frac{1}{p_i^4} + ... \right] \tag{11.34}$$

and then introduce the zetafunction again, making use of (11.16). This yields

$$\sum_{p_i=2}^{\infty} \frac{1}{p_i(p_i-1)} = \ln \zeta(2) + \ln \zeta(3)$$

$$+ \frac{1}{2} \ln \zeta(4) + \ln \zeta(5)$$

$$+ \frac{1}{6} \ln \zeta(6) + ... \approx 0.77317 \qquad \text{or} \tag{11.35}$$

$$\overline{\Omega}(n) \approx \overline{\omega}(n) + 0.77317 \approx \ln(\ln n) + 1.0346 . \tag{11.36}$$

What do Abramowitz and Stegun have to say? In their table of prime factors for n in the range 9000 to 9499 [Ref. 11.2, p. 862], I counted a total of 1650 prime factors, including multiplicity, yielding $\overline{\Omega} = 3.30$. Our theoretical value $\ln(\ln 9250) + 1.0346 \approx 3.25$, which is as similar as could be expected.

Incidentally, sums taken over all primes, with primes appearing in the denominator as in (11.13) and (11.34), need not always lead to irrational results. A noteworthy counterexample (from an entire family of like-fashioned expressions) is

$$\prod_{p=2}^{\infty} \frac{p^2+1}{p^2-1} = \frac{5}{2} . \tag{11.37}$$

This seems preposterous, but a quick numerical check indicates that the product certainly could not deviate much from 5/2, and in fact, the infinite product *does* equal 5/2. This is actually not too difficult to see, because

$$\Pi \; \frac{p^2+1}{p^2-1} = \Pi \; \frac{p^4-1}{(p^2-1)^2} = \Pi \; \frac{1 - \dfrac{1}{p^4}}{\left(1 - \dfrac{1}{p^2}\right)^2} \; ,$$

or, expanding into geometric series:

$$\Pi \; \frac{p^2+1}{p^2-1} = \frac{\Pi\left(1 + \dfrac{1}{p^2} + \dfrac{1}{p^4} + ...\right)^2}{\Pi\left(1 + \dfrac{1}{p^4} + \dfrac{1}{p^8} + ...\right)} = \frac{\left(\displaystyle\sum_{n=1}^{\infty} \dfrac{1}{n^2}\right)^2}{\displaystyle\sum_{n=1}^{\infty} \dfrac{1}{n^4}} \; . \tag{11.38}$$

We have encountered the sum in the numerator several times before (Chaps. 4, 8), and found it to equal $\pi^2/6$. The sum in the denominator equals $\zeta(4) = \pi^4/90$, and if we had not heard of the zetafunction, we could find out by calculating a certain definite integral over the Fourier series

$$\sin x - \sin 3x + \sin 5x - ... .$$

(The reader may want to try this.) The result is

$$\prod_{p=2}^{\infty} \frac{p^2+1}{p^2-1} = \frac{\dfrac{\pi^4}{36}}{\dfrac{\pi^4}{90}} = \frac{5}{2} \; .$$

Consideration of this product also leads to some rather unexpected relations for $\Omega(n)$. Expanding

$$\Pi \; \frac{p^2+1}{p^2-1} = \frac{\Pi\left(1 + \dfrac{1}{p^2} + \dfrac{1}{p^4} + ...\right)}{\Pi\left(1 - \dfrac{1}{p^2} + \dfrac{1}{p^4} - ...\right)} \tag{11.39}$$

and multiplying out, one obtains, in the denominator, a sum of each reciprocal square $1/n^2$ exactly once, with a sign that depends on the parity (odd or even) of the total numbers of prime factors of n. Thus, with (11.37), remembering that $\Omega(1) = 0$:

$$\sum_{n=1}^{\infty} \frac{(-1)^{\Omega(n)}}{n^2} = \frac{\pi^2}{15} \; , \tag{11.40}$$

or

$$\sum_{\Omega(n)\,\text{odd}} \frac{1}{n^2} = \frac{\pi^2}{20} \, , \tag{11.41}$$

two noteworthy results.

Similar procedures give the equally remarkable

$$\sum_{n=1}^{\infty} \frac{(\pm 1)^{\Omega(n)}}{n^2} \, 2^{\omega(n)} = \left[\frac{5}{2}\right]^{\pm 1} \, , \tag{11.42}$$

or

$$\sum_{\Omega(n)\,\text{odd}} \frac{2^{\omega(n)}}{n^2} = \frac{21}{20} \, . \tag{11.43}$$

11.4 The Harmonic Mean of $\Omega(n)$

In order to estimate, as we would like to, the geometric mean of the prime factors of n, we need the *harmonic* mean of $\Omega(n)$. If we designate geometric means by a tilde, then the desired mean is given by

$$\tilde{p}(n) := n^{1/\Omega(n)} \, . \tag{11.44}$$

Now if we average over several (similar) values of n, we are led to the harmonic mean of $\Omega(n)$, which we identify by a "hat":

$$\hat{\Omega}(n) := (\overline{1/\Omega(n)})^{-1} \, . \tag{11.45}$$

With this notation, we have

$$\tilde{p}(n) \approx n^{1/\hat{\Omega}(n)} \, . \tag{11.46}$$

Of course, like any harmonic mean of a fluctuating quantity, $\hat{\Omega}(n)$ is smaller than the previously computed arithmetic mean $\overline{\Omega}(n) \approx \ln(\ln n) + 1.035$. By how much? To answer this question, we have to find out about the *distribution* of $\Omega(n)$. Reverting to our earlier "unaveraged" estimate of $\Omega(n)$:

$$\Omega(n) \approx \sum_{p_i \leq n} e_i \, \frac{1}{p_i^{e_i}} \left[1 - \frac{1}{p_i}\right] \, , \tag{11.47}$$

we recognize geometric distributions[1] in the exponents e_i. The mean value m_i for each term of the sum is

$$m_i = \frac{1}{p_i - 1} , \tag{11.48}$$

a result we used before in estimating $\overline{\Omega}(n)$.

Now we also want the *variance* σ_i^2 of each term, which for a geometric distribution is given in terms of the mean m_i by the following well-known formula[2]:

$$\sigma_i^2 = m_i + m_i^2 . \tag{11.49}$$

By summing over the index i, assuming independence of the p_i, we obtain the variance of $\Omega(n)$:

$$\sigma_\Omega^2 = \overline{\Omega}(n) + \sum_{p_i-2} \frac{1}{(p_i-1)^2} . \tag{11.50}$$

Using the expansion (11.16) again, we can write the sum here as

$$\sum_{p_i} \left[\frac{1}{p_i^2} + \frac{2}{p_i^3} + \frac{3}{p_i^4} + ... \right] = \ln \zeta(2) + 2\ln \zeta(3)$$

$$+ \frac{5}{2} \ln \zeta(4) + ... \approx 1.3751 . \tag{11.51}$$

Again, $\sigma_\Omega^2 \approx \overline{\Omega}$, and we also expect a shifted Poisson distribution for Ω:

$$\text{Prob}\{\Omega(n) = k\} \approx \frac{(\overline{\Omega}-1)^{k-1}}{(k-1)!} e^{-\overline{\Omega}+1} , \quad k = 1,2, ..., \overline{\Omega} > 1 , \tag{11.52}$$

with $\overline{\Omega}$ from (11.36).

---

[1] Physicists call a related distribution "Bose-Einstein" in honor of Bose, the Indian scientist who discovered its significance for photons and other "bosons", and Einstein, who publicized it when people would not believe it.

[2] This formula played a role in physics that can hardly be overestimated. According to Maxwell's equations, the intensity fluctuations σ_i^2 in "black-body" radiation should equal the squared intensity m_i^2. It was Einstein who discovered, from deep considerations of entropy, that the actual fluctuations exceeded m_i^2 by m_i, recognizing the additional term m_i as stemming from a non-Maxwellian "granularity" of the field. This observation led him to the *photon* concept for electromagnetic radiation on much more persuasive grounds than Planck's inherently contradictory discretization of the energies of harmonic oscillators. As a result, Einstein believed in the reality of the photons from 1905 on (and he received his Nobel prize in physics for this work and not for his theory of relativity), while Planck continued to doubt the meaningfulness of his "ad hoc" trick.

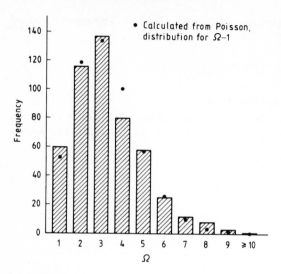

Fig. 11.2. The distribution of the number of prime factors (*bars*) in the interval $9000 \leqslant n \leqslant 9499$ and the Poisson distribution (*dots*) for the theoretical mean

This theoretical distribution is shown by dots in Fig. 11.2 for $\overline{\Omega} = 3.25$ ($n \approx 9500$). The shaded bars are from actual prime factor counts between $n = 9000$ and 9499. The agreement is remarkably good.

For the shifted Poisson distribution, the harmonic mean $\hat{\Omega}$ is easily evaluated:

$$\hat{\Omega} = \frac{\overline{\Omega}-1}{1 - e^{-\overline{\Omega}+1}} , \quad \text{or} \tag{11.53}$$

$$\hat{\Omega} \approx \frac{\ln(1.035 \ln n)}{1 - 1/1.035 \ln n} . \tag{11.54}$$

For $n = 9250$, we obtain $\hat{\Omega} \approx 2.51$. The "experimental" value for the range $n = 9000$ to 9499 is $\hat{\Omega} = 2.47$ — as close as one can hope.

The geometric mean of the prime factors, calculated with the theoretical value of $\hat{\Omega}$, becomes

$$\bar{p}(9250) \approx 38 ,$$

while the actual value in the range $n = 9000$ to 9499 is $\bar{p} = 40$.

For $n = 10^{50}$, a range of interest for public-key encryption, $\hat{\Omega} \approx 4.8$, and the geometric mean $\bar{p} \approx 2.4 \cdot 10^{10}$ — 40 orders of magnitude smaller than n.

11.5 Medians and Percentiles of $\Omega(n)$

With (11.36), the probability that the integer n equals a prime that divides N is given approximately by

$$w(n = p \mid N) = \frac{1}{n \ln n \, (\ln(\ln N) + 1.035)} \, . \qquad (11.55)$$

Thus, the cumulative distribution for a prime divisor of N to be smaller than n is approximated by

$$W(n;N) = \frac{\ln(\ln n) + 1.035}{\ln(\ln N) + 1.035} \, . \qquad (11.56)$$

From this expression the median value $n_{0.5}$ follows directly:

$$n_{0.5} = e^{\sqrt{\ln N / 2.81}} \, . \qquad (11.57)$$

Example: $N = 9250$, $n_{0.5} = 6$. Thus, the primes 2, 3, and 5 should account for roughly half the prime factors around $N = 9250$. The actual count in the interval 9000 to 9499 is as follows (with the theoretical value, $500/(p_i - 1)$, in parenthesis):

$p_i = 2$: 500 times   (500)

$p_i = 3$: 250 times   (250)

$p_i = 5$: 126 times   (125)

Thus, the total number of occurrences of 2, 3, and 5 is 876 times, or 53% of the total of 1650 prime factors in that interval — in very good agreement with our theoretical prediction.

For $N = 10^{50}$, $n_{0.5} = 600$, a remarkably small value.

The above distribution formula gives the following percentile values n_f, defined by $W(n_f) = f$:

$$n_f(N) = \exp[(\ln N)^f 2.81^{f-1}] \, . \qquad (11.58)$$

The lower-quartile value $n_{0.25}$ (for $N = 9250$) becomes 2.2, which compares well with the count of 30% (500 out of 1650) for the factor 2 in the interval 9000 to 9499. In fact, according to (11.56), 29% of the prime factors should be below 2.5.

The theoretical upper-quartile value $n_{0.75} = 57.8$ is in very good agreement with the count of 75%. (1231 out of 1650) prime factors up to and including 59. But the median of the largest prime factor of N is about $N^{0.6}$ (!).

11.6 Implications for Public-Key Encryption

For $N = 10^{50}$, the theoretical lower- and upper-quartile values for the prime factors are 4.5 and $6 \cdot 10^{11}$, respectively. Thus, in three out of four cases of integers around 10^{50}, one will encounter prime factors not exceeding $6 \cdot 10^{11}$. If one assumes that *rapid* factoring of such integers is no problem, then 75% of such large, *randomly* selected integers can be easily factored.

This conclusion is in stark contrast to the (correct) assertion that sufficiently large integers *constructed so as to contain only two very large prime factors* cannot be easily factored.

Additional results on large prime factors in a given interval can be found in [11.4]. As an introduction to the art of generating large primes, see [11.5, 6].

12. Certified Signatures

Here we learn how certified signatures can be attached to secret messages in the context of public-key encryption. The degree of certitude (in the sense of avoiding random confusions) achievable by this method, which is based on modular arithmetic, appears to exceed by far that of notarized signatures, fingerprinting or, conceivably, even genetic analysis.

Certified signatures are also important in protecting computer systems against illicit entry and manipulation, and safeguarding data files from unauthorized "readers," falsification or destruction.

12.1 A Story of Creative Financing

Baron von Münchhausen, a close relative of the fabulous liar of the same name, and founder of the Georg-August University at Göttingen under the auspices of his King in Hanover, Georg August[1], received a secret message in (say) 1743 saying (in part):

"SPEND ALL EXCESS FUNDS OF KINGDOM ON NEW UNIVERSITY IN GOTTINGEN." signed "GEORGE."

How does von Münchhausen know that it was really King George who sent that generous but unlikely message? George is about to establish two more institutions of higher learning in his American colonies: King's College on an island called Manhattan (later to be known as Columbia University) and the College of New Jersey (now Princeton University) and the royal treasury has few, if any, "excess funds" to throw in the direction of Göttingen. The signature looks fine, but it could have been faked.

12.2 Certified Signature for Public-Key Encryption

In one of the great advances of modern secure and reliable communication (apart from public-key encryption itself), *certified signatures* can now be attached to public-key encryption messages in such a manner as to remove any doubt about the sender [12.1]. This works as follows: The sender, call

---

[1]   Also known in London as George II, King of England, etc., etc.

him N, encrypts his name, address, etc., by his *de*crypting key t_o (which only he knows!). Thus, he forms

$$S \equiv N^{t_o} \pmod{r_o} ,\tag{12.1}$$

which he appends to his message M (which includes his name) and encrypts both M and S by the (public!) encrypting key of the receiver s_1, r_1.

The receiver decrypts using his secret key t_1, and reads the message M followed by a string of "garbled" symbols S, which must be the certified signature, because the message was identified as carrying such a signature. The message also purports to have been sent by N. Thus, knowing the protocol, the receiver applies the publicly known encrypting parameters of N, namely s_o and r_o, to S and obtains

$$S^{s_o} \equiv N^{t_o s_o} \equiv N \pmod{r_o} ,\tag{12.2}$$

i.e., the name and address, etc., of the sender. And no one, but no one, who did not know t_o, could have constructed S so that with the above operation it would yield N. A certified signature to put all other "certified" signatures — including fingerprinting and (present-day) genetic analysis — to shame!

The reader can find further information on digital signatures and authentications to counteract potential threats[2] in financial, diplomatic and military "transactions" in [12.2-4].

---

[2]
reneging	the *originator* subsequently disowns a transaction
forgery	the *recipient* fabricates a transaction
alteration	the *recipient* alters a previous valid transaction
masquerading	an *originator* attempts to masquerade as another

13. Primitive Roots

In this chapter we introduce the concepts of *order* and the *primitive root*, two of the more fascinating and useful ideas in number theory. On the fundamental side, they helped the young Gauss to reduce the equation $x^{16} + x^{15} + \ldots + x + 1 = 0$ to several quadratic equations leading to the construction of the regular 17-gon. These same concepts also allow us to see why the decimal fraction of $1/7$ has a period of length 6, while the decimal fraction for $1/11$ has a period of only 2. And why does $1/99007599$, written as a binary fraction, have a period of nearly 50 million 0's and 1's? We shall see!

Closely related to the primitive root is the concept of *index*, a kind of number-theoretic logarithm that permits us to solve exponential Diophantine equations and even show that $2^n = 3^m - 1$ has only two, and precisely two solutions ($n = 1$, $m = 1$ and $n = 3$, $m = 2$).

Periodic sequences constructed from primitive roots also have an interesting Fourier-transform property that permits the construction of wave-scattering surfaces with very broad scatter and little specular reflection. Such surfaces can be useful in improving concert hall acoustics, in noise abatement measures, and in making ships and planes more difficult to see by sonar or radar. And, of course, there are applications to our main theme: digital encryption and electronic contracting (Sect. 19.3).

13.1 Orders

Some of the things we want to accomplish by electronic mail — other than public key encryption and certified signatures — have to do with certifiable "coin tossing," *registered mail* with or without *receipt*, and *signed contracts*. For these options we need the number-theoretic concepts of a *primitive root* and a *quadratic residue*, both delightful entities in their own right.

Let us look at increasing powers of 2 modulo 7:

$n = 1$	2	3	4	5	6	
$2^n \equiv 2$	4	1	2	4	1	etc.

Here the period after which the sequence repeats for the first time is obviously 3. One therefore says that the integer 2 has *order* 3 modulo 7:

$$\text{ord}_7 2 = 3 \ . \tag{13.1}$$

Of course, the order of any integer must be divisor of $p-1$, where p is the prime modulus (7 in our example). This is so because of Fermat's theorem, which requires of any integer b coprime to the modulus p that the congruence $b^{p-1} \equiv 1 \pmod{p}$ must hold. Obviously, for $p = 7$, the order could never be 4, for example, because then the sequence of powers would repeat after 4 and 8 steps, etc., and not, as required by Fermat, after $p-1 = 6$ steps.

What is the largest order of any integer modulo a prime p? Certainly it cannot be larger than $p-1$, because there are only $p-1$ values of least positive residues such that $(m,p) = 1$, and once all residues have appeared they *must* repeat.

What is the order of 3 modulo 7? The following table will tell us.

$n =$	1	2	3	4	5	6	7	8	
$3^n \equiv$	3	2	6	4	5	1	3	2	etc.

Thus, the order of 3 modulo 7 is 6, the highest possible value. Therefore 3 is called a *primitive root* modulo 7. A primitive root is also called a generating element, or generator, because it generates a complete residue system (in our example the integers from 1 to 6) in some permutation.

Once we have found a primitive root g, we can immediately find another one, its inverse modulo p:

$$g_2 \equiv g_1^{\phi(p)-1} \pmod{p} \ , \tag{13.2}$$

or, since $\phi(p) = p-1$ for a prime,

$$g_2 \equiv g_1^{p-2} \ . \tag{13.3}$$

In our example, with $g_1 = 3$, we get $g_2 \equiv 3^5 = 243 \equiv 5 \pmod{7}$. Check: $5 \cdot 3 = 15 \equiv 1 \pmod{7}$. Check! And 5 raised to successive powers yields

$n =$	1	2	3	4	5	6	7	8	
$5^n \equiv$	5	4	6	2	3	1	5	4	etc.

Thus 5, too, has order $p-1 = 6$ and is therefore another primitive root.

How many primitive roots are there? If we raise a given primitive root g to the power $m > 1$, where $(m,\phi(p)) = 1$, then g^m must be another primitive root. Thus there are $\phi(\phi(p))$ primitive roots. (For $p = 7$, the number is $\phi(6) = 2$, both of which we have already found: 3 and 5.)

If, by contrast, the greatest common divisor (GCD) d of m and $\phi(p)$ is greater than 1, $(m,\phi(p)) = d > 1$, then the order of $g = g_1^m$ is only $\phi(p)/d$. To show this, we observe first that $\phi(p)/d$ is a period of g:

$$g^{\phi(p)/d} = g_1^{\phi(p)m/d} \equiv 1^{m/d} = 1 \pmod{p} , \tag{13.4}$$

and second that it is the *shortest* period, because by introducing the least common multiple $[\phi(p), m]$, we can write

$$g^{\phi(p)/d} = g_1^{[\phi(p),m]} = g_1^{\phi(p)\cdot k} \equiv 1^k = 1 \pmod{p} . \tag{13.5}$$

Thus, $\phi(p)/d$ is the smallest exponent for which $g^{\phi(p)/d}$ is congruent 1 modulo p.

How many positive $m < p$ are there that have order $T = \phi(p)/d$? As we saw in Chap. 7, there are exactly $\phi\left\lfloor\phi(p)/d\right\rfloor$ values of m that have d as the GCD with $\phi(d)$. Hence, the number of residue classes that have order T equals $\phi(T)$. This is illustrated by the following table for $p = 7$, $\phi(p) = 6$:

m	m^2	m^3	m^4	m^5	m^6	T
1	1	1	1	1	1	1
2	4	1	2	4	1	3
3	2	6	4	5	1	6
4	2	1	4	2	1	3
5	4	6	2	3	1	6
6	1	6	1	6	1	2

Indeed, there are exactly $\phi(1) = 1$ order $T = 1$, $\phi(2) = 1$ order $T = 2$, $\phi(3) = 2$ orders $T = 3$ and $\phi(6) = 2$ orders $T = 6$. Further, all T divide $\phi(7) = 6$.

Primitive roots are possessed by the integers 1, 2, 4, p^k and $2 \cdot p^k$ (where p is an *odd* prime and $k > 0$). All that has been said about primitive roots for a prime modulus transfers, *mutatis mutandi,* to these other cases.

The smallest integer not having a primitive root is 8. A prime residue system modulo 8 is given by 1, 3, 5, and 7, and all of these have order 1 or 2: $1^1 \equiv 3^2 \equiv 5^2 \equiv 7^2 \equiv 1 \pmod{8}$. There is no residue that has order $\phi(8) = 4$.

Why are 3 and 5 primitive roots modulo 7 and not, say, 4? How are the primitive roots distributed within a residue system? For example, 71 and 73 both have 24 primitive roots, of which they share exactly one half, namely

11, 13, 28, 31, 33, 42, 44, 47, 53, 59, 62, 68 .

What distinguishes these numbers?

Gauss said in his *Disquisitiones* [13.1] that the distribution of primitive roots is a deep mystery; there is no way to predict where they will occur — only their total number is known. But Gauss *did* give some fast algorithms for ferreting them out.

13.2 Periods of Decimal and Binary Fractions

As every high-school student knows, 1/2 written as a decimal fraction is 0.5 and 1/50 becomes 0.02. Both 0.5 and 0.02 are *terminating* decimal fractions. By contrast 1/3 becomes a nonterminating decimal fraction, namely 0.3333..., and so does 1/7:

0.142857142857 ... .

Both 1/3 and 1/7 lead to *periodic* decimal fractions. By contrast, $\sqrt{2} = 1.41421356...$ and $\pi = 3.14159265...$ are irrational and have nonterminating aperiodic decimal representation.

What reduced rational fractions m/n, where $(m,n) = 1$, have terminating decimal representation? The answer is very simple and devolves directly from the prime factor decomposition of the denominator n:

$$n = \prod_{p_i \mid n} p_i^{e_i} , \tag{13.6}$$

where the product is over all prime p_i that divide n. Now, if the only p_i in (13.6) are 2 and 5, then the fraction terminates because 2 and 5 are the only prime factors of 10.

Specifically, if $n = 2^a 5^b$ and, for example, $a > b$, then $n = 2^{a-b} 10^b$ and for $a - b = 2$, say,

$$\frac{1}{n} = 0.0 ... 025 ,$$

where the number of zeroes to the right of the decimal point equals b.

Example: $n = 80 = 2^4 \cdot 5 = 2^3 \cdot 10$. Thus, $1/80 = 2^{-3} \cdot 10^{-1} = 0.0125$.

The numerator m of the fraction m/n simply converts a terminating decimal fraction into another terminating decimal fraction.

Things become more interesting if the denominator n is divisible by prime factors other than 2 or 5. Let us begin with the prime factor 3 and write

$$\frac{1}{3} = \frac{3}{9} = \frac{3}{10-1} = \frac{3}{10} \cdot \frac{1}{1 - \dfrac{1}{10}} = \frac{3}{10} \cdot \left(1 + \frac{1}{10} + \frac{1}{100} \cdots \right).$$

This brings the periodic nature of the decimal fraction for 1/3 into direct evidence. The fraction 3/10 is, of course, 0.3 and the 3 has to be repeated over and over again with increasing right shifts:

$$\frac{1}{3} = 0.3 + 0.03 + 0.003 + \cdots = 0.333... \, .$$

To save ink, periodic decimal fractions are usually written with a bar over a single period. Thus, $1/3 = 0.\overline{3}$ and $1/7 = 0.\overline{142857}$.

But why does 1/7 have a period length of 6? Modeling 1/7 on what we did to 1/3, we might try to express 1/7 as a rational fraction with a denominator that is one less than a power of 10. Thus, we are looking for the smallest positive factor f such that

$$7f = 10^k - 1 \, , \tag{13.7}$$

or, equivalently, we want to know the smallest k for which

$$10^k \equiv 1 \pmod 7 \, . \tag{13.8}$$

That is, of course, just the definition of *order* (in the arithmetic sense!) that we encountered in Sect. 13.1. Thus,

$$k = \mathrm{ord}_7 \, 10 = 6 \, . \tag{13.9}$$

Check: $10^6 = 142857 \cdot 7 + 1$, and no lower power of 10 exceeds a multiple of 7 by 1. Check!

Hence, 1/7 has a decimal period of length 6 with the digits

$$f = (10^6 - 1)/7 = 142857 \, , \qquad \text{or} \tag{13.10}$$

$$\frac{1}{7} = 0.\overline{142857} \, .$$

More generally, $1/p$, $p \neq 2$ or 5, has a period length

$$k = \mathrm{ord}_p \, 10 \, . \tag{13.11}$$

Example: For $p = 11$, $k = 2$, $f = 99/11 = 9$; hence $1/11 = 0.\overline{09}$.   For $p = 13$, $k = 6$, $f = 999999/13 = 76923$; hence $1/13 = 0.\overline{076923}$.

It is also clear that the period cannot be longer than $p-1$, because in carrying out the long division $1/p$, there are at most $p-1$ possible remainders, namely 1,2, ..., $p-1$, after which the remainders and therefore also the decimal digits *must* repeat.

In fact, $\mathrm{ord}_p 10$ is always less than p, because according to Fermat's theorem, for $(p,10) = 1$:

$$10^{p-1} \equiv 1 \ (\mathrm{mod}\ p) \ . \tag{13.12}$$

Thus $\mathrm{ord}_p 10$ is either $p-1$ (as in the case of $p = 7$) or a proper divisor of $p-1$ (as in the cases $p = 11$ and $p = 13$).

The longest possible period $p-1$ occurs whenever 10 is a primitive root of p. According to *Abramowitz* and *Stegun* [13.2], 10 is a primitive root of $p = 7, 17, 19, 23, 29, 47, 59, 61, 97$, etc.

Example: $\dfrac{1}{17} = 0.\overline{0588235294117647}$, which indeed has period length 16.

Of course, pocket calculators (and even big computers) are not accurate enough to determine the 96 digits of the decimal period of 1/97 directly. However, there is a trick that allows us to get the desired digits nevertheless. We shall illustrate this with the 16 digits of the period of 1/17. A 10-digit pocket calculator shows that

$$\frac{100}{17} = 5.88235294(1) \ ,$$

where the last digit may have been rounded off and is therefore uncertain. We have thus found 9 of the 16 digits. The next digits are obtained by calculating, say,

$$\frac{160}{17} = 9.41176470(6) \ .$$

Thus we have found all 16 digits of 1/17. By merging the two digit strings, we obtain

$$\frac{1}{17} = 0.\overline{0588235294117647} \ .$$

We leave it to the reader to discover a general and efficient algorithm for generating the digits of periodic fractions with a calculator of limited accuracy. In [13.3] all primes with period lengths less than 17 are listed.

Surprisingly, 37 is the only prime with period length 3.

Without derivation or proof we also state that for

$$n = \prod_{p_i \neq 2,5} p_i^{n_i} , \tag{13.13}$$

the decimal expression has a period length T equal to the least common multiple of the orders of 10 with respect to the different p_i. Thus, with

$$k_i := \text{ord}_{p_i^{n_i}} 10 , \tag{13.14}$$

$$T = [k_1, k_2, ...] . \tag{13.15}$$

Proving this is a nice exercise. If n also contains factors 2 or 5, the decimal fractions are mixed, meaning they have a nonperiodic "head."

Examples: $1/119 = 1/(7 \cdot 17)$ has period length $T = [6,16] = 48$. And for $1/2737 = 1/(7 \cdot 17 \cdot 23)$, $T = [6,16,22] = 528$.

Nonunitary fractions have a cyclically shifted period with respect to the corresponding unitary fractions, provided $T = \phi(n)$. Otherwise there are $\phi(n)/T$ different cycles, all of length T.

Example: $1/7 = 0.\overline{142857}$ and $6/7 = 0.\overline{857142}$. But for 13 we have $\phi(13) = \underline{12}$ and $\text{ord}_{13} 10 = 6$; thus there are $12/6 = 2$ different cycles: $1/13 = 0.\overline{076923}$ and $2/13 = 0.\overline{153846}$.

Everything that has been said here about decimal fractions carries over to other number bases. For example, 1/3 in binary notation has period length $T = \text{ord}_3 2 = 2$. In fact, $1/3 = 0.\overline{01}$. For 1/5 in binary, $T = \text{ord}_5 2 = 4$, and indeed, $1/5 = 0.\overline{0011}$. The prime 9949 has 2 as a primitive root [13.2]. Therefore, $\text{ord}_{9949} 2 = 9948$ and 1/9949 will generate a sequence of 0's and 1's with period lengths $9948 = 2^2 \cdot 3 \cdot 829$.

Another prime in the same range having 2 as a primitive root is 9851. With $9850 = 2 \cdot 5^2 \cdot 197$, the binary expansion of $1/99007599 = 1/(9949 \cdot 9851)$ has a period length $T = [2^2 \cdot 3 \cdot 829, 2 \cdot 5^2 \cdot 197] = 48993900$. Here is a method of generating long pseudorandom binary sequences! What are their spectral properties?

13.3 A Primitive Proof of Wilson's Theorem

Because the sequence of least positive residues $g^k \pmod{p}$, $k = 1, 2, ..., p-1$ (where g is a primitive root of the prime p), is a permutation of the integers $1, 2, ..., p-1$, one has

$$(p-1)! = 1 \cdot 2 \cdots (p-1) \equiv g \cdot g^2 \cdots g^{p-1} = g^{p\frac{p-1}{2}} \equiv g^{\frac{p-1}{2}} \pmod{p} . \quad (13.16)$$

Now according to Fermat,

$$g^{p-1} \equiv 1 \pmod{p} . \tag{13.17}$$

Therefore

$$g^{\frac{p-1}{2}} \equiv \pm 1 \pmod{p} . \tag{13.18}$$

However the plus sign is impossible, because g is a primitive root and $p-1$ is the smallest exponent m for which g^m is congruent to 1. Thus,

$$g^{\frac{p-1}{2}} \equiv -1 \pmod{p} , \qquad \text{i.e.,} \tag{13.19}$$

$$(p-1)! \equiv -1 \pmod{p} , \tag{13.20}$$

which is Wilson's theorem. (However, note that we had to assume the *existence* of a primitive root!)

13.4 The Index — A Number-Theoretic Logarithm

Let m have the primitive root g. For the prime residue system $(k,m) = 1$, one defines the *index* of k modulo m as the smallest positive t for which

$$g^t \equiv k \pmod{m} , \tag{13.21}$$

and writes

$$t = \text{ind}_g \, k .$$

Read: t equals the index to the base g of k.

Example: for $m = 5$ and $g = 2$:

$$\text{ind}_2 \, 1 = 0, \quad \text{ind}_2 \, 2 = 1, \quad \text{ind}_2 \, 3 = 3, \quad \text{ind}_2 \, 4 = 2 .$$

It is easy to see that

$$\text{ind}_g(ab) \equiv \text{ind}_g a + \text{ind}_g b \pmod{\phi(m)} , \tag{13.22}$$

a property the index shares with the logarithm. And in fact, the index is used much like a logarithm in numerical calculations in a prime residue system. For example, the congruence

$$3x \equiv 2 \pmod 5$$

is converted to

$$\mathrm{ind}_2\, 3 + \mathrm{ind}_2\, x \equiv \mathrm{ind}_2\, 2 \pmod 4 \, ,$$

or with the above "index table "

$$\mathrm{ind}_2\, x \equiv 1 - 3 = -2 \equiv 2 \pmod 4 \, .$$

Thus,

$$x = 4 \, .$$

Check: $3 \cdot 4 = 12 \equiv 2 \pmod 5$. Check!

A rule that is handy for base conversion is

$$\mathrm{ind}_a b \cdot \mathrm{ind}_b a \equiv 1 \pmod{\phi(m)} \, , \tag{13.23}$$

which is reminiscent of $\log_a b \cdot \log_b a = 1$ for logarithms.

In preparing index tables, it is only necessary to list values for primes, because index values of composites are obtained by addition. Also, one-half of a complete index table is redundant on account of the following symmetry relation:

$$\mathrm{ind}(m-a) \equiv \mathrm{ind}\, a + \frac{1}{2}\, \phi(m) \pmod{\phi(m)} \, , \tag{13.24}$$

which is a consequence of

$$g^{\frac{1}{2}\, \phi(m)} \equiv -1 \pmod m \, . \tag{13.25}$$

13.5 Solution of Exponential Congruences

The exponential congruence

$$a^x \equiv b \pmod m \, , \tag{13.26}$$

if m has a primitive root, can be solved by index-taking

$$x \cdot \mathrm{ind}\, a \equiv \mathrm{ind}\, b \pmod{\phi(m)} \, , \tag{13.27}$$

which has a solution *iff*

$$(\text{ind } a, \phi(m)) \mid \text{ind } b \ . \tag{13.28}$$

In fact, in that case, there are $(\text{ind } a, \phi(m))$ incongruent solutions.

Example: $7^x \equiv 5 \pmod{17}$; with the primitive root $g = 3$ as a base, we have

$$x \cdot 11 \equiv 5 \pmod{16} \ .$$

Since $(11,16) = 1$ divides 5, there is one (and only one) incongruent solution. Using Gauss's recipe,

$$x \equiv \frac{5}{11} \equiv \frac{5}{-5} = -1 \equiv 15 \pmod{16} \ .$$

Check: $7^{15} = 7^{16}/7 \equiv 1/7 \equiv 5 \pmod{17}$. Check!

In Sect. 7.6 on exponential Diophantine equations we considered the equation

$$2^n = 3^m - 1 \tag{13.29}$$

and asked whether there were solutions other than $2 = 3 - 1$ and $8 = 9 - 1$. Unfortunately, the answer was negative, otherwise we could have used ternary maximum-length sequences for precision measurements whose period was a power of 2, making them amenable to Fast Fourier Transformation (FFT) algorithms.

Now we consider another equation and ask: does

$$3^n = 2^m - 1$$

have any solutions other than $n = 1$, $m = 2$? If so, we could use binary maximum-length sequences whose period is a power of 3, making only slightly less efficient FFT algorithms based on the factor 3 (rather than 2) applicable.

We shall answer the above question using the concept of the *order* of an integer [13.4]. We ask: is there a solution of the above equation for $n > 1$? If there were, then $3^n = 9 \cdot k$ for some integer $k \geqslant 1$. Thus,

$$2^m \equiv 1 \pmod{9} \ .$$

Now the order of 2 modulo 9 is 6:

$$\text{ord}_9 2 = 6 \ .$$

Check: $2^r \equiv 2, 4, 8, 7, 5, 1 \pmod{9}$. Check! This means that 6 must divide the exponent m in the above congruence:

$$m = 6b \qquad \textit{for some integer } b \ .$$

Hence,

$$2^m = 2^{6b} = (2^3)^{2b} = 8^{2b} \equiv 1^{2b} = 1 \pmod 7 .$$

In other words, 7 divides $2^m - 1 = 3^n$, a contradiction because by the fundamental theorem, 3^n cannot be divisible by 7. Consequently, $n \leqslant 1$ and $3^1 = 2^2 - 1$ is the only solution. Too bad for our intended application!

In a similar vein [13.4], we prove that there are no solutions to

$$2^a = 3^b - 1 \tag{13.30}$$

for $a > 3$ or $2^a = 16 \cdot k$ for some integer $k \geqslant 1$. Thus, for $a > 3$:

$$3^b \equiv 1 \pmod{16} .$$

Since $\mathrm{ord}_{16}\, 3 = 4$, b must be some multiple of 4:

$$b = 4 \cdot r$$

for some integer $r \geqslant 0$. Thus,

$$3^{4r} \equiv 1 \pmod{16} .$$

Now, note that $\mathrm{ord}_5 3 = 4$, i.e.,

$$3^4 \equiv 1 \pmod 5 ,$$

and therefore also

$$3^{4r} = 3^b \equiv 1 \pmod 5 ,$$

or, equivalently,

$$5 | (3^b - 1) = 2^a ,$$

a contradiction because 5 cannot divide a power of 2! Thus, $a = 1$, $b = 1$ and $a = 3$, $b = 2$ are the sole solutions of $2^a = 3^b - 1$.

13.6 What is the Order T_m of an Integer m Modulo a Prime p?

As another example of solving exponential congruences we shall consider the congruence

$$m^{T_m} \equiv 1 \pmod{p} \qquad \text{or} \tag{13.31}$$

$$T_m \cdot \text{ind } m \equiv 0 \pmod{\phi(p)}, \qquad \text{i.e.,} \tag{13.32}$$

$$T_m \cdot \text{ind } m = k \; \phi(p). \tag{13.33}$$

Here the left-hand side must be both a multiple of $\phi(p)$ and ind m and, because of the definition of T_m as the *smallest* solution, $T_m \cdot$ ind m must be the *least* common multiple [ind m, $\phi(p)$]:

$$T_m \cdot \text{ind } m = [\text{ind } m, \phi(p)] = \frac{\text{ind } m \cdot \phi(p)}{d}, \tag{13.34}$$

where d is the greatest common divider of ind m and $\phi(p)$. Thus,

$$T_m = \frac{\phi(p)}{d}. \tag{13.35}$$

For example, for $p = 7$ and $m = 2$ and using 3 as the index base: $\text{ind}_3 2 = 2$ and $T_2 = 6/(2,6) = 3$, i.e., the order of 2 modulo 7 is 3. Check: $2^3 = 8 \equiv 1 \pmod 7$ and $2^k \not\equiv 1 \pmod 7$ for $k < 3$. Check!

If we had taken 5 as the index base, the answer would have been the same: $\text{ind}_5 2 = 4$ and $T_2 = 6/(4,6) = 3$. Check!

13.7 Index "Encryption"

The public-key encryption method described earlier is based on the fact that exponentiation modulo a large composite number whose factors are not known is apparently a "trap-door function," i.e., it is easy to exponentiate with a known exponent and to calculate a remainder, but it is very difficult to go in the opposite direction, i.e., to determine which number has to be exponentiated to yield a known remainder.

Another way to describe this situation, for the case that the modulus has a primitive root, is to say that taking logarithms in number theory (i.e., determining an index) is a difficult operation. While the encrypted Message E is given by

$$E \equiv M^s \pmod r, \tag{13.36}$$

the original message M can be obtained, at least formally, by taking the index to the base g, where g is a primitive root of r:

$$\text{ind}_g E \equiv s \cdot \text{ind}_g M \pmod{\phi(r)} \qquad \text{or} \tag{13.37}$$

$$M \equiv g^{(\text{ind } E)/s} \pmod{r} . \tag{13.38}$$

Example: $r = 17$, $g = 3$, $s = 5$. Say the cryptogram is $E = 7$. Then, with

$$\text{ind}_3 7 = 11 \pmod{16} \qquad \text{and}$$

$$\frac{\text{ind } E}{s} = \frac{11}{5} \equiv \frac{-5}{5} = -1 \equiv 15 \pmod{16} ,$$

the original message is

$$M \equiv 3^{15} \equiv 6 \pmod{17} .$$

Check: $6^5 \equiv 7 \pmod{17}$. Check!

The disadvantage in serious applications of index encryption is that the modulus r is limited to integers that have primitive roots, i.e., primes, odd prime powers and twice odd prime powers (apart from 1, 2 and 4).

13.8 A Fourier Property of Primitive Roots and Concert Hall Acoustics

Consider the sequence

$$a_n = \exp(i2\pi g^n/p), \tag{13.39}$$

where g is a primitive root of the prime p. This sequence is periodic, with period $\phi(p) = p-1$. Also, the a_n have magnitude 1.

The *periodic correlation sequence* is defined by

$$c_m := \sum_{n=0}^{p-2} a_n a_{n+m}^* , \tag{13.40}$$

where a^* stands for the conjugate complex of a. Obviously,

$$c_0 = p-1 , \tag{13.41}$$

or, more generally, $c_m = p-1$ for $m \equiv 0 \pmod{p-1}$.

On the other hand, for $m \not\equiv 0 \pmod{p-1}$,

$$c_m = \sum_{n=0}^{p-2} \exp[i2\pi g^n(1-g^m)/p] . \tag{13.42}$$

Here the factor $1-g^m \not\equiv 0 \pmod{p}$, and $g^n(1-g^m)$ therefore runs through a complete prime residue system except 0 as n goes from 0 to $p-2$. Thus, c_m is the sum over a complete set of pth roots of 1, except 1 itself. Since the "complete" sum equals 0, we have

$$c_m = -1, \quad \text{for } m \not\equiv 0 \pmod{p-1}. \tag{13.43}$$

Now, a periodic correlation function that has only two distinct values ($p-1$ and -1 in our case) has a *power spectrum* with only two distinct values [13.5]. By power spectrum we mean the absolute square of the Discrete Fourier Transform (DFT) defined by

$$A_m := \sum_{n=0}^{p-2} a_n e^{-2\pi i nm/(p-1)}. \tag{13.44}$$

It is easy to show that the power spectrum is given by

$$|A_m|^2 = \sum_{k=0}^{p-2} c_k e^{-2\pi i km/(p-1)}, \tag{13.45}$$

i.e., by the DFT of the correlation sequence. This is reminiscent of the well-known Wiener-Khinchin theorem [13.6].

For $m = 0$, or more generally $m \equiv 0 \pmod{p-1}$, we have, with the above two values for c_k,

$$|A_0|^2 = 1. \tag{13.46}$$

For $m \not\equiv 0 \pmod{p-1}$, with $c_0 = p-1$ and $c_k = -1$, we get

$$|A_m|^2 = p-1 - \sum_{k=1}^{p-2} e^{-2\pi i km/(p-1)}, \tag{13.47}$$

where the sum is again over a complete set of roots of 1, except 1 itself. Thus,

$$|A_m|^2 = p \quad \text{for all } m \not\equiv 0 \pmod{p-1}. \tag{13.48}$$

Such a constant power spectrum is called "flat" or "white" (from "white light," except that white light has a flat spectrum only in the *statistical* sense).

13.9 More Spacious-Sounding Sound

Flat power spectra are important in physics and other fields. (For example, a good loud-speaker is supposed to radiate a flat power spectrum when driven by a short electrical impulse.) Here, in addition, the original sequence a_n, whose spectrum is flat, has constant magnitude 1. This leads to an interesting application in concert hall acoustics.

It has been shown that concert halls with laterally traveling sound waves, all else being equal, have a superior sound [13.7] to those halls that furnish short-path sound arriving only from the front direction — as is the case in many modern halls with low ceilings (dictated by high building costs and made possible by modern air conditioning). To get more sound energy to arrive at the listeners' ears from the sides (laterally), the author [13.8] proposed scattering, or diffusing, the sound which emanates from the stage and is reflected from the ceiling in all directions except the specular direction [13.9]. Also, the ceiling should not absorb any sound: in a large modern hall every "phonon," so to speak, is valuable; otherwise the overall sound level (loudness) will be too low.

Thus, what is called for on the ceiling is something the physicist calls a *reflection phase-grating* that scatters equal sound intensities into all diffraction orders except the zero order. Here "order" is used not as defined in mathematics but as in physics. Zero-order diffraction corresponds to the specular direction, i.e., straight downward.

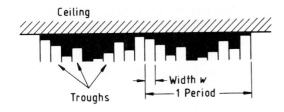

Fig. 13.1 Concert hall ceiling designed as a reflection phase-grating (based on the primitive root 2 of the prime 11)

Reflection phase-gratings can be realized by a hard surface with "wells" of different depths d_n, as shown in Fig. 13.1. Upon reflection, the phase of a normally incident wave is changed by $2d_n 2\pi/\lambda$, where λ is the wavelength. Now, if the depths d_n are chosen according to

$$d_n = \frac{1}{2} \lambda \, g^n/p \, , \tag{13.49}$$

where g is a primitive root of the prime p, and g^n can be the least residue modulo p, then the reflected wave has complex amplitudes[1] on its "wavefront" according to

---

[1]   This is taking an approximate ("Kirchhoff") view of diffraction. In reality, the complex amplitude cannot change abruptly. For an exact treatment, see [13.10].

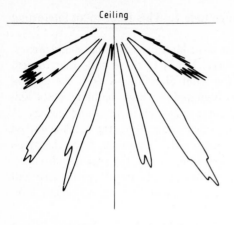
Ceiling

Fig. 13.2. Backscatter from primitive root ceiling. Note low specular reflection (vertically downward). $p = 7$, $g = 3$

$$a_n = e^{2\pi i g^n/p} \; ,$$

just like the periodic sequence that we considered above and that had a flat power spectrum.

Now, if the spatial distribution of wave amplitudes along a plane surface has a *flat* power spectrum, then the intensities of the wavelets scattered into the different diffraction orders will all be equal. Hence we expect a ceiling constructed according to this principle, as shown in Fig. 13.1 for $p = 11$ and $g = 2$, to scatter sound widely except in the specular direction (downward). That this is indeed so is illustrated in Fig. 13.2, which shows the result of actual measurements on a "primitive root" ceiling designed for improving concert hall acoustics. Such ceilings can be expected to increase the feeling of spaciousness, i.e., of being surrounded by or "bathed" in sound.

In order to form a *two*-dimensional array that scatters equal intensities into all diffraction orders (except the zeroth) over the *solid* angle, the prime p must be so chosen that $p-1$ has two coprime factors. For example, for $p = 11$, $p-1 = 10 = 2 \cdot 5$ and the ten numbers a_n can be used to fill a 2-by-5 array in "Chinese remainder" fashion (Chap. 16), for example as follows:

$$a_1 \quad a_7 \quad a_3 \quad a_9 \quad a_5$$
$$a_6 \quad a_2 \quad a_8 \quad a_4 \quad a_{10} \; , \tag{13.50}$$

i.e., the horizontal (left-to-right) location of a_n in the array corresponds to $\langle n \rangle_5$ and the vertical (up-down) location of a_n is given by $\langle n \rangle_2$. More generally, the array locations can be given by $\langle k \cdot n \rangle_5$, with any k for which $(k,5) = 1$, and $\langle m \cdot n \rangle_2$, with $(m,2) = 1$ (i.e., m must be odd). Here the acute brackets signify least remainders (see Sect. 16.2 on Sino-representation).

In the most general case, an r-dimensional array with the desired r-dimensional correlation and Fourier properties can be formed if $p-1$ has r

pairwise coprime factors q_1, q_2, ..., q_r and the location in the array of a_n has the coordinates

$$\langle k_i \cdot n \rangle_{q_i}, \quad \text{with } (k_i, q_i) = 1, \; i = 1, 2, ..., r \; . \tag{13.51}$$

For three-dimensional arrays, the smallest prime p such that $p-1$ has three coprime factors is $31 : 30 = 2 \cdot 3 \cdot 5$, giving a 2-by-3-by-5 array.

Another important principle that can be employed to diffuse sound involves *quadratic residues* (Chap. 15), and an interesting application of primitive roots, to the splicing of telephone cables, is given in [13.11].

13.10 A Negative Property of the Fermat Primes

The Fermat primes are, as Gauss discovered, precisely those primes p for which a "Euclidean" construction of the regular p-gon is possible (see Sect. 3.9). Thus, being a Fermat prime makes something possible.

Curiously, being a Fermat prime also makes something *impossible,* namely the construction of two- or higher-dimensional primitive-root arrays, as described in the preceding section. For such arrays, $p-1$ must be factored into two or more coprime factors, but $p-1$ has only *one* prime factor, namely 2. Thus, the same circumstance that allows the Euclidean construction forbids the construction of primitive-root arrays with more than one dimension (a form of mathematical justice?).

14. Knapsack Encryption

As a diversion we return in this chapter to another (once) promising public-key encryption scheme using a trap-door function: *Knapsack encryption*. It, too, is based on residue arithmetic, but uses multiplication rather than exponentiation, making it easier to instrument and theoretically more transparent.

The required trapdoor is obtained from the ancient knapsack puzzle: given the total weight of a knapsack and the weight of individual objects, which objects are in the bag? As it turns out, this problem can be made quite difficult to solve for someone who doesn't have the proper "key."

14.1 An Easy Knapsack

Number-theoretic exponentiation is not the only trap-door function potentially suitable for public-key encryption. Another trap door is opened (and kept almost closed) by *knapsack problems*.

Suppose we have a set of six stones, weighing 1, 2, 4, 8, 16 and 32 kilograms, respectively. Now, if a knapsack containing some of these stones weighs 23 kilograms more than its empty weight, then which stones are in the knapsack? The answer is given by the binary decomposition of 23:

$$23 = 16 + 4 + 2 + 1 ,$$

i.e., the stones weighing 1, 2, 4 and 16 kilograms are in the knapsack and no others.

This is an example of an *easy* knapsack problem. The problem remains easy if each weight exceeds the sum of the lower weights by at least one measurable unit. The binary sequence of weights 1, 2, 4, 8, etc., fulfills this condition, because $2 > 1$, $4 > 2 + 1$, $8 > 4 + 2 + 1$, etc. Such sequences are called "superincreasing." In fact, the binary sequence is the smallest positive super-increasing sequence if 1 is the just discriminable weight difference [14.1].

Table 14.1 shows such an easy knapsack embedded in leading and trailing digits. If the clear message in binary form is 110101, then the encrypted message is 21,853,232. It is obtained by multiplying each knapsack row by 0 or 1 in accordance with the binary message and adding as shown in

Table 14.1. Encryption with an easy knapsack using binary "weights" 1, 2, 4, 8, 16, and 32 (see fourth and third columns in center)

MESSAGE	EASY KNAPSACK	ENCRYPT
1	8 3 0 1 0 5 1	8 3 0 1 0 5 1
0	2 0 2 0 6 1	
1	7 8 0 4 0 9 0	7 8 0 4 0 9 0
0	3 5 0 8 0 4 9	
1	2 4 1 6 0 1 3	2 4 1 6 0 1 3
1	3 3 3 2 0 7 8	3 3 3 2 0 7 8
		2 1 8 5 3 2 3 2

53 = 110101 IN BINARY
WHICH IS THE MESSAGE

Table 14.1. To decrypt this number we only need the two digits 53 and its binary expansion

$$53 = 32 + 16 + 4 + 1,$$

corresponding to 110101 in binary notation, the original clear message.

14.2 A Hard Knapsack

Adding extra digits has not made the knapsack one bit harder to solve — so far. But now look at Table 14.2, in which each row of the simple knapsack is

Table 14.2. Encryption with a hard knapsack, obtained from the easy knapsack at left by multiplication and residue reduction (After *N. J. A. Sloane* [14.1])

MESSAGE	EASY KNAPSACK	HARD KNAPSACK	ENCRYPT
1	8 3 0 1 0 5 1	5 1 5 8 6 2 9 6 7	5 1 5 8 6 2 9 6 7
0	2 0 2 0 6 1	6 7 9 2 7 3 3 9 7	
1	7 8 0 4 0 9 0	5 1 3 4 3 8 3 0 5	5 1 3 4 3 8 3 0 5
0	3 5 0 8 0 4 9	4 0 2 1 6 1 7 8 3	
1	2 4 1 6 0 1 3	4 2 7 5 3 1 7 9 1	4 2 7 5 3 1 7 9 1
1	3 3 3 2 0 7 8	3 7 6 0 5 2 6 4 1	3 7 6 0 5 2 6 4 1
			1 8 3 2 8 8 5 7 0 4

CIPHER

MULTIPLY BY s = 324358647
AND REDUCE MOD r = 786053315

multiplied by $s = 324{,}358{,}647$ and reduced modulo $r = 786{,}053{,}315$ to form a *hard* knapsack.

Now, if encryption proceeds as before by multiplying the binary message with the rows of the knapsack (the hard knapsack!), the result, after the multiplied rows have been summed, is $1{,}832{,}885{,}704$, which is "a little" harder to decrypt if only the public encryption key is available.

To design a trap-door function based on a hard knapsack, proceed as follows [14.2]:

1) Pick a set of easy knapsack weights a_i, i.e., one that forms a superincreasing sequence:

$$a_{i+1} > \sum_{k=1}^{i} a_k .$$ (14.1)

2) Pick a modulus r and a coprime multiplier s, i.e., $(s,r) = 1$.

3) Calculate the hard knapsack

$$b_k \equiv sa_k \pmod{r}$$ (14.2)

and publish *only* the b_k.

4) Calculate a decrypting multiplier t such that

$$st \equiv 1 \pmod{r}$$ (14.3)

and *do not* publish it.

Suppose M is a message and its binary digits are m_k:

$$M = \sum_{k=0}^{K} m_k 2^k .$$ (14.4)

The encrypted message E is now calculated as follows:

$$E = \sum_{k=0}^{K} m_k b_k .$$ (14.5)

To decrypt, the legitimate receiver, knowing t, now forms

$$tE = \sum_{k} m_k t b_k .$$ (14.6)

But

$$tb_k \equiv tsa_k \equiv a_k \pmod{r} .$$ (14.7)

Thus,

$$tE \equiv \sum_k m_k a_k \pmod{r} , \tag{14.8}$$

and to recover the message bits m_k is an easy knapsack problem, because the a_k form a super-increasing sequence. Specifically, for $a_k = 2^k$, tE is the original message:

$$tE \equiv M \pmod{r} . \tag{14.9}$$

One of the advantages of knapsack encryption is that it does not rely on the supposed difficulty of factoring specially constructed very large numbers. On the other hand, knapsacks are presently under attack because *Shamir* [14.3] and others [14.4] have shown the equivalence of the knapsack problem to a problem in integer programming for which a "fast" algorithm was recently invented by H. W. Lenstra of the University of Amsterdam. Further progress in knapsack ripping has been made by L. Adleman, and by J. C. Lagarias and A. M. Odlyzko [14.5]. Thus knapsacks, as described here, have developed holes through which our "secret" weights can fall for everyone to see. Who will darn the ripped knapsack?

15. Quadratic Residues

Here we will acquaint ourselves with the fundamentals of *quadratic residues* and some of their applications, and learn how to solve *quadratic congruences* (or perhaps see when there is no solution).

Certain periodic sequences based on quadratic residues have useful Fourier-transform and correlation properties. In fact, from the related *Legendre symbol,* we can construct binary-valued sequences that equal their own Fourier transform, up to a constant factor — a property reminiscent of the Gauss function $\exp(-x^2/2)$ which reproduces itself upon Fourier transformation. These Fourier properties are intimately related to the fact that quadratic-residue sequences generate perfect difference sets.

To demonstrate these Fourier properties, we will introduce the so-called *Gauss sums,* which added up to so many headaches for Carl Friedrich. We will also see (literally) that incomplete Gauss sums make pretty pictures.

An interesting physical application of quadratic-residue *sequences,* stemming from their Fourier property, is the design of broadly scattering diffraction gratings.

Another application, based on the auto- and cross-correlation properties of families of quadratic-residue sequences, occurs in *spread-spectrum* communications (an important technique for combating jamming and other interference).

And, of course, we will mention *quadratic reciprocity* and Gauss's famous *Theorema Fundamentale.*

15.1 Quadratic Congruences

Consider the quadratic congruence [15.1]

$$Ax^2 + Bx + C \equiv 0 \pmod{m} .\tag{15.1}$$

For m a prime p, there is an inverse A' of A:

$$A'A \equiv 1 \pmod{p} ,\tag{15.2}$$

and we can write

$$x^2 + A'Bx + A'C \equiv 0 \pmod{p} .\tag{15.3}$$

If $A'B$ is even, then with

$$y = x + \frac{1}{2} A'B , \tag{15.4}$$

the original congruence can be written

$$y^2 \equiv b \pmod{p} , \qquad \text{where} \tag{15.5}$$

$$b = (\frac{1}{2} A'B)^2 - A'C . \tag{15.6}$$

(For $A'B$ odd and p odd, $A'B$ is replaced by $A'B + p$ in the above substitution of y for x.)

Thus, to solve quadratic congruences, all we have to learn is how to extract square roots. For example, for $p = 7$ and $b = 2$, the solutions are $y = 3$ or $y = 4$. Of course, there may be no solution at all. For example, there is no y for which $y^2 \equiv 3 \pmod{7}$. The same is true for $y^2 \equiv 5$ or $y^2 \equiv 6 \pmod{7}$.

For a solution to $y^2 \equiv b \pmod{p}$ to exist, we say that b has to be a *quadratic residue* (R) modulo p. If b is a quadratic *nonresidue* (N), then there is no solution. For $p = 7$, the least positive residues are

$$R = 1, 2, 4 ;$$

and the nonresidues are

$$N = 3, 5, 6 .$$

Not counting 0, there are exactly $(p-1)/2$ residues and also $(p-1)/2$ nonresidues. The properties R and N obey the rules for the multiplication of signs in ordinary arithmetic with R corresponding to $+$ and N to $-$:

$$R \cdot R \equiv R, \; R \cdot N \equiv N \cdot R \equiv N, \; N \cdot N \equiv R . \tag{15.7}$$

Example: for $N \cdot N \equiv R$:

$$5 \cdot 6 = 30 \equiv 2 \pmod{7} .$$

15.2 Euler's Criterion

The integer a, $(a,p) = 1$, is a quadratic residue modulo p, p odd, *iff*

$$a^{\frac{1}{2}(p-1)} \equiv 1 \pmod{p} ; \tag{15.8}$$

it is a quadratic nonresidue *iff*

$$a^{\frac{1}{2}(p-1)} \equiv -1 \pmod{p} .$$
(15.9)

Proof: According to Fermat, for $(a,p) = 1$:

$$a^{p-1} \equiv 1 \pmod{p} .$$
(15.10)

Thus, since in a prime residue system there are only two values for the square root:

$$a^{\frac{1}{2}(p-1)} \equiv \pm 1 \pmod{p} .$$
(15.11)

Now, if g is a primitive root modulo p and if a can be written as follows:

$$a \equiv g^{2n} \pmod{p} ,$$
(15.12)

then

$$a^{\frac{1}{2}(p-1)} = g^{(p-1)n} \equiv 1^n = 1 \pmod{p} ,$$
(15.13)

and the solution of

$$y^2 \equiv a \pmod{p}$$
(15.14)

is

$$y \equiv g^n \pmod{p} ,$$
(15.15)

i.e., a is quadratic residue.

For a equal to an odd power of a primitive root,

$$a^{\frac{1}{2}(p-1)} \equiv g^{\frac{1}{2}(p-1)} \equiv -1 \pmod{p} ,$$
(15.16)

and a is a nonresidue. The same conclusion also follows from the fact that the formula g^{2n}, $n = 1,2, \ldots$ generates $(p-1)/2$ distinct quadratic residues, so that the remaining $(p-1)/2$ residue classes must be quadratic nonresidues.

In a sequence of successive powers of a primitive root, R's and N's must therefore alternate.

Example: $p = 7$, $g = 3$, $n = 1,2, \ldots$; $g^n \equiv 3, \underline{2}, 6, \underline{4}, 5, \underline{1} \pmod{7}$. Here the R's, appearing for n even, are underlined.

15.3 The Legendre Symbol

The Legendre symbol (a/p) is a shorthand notation for expressing whether a is or is not a quadratic residue modulo p:

$$(a/p) := \begin{Bmatrix} 1 & \text{for} & a = R \\ -1 & \text{for} & a = N \\ 0 & \text{for} & a \equiv 0 \end{Bmatrix} \pmod{p} \, . \tag{15.17}$$

Because of the multiplication rule for residues, one has

$$(a/p) \cdot (b/p) = (ab/p) \, .$$

Let us look at an interesting property of what may be called *Legendre sequences:*

$$b_n = (n/p) \, .$$

Example: $p = 7$:

$$b_n = 0, \, 1, \, 1, \, -1, \, 1, \, -1, \, -1; \; 0, \, ... \, ,$$

etc., repeated periodically with period p. The sequence b_n is three-valued and its average is 0. Also, it reproduces itself by "decimating" with a quadratic residue. If we take every ath term, where a equals some R, then, because of

$$b_{na} = (an/p) = (a/p) \cdot (n/p) = (n/p) \, , \tag{15.18}$$

we get

$$b_{na} = b_n \, . \tag{15.19}$$

For example, in the above sequence, with $a = 2$:

$$b_{2n} = 0, \, 1, \, 1, \, -1, \, 1, -1, \, -1; \; 0,... \, ,$$

which exactly equals b_n.

For a equal to a *non*residue, we get $b_{an} = -b_n$. This follows from the above derivation with $(a/p) = -1$ for a nonresidue. For example, with $a = 6 \equiv -1$, we obtain the backward-running sequence:

$$b_{-n} = 0, \, -1, \, -1, \, 1, -1, \, 1, \, 1; \; 0, \, ... = -b_n \, ,$$

which is also the original sequence multiplied by -1.

More generally, we can say that when $p-1$ is a quadratic nonresidue of p, as in our example for $p = 7$, the sequence b_n is *antisymmetric*, i.e., $b_{-n} = -b_n$. From this follows that its discrete Fourier transform (DFT) is purely imaginary. For $p \equiv 3 \pmod 4$, $p-1$ is a quadratic nonresidue. On the other hand, for $p \equiv 1 \pmod 4$, $p-1$ *is* a residue.

Examples: for $p = 5$: $p-1 = 4 = 2^2$; similarly for $p = 13$: $p-1 = 12 \equiv 25 = 5^2 \pmod{13}$.

It is easy to see that this is so, because if $p \equiv 1 \pmod 4$, then $p-1 \equiv 0 \pmod 4$ and the congruence $y^2 \equiv 0 \pmod 4$ always has a solution. In fact, it has two incongruent solutions, namely $y = 0$ and $y = 2$.

By contrast, if $p \equiv 3 \pmod 4$, then $p-1 \equiv 2 \pmod 4$, and there are *no* solutions to $y^2 \equiv 2 \pmod 4$. In fact, $0^2 = 0$, $1^2 = 1$, $2^2 \equiv 0$, $3^2 \equiv 1 \pmod 4$.

A convenient way to generate a Legendre sequence is by means of Euler's criterion

$$b_n := (n/p) \equiv n^{\frac{1}{2}(p-1)} \pmod p , \tag{15.20}$$

making use of the binary expansion of $(p-1)/2$ and reducing to least residues after each multiplication, as discussed in Sect. 6.3.

15.4 A Fourier Property of Legendre Sequences

The DFT of a Legendre sequence is given by

$$B_m = \sum_{n=0}^{p-1} b_n e^{-2\pi i nm/p} . \tag{15.21}$$

For $m \equiv 0 \pmod p$, we get

$$B_0 = 0 . \tag{15.22}$$

For any m, we may replace b_n by $b_{nm}(m/p) = b_{nm}b_m$ yielding

$$B_m = b_m \sum_{n=0}^{p-1} b_{nm} e^{-2\pi i nm/p} , \tag{15.23}$$

or with $nm = k$, $m \not\equiv 0 \pmod p$:

$$B_m = b_m \sum_{k=0}^{p-1} b_k e^{-2\pi i k/p} = b_m B_1 , \tag{15.24}$$

which, because of $b_0 = 0$, also holds for $m \equiv 0 \pmod{p}$. Thus, B_m, the DFT of the Legendre sequence, equals b_m times a constant (B_1) — a remarkable property, reminiscent of the fact that the Fourier integral of the Gauss distribution function $\exp(-x^2/2)$ reproduces the function to within a constant factor.[1]

15.5 Gauss Sums

To determine the multiplier B_1, we consider the sum

$$S(p) := \sum_{k=0}^{p-1} (b_k+1)\, e^{2\pi i k/p} . \tag{15.25}$$

Here $b_k + 1 = 2$ if $k \neq 0$ is a quadratic residue modulo p and $b_k + 1 = 0$ if k is a nonresidue. Thus, only those terms contribute to $S(p)$ for which $k \equiv n^2$ $\pmod{p}$ for some n. As n goes from 1 to $p-1$, every such k is "touched" exactly twice. Hence

$$S(p) = \sum_{n=0}^{p-1} e^{2\pi i n^2/p} , \tag{15.26}$$

a type of sum studied by Gauss.

Now, it is easy to show that

$$S^2(p) = |S(p)|^2 = p \quad \text{for } p \equiv 1 \pmod{4}$$
$$S^2(p) = -\,|S(p)|^2 = -p \quad \text{for } p \equiv 3 \pmod{4} , \tag{15.27}$$

but it took Gauss (somewhat inexplicably) several painful years to determine the sign of $S(p)$, i.e., to prove that

$$S(p) = \sqrt{p} \quad \text{for } p \equiv 1 \pmod{4} \qquad \text{and}$$
$$S(p) = i\sqrt{p} \quad \text{for } p \equiv 3 \pmod{4} , \tag{15.28}$$

where in both cases the positive sign of the square root is to be taken.

For p being replaced by the product of two primes, p and q, one has

$$S(pq) = (-1)^{(p-1)(q-1)/4} S(p)\, S(q) . \tag{15.29}$$

---

[1] More generally, this is true for the Gauss function multiplied by a Hermite polynomial, a circumstance that plays an important role in quantum mechanics, because of the manifest symmetry in the phase space of the harmonic oscillator.

The sums $S(p)$, or more generally $S(n)$ for any n, are called Gauss sums, and Gauss was eventually able to show that

$$
S(n) = \begin{cases}
(1+i)\sqrt{n} & \text{for } n \equiv 0 \\
\sqrt{n} & \text{for } n \equiv 1 \\
0 & \text{for } n \equiv 2 \\
i\sqrt{n} & \text{for } n \equiv 3
\end{cases} \pmod 4 . \tag{15.30}
$$

Applying the above results to the DFT of our Legendre sequences, $b_m = (m/p)$, we obtain:

$$
B_m = \sqrt{p}\, b_m \qquad \text{for } p \equiv 1 \pmod 4
$$

$$
B_m = -i\sqrt{p}\, b_m \quad \text{for } p \equiv 3 \pmod 4 . \tag{15.31}
$$

If we replace $b_m = 0$ by $\tilde{b}_m = 1$, i.e., if we make our sequence binary-valued, then for $p \equiv 3 \pmod 4$, the new DFT is

$$
\tilde{B}_m = 1 - i\sqrt{p}\, b_m . \tag{15.32}
$$

Thus, we have succeeded in constructing a *binary* sequence ($\tilde{b}_m = \pm 1$) with a *binary*-valued power spectrum

$$
|\tilde{B}_m|^2 = 1 + p \quad \text{for } m \not\equiv 0 \pmod p , \tag{15.33}
$$

and

$$
|B_0|^2 = 1 . \tag{15.34}
$$

Such binary sequences with a binary power spectrum are extremely useful in several applications requiring Fourier transforms. If we store b_m in a computer memory, we need not also store the DFT, because a simple "multiply-and-add" operation converts the b_m into the $\tilde{B}_m$.

In Chap. 25 we will encounter another type of binary-valued periodic sequences with a binary power spectrum, the so-called pseudorandom or *maximum-length sequences,* with periods $2^n - 1$. Their DFT, by contrast, has $\phi(2^n - 1)/n$ different phase angles. For a typical value (for concert hall measurements), $n = 16$ and $\phi(2^n - 1)/n = 2048$, a sizable number when memory is expensive.

15.6 Pretty Diffraction

"Incomplete" Gauss sums defined by

$$S(m,k) := \sum_{n=0}^{k} e^{2\pi i n^2/m}, \; k = 0,1, \ldots, m-1 \tag{15.35}$$

make pretty pictures. Figure 15.1 shows $S(m,k)$ for $m \equiv 2$ (mod 4), for which the complete sum equals 0, i.e., the incomplete sum "returns to the origin" (marked by 0) after first having gone into and out of two spirals. For $m \to \infty$, these spirals, properly scaled, will approach Cornu's spiral [15.2], the visual counterpart of the Fresnel integrals, which are so important in diffraction theory. See also [15.3].

Figure 15.2 illustrates the case $m \equiv 3$ (mod 4), where again the partial sum goes into and out of two spirals but "gets stuck" at $\sqrt{m}$ instead of returning to the origin (0), as predicted by Gauss.

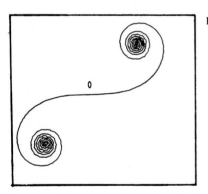

Fig. 15.1

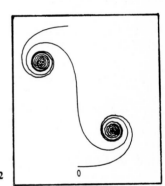

Fig. 15.2

Fig. 15.1. Partial Gauss sums forms spirals. Here the case $m \equiv 2$ (mod 4), for which the complete sum returns to the origin (marked by 0)

Fig. 15.2. Spiral representing partial Gauss sums for the case $m \equiv 3$ (mod 4). The origin is marked by 0. The complete sum equals $+1$

15.7 Quadratic Reciprocity

If p and q are odd primes, then using the Legendre symbol, the following law of *quadratic reciprocity* holds:

$$(p/q) \cdot (q/p) = (-1)^{\frac{(p-1)(q-1)}{4}}. \tag{15.36}$$

Another way to state this *Theorema Fundamentale,* so called by Gauss, who gave the first hole-free proof of this mathematical "gem" at age 18, is the following:

for $p \equiv q \equiv 3$ (mod 4)

$(p/q) \cdot (q/p) = -1$. Otherwise the product equals $+1$.

Example: $p = 3, q = 7$; $(3/7) = -1$, $(7/3) = (1/3) = 1$. Check!

The law of quadratic reciprocity can be very convenient when calculating quadratic residues.

Example: $(11/97) = (97/11) = (9/11) = (3/11)^2 = 1$.

Of particular interest are the following special cases:

$$(-1/p) = (-1)^{\frac{1}{2}(p-1)} , \tag{15.37}$$

which we have already encountered, in a more general form, as Euler's criterion, and

$$(2/p) = (-1)^{(p^2-1)/8} , \tag{15.38}$$

i.e., 2 is a quadratic residue *iff* $p \equiv \pm 1$ (mod 8).

15.8 A Fourier Property of Quadratic-Residue Sequences

Consider the complex sequence

$$r_n = e^{2\pi i n^2/p} . \tag{15.39}$$

The magnitude of its individual terms equals 1, and it is periodic with period p. Its periodic correlation sequence is given by

$$c_k := \sum_{n=0}^{p-1} r_n r_{n+k}^* = e^{-2\pi i k^2/p} \sum_{n=0}^{p-1} e^{-4\pi i n k/p} . \tag{15.40}$$

Now, for any $k \not\equiv 0$ (mod p), the sum, being over a complete set of pth roots of 1, vanishes. Thus,

$$c_k = 0 \quad \text{for } k \not\equiv 0 \pmod{p} . \tag{15.41}$$

On the other hand, for $k \equiv 0 \pmod{p}$, we obtain

$$c_0 = p . \tag{15.42}$$

Thus, we have found another constant-magnitude periodic sequence whose periodic correlation sequence has only two distinct values, one of which equals zero. As a direct consequence, the squared magnitude of the DFT, i.e., the power spectrum of the sequence r_n, has only *one* value:

$$|R_m|^2 = p, \quad \text{for } all \ m . \tag{15.43}$$

Again there are applications to broad scattering of waves [15.4], such as diffusing sound waves from concert hall walls and ceilings to improve acoustic quality. Figure 15.3 shows a "quadratic-residue" ceiling for $p = 17$.

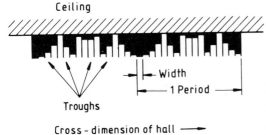

Fig. 15.3. Proposed concert hall ceiling design, a reflection phase-grating based on a quadratic-residue sequence for the prime 17

Such surface structures have the highly diffusing property not only at the "fundamental" frequency, but at a set of $p-1$ discrete frequencies. The reason is that for a frequency m times higher than the fundamental frequency, the phase shift suffered upon reflection from the "wells" is m times greater, i.e., the effective reflection coefficient is raised to the mth power. It is not difficult to show that the correlation property derived above holds not only for the sequence r_n, but also for the sequence r_n^m, as long as $m \not\equiv 0 \pmod{p}$. Thus, the ceiling shown in Fig. 15.3 is good for $p - 1 = 16$ frequencies, spanning a range of 4 musical octaves, see Fig. 15.4 for $m = 3$.

Fortunately, the good diffusing properties of such a quadratic-residue surface are also maintained at nonrational frequencies. Figure 15.5 shows a scatter diagram for $2\sqrt{2}$ times the fundamental frequency.

Such *reflection phase-gratings,* to use the physicist's term for such structures, also scatter well for oblique incidence. To guarantee good backscatter even at grazing incidence, the grating constant (width of the "wells") should be a *quarter*-wavelength, instead of the half-wavelength sufficient for normal incidence.

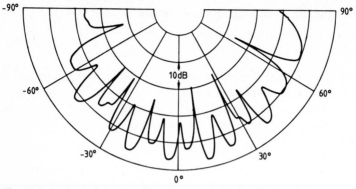

Fig. 15.4. Backscatter from model of concert hall ceiling illustrated in Fig. 15.3 for 3 times the fundamental design frequency. Note broad scatter, including strong specular reflection (vertically downward)

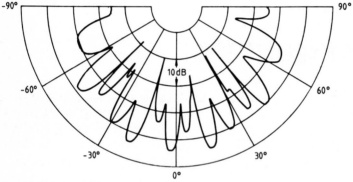

Fig. 15.5. Same as Fig. 15.4, for an irrational multiple of the fundamental design frequency

Quadratic-residue sequences are easily adapted to *two*-dimensional arrays a_{nm}, scattering equal intensities into all diffraction orders over the *solid* angle. We simply take *two* sequences r_n and r_m (based on the same prime p or different primes p_1 and p_2) and multiply them together:

$$a_{nm} = r_n \cdot r_m = \exp\left[2\pi i\left(\frac{n^2}{p_1} + \frac{m^2}{p_2}\right)\right]. \tag{15.44}$$

Equation (15.44) is useful in the design of surfaces that scatter incident radiation widely with little of the total reflected energy going in any given direction. Potential applications are in radar and sonar camouflage and noise abatement.

Equation (15.44) obviously generalizes to arrays in three or more dimensions having the desired multidimensional correlation and Fourier properties.

Quadratic-residue sequences are also useful in the realization of broad (antenna) radiation patterns [15.5], which are discussed in Chap. 28.

15.9 Spread Spectrum Communication

In order to minimize interference, and especially *jamming,* in wireless ("radio") communication, the use of broad spectra is recommended [15.6]. Since frequency space in the "ether" is scarce, several users must therefore share the same frequency band. How should they signal among themselves to keep interference with other simultaneous users at a minimum?

Here again, quadratic residues offer a solution, and one that is especially transparent. Each user is assigned a particular quadratic-residue sequence as his "signature." By generalizing (15.39) we can construct $p-1$ such sequences:

$$r_n^{(m)} = e^{2\pi i m n^2/p} , \quad m = 1, 2, \dots , p-1 . \tag{15.45}$$

Each of these sequences has, of course, the desired *auto*correlation property (15.41). But in addition, the cross-correlations with all other sequences is small. Thus, if each receiver is "matched" to his own sequence, he will only receive messages addressed to him — *without interference from any other simultaneously active user*, although they all share the same frequency band! (See Sect. 25.9 and [15.7, 8] for further details and additional references.)

15.10 Generalized Legendre Sequences Obtained Through Complexification of the Euler Criterion

Consider the Legendre sequence

$$b_n = \left(\frac{n}{p}\right) , \tag{15.47}$$

where (n/p) is the Legendre symbol, see (15.17). Using Euler's criterion (15.8), (15.47) may be written as follows

$$b_n \equiv n^{(p-1)/2} \bmod p , \tag{15.48}$$

or with a primitive root g of p:

$$b_0 = 0, \quad b_n = g^{(\mathrm{ind}_g n)(p-1)/2}, \tag{15.49}$$

where n is taken modulo p and $b_n = \pm 1$ for $n \not\equiv 0$. In (15.49) $\mathrm{ind}_g n$ is the index function (see Sect. 13.4) or "number-theoretic logarithm" defined by

$$g^{\mathrm{ind}_g n} \equiv n \bmod p . \tag{15.50}$$

Since

$$g^{(p-1)/2} \equiv -1 \bmod p , \tag{15.51}$$

(15.49) can also be written as

$$b_0 = 0, \quad b_n = e^{i\pi \mathrm{ind}_g n} . \tag{15.52}$$

Now, instead of admitting as possible values for b_n only two roots of unity (± 1), we can generalize (15.52) by allowing each b_n to equal one of the $p-1$ different $(p-1)$-th roots corresponding to the $p-1$ different values of $\mathrm{ind}_g n$. Thus, we define a *generalized Legendre sequence*, see also *Lerner* [15.9]:[2]

$$a_0 = 0, \quad a_n = \exp[2\pi i(\mathrm{ind}_g n)/(p-1)], \quad n \not\equiv 0 \bmod p . \tag{15.53}$$

Note that $a_n \cdot a_m = a_{nm}$, as for the original Legendre sequence.

What are the spectral properties of the so defined periodic sequence? The (periodic) autocorrelation sequence

$$c_k = \sum_{n=1}^{p-1} a_n a_{n+k}^* , \tag{15.54}$$

equals $p-1$ for $k \equiv 0 \bmod p$. For $k \not\equiv 0$, (15.54) gives

$$c_k = \sum_{\substack{n=1 \\ n+k \not\equiv 0}}^{p-1} \exp\{2\pi i[\mathrm{ind}_g n - \mathrm{ind}_g(n+k)]/(p-1)\} , \tag{15.55}$$

where the difference inside the brackets in the exponent assumes all integer values between 1 and $p-2$, modulo $p-1$, exactly once. Thus,

$$c_0 = p-1 \quad \text{and} \quad c_k = -1 \quad \text{for} \quad k \not\equiv 0 \bmod p , \tag{15.56}$$

again as for the original Legendre sequence, see (28.22).

---

[2] I am grateful to C.M. Rader for alerting me to Lerner's work.

As a consequence of the correlation function being two-valued, the magnitude of the discrete Fourier transform (DFT), see Sect. 13.8 and [15.10], is also two-valued:

$$A_0 = 0 \quad \text{and} \quad |A_m|^2 = p - 1 \quad \text{for} \quad m \not\equiv 0 \bmod p . \tag{15.57}$$

In fact, the DFT A_m of the generalized Legendre sequence is *proportional* to the conjugate complex or reciprocal sequence:

$$A_m = A_1 a_m^{-1}, \tag{15.58}$$

a most remarkable property which also holds for the original Legendre sequence b_m, see (15.24).

In a further generalization, a scaling factor, $r = 1, 2, \ldots, p - 1$, can be introduced in the definition (15.53), yielding

$$a_0 = 0 \quad \text{and} \quad a_n = \exp[2\pi i \, r(\mathrm{ind}_g n)/(p-1)], \quad r \not\equiv 0, n \not\equiv 0 \bmod p . \tag{15.59}$$

For $r = (p-1)/2$ the original three-valued Legendre sequence is recovered. The generalized Legendre sequence can be used in the design of phase arrays and gratings and for spread-spectrum communication (see Sect. 15.9).

In the latter application, because of the low crosscorrelation between sequences with different values of r, the possibility of choosing $p - 1$ different values of r allows the specification of as many different "signature codes" for up to $p - 1$ different users sharing the same frequency channel (see also Sect. 25.9).

16. The Chinese Remainder Theorem and Simultaneous Congruences

One of the most useful and delightful entities in number theory is the *Chinese Remainder Theorem* (CRT). The CRT says that it is possible to reconstruct integers in a certain range from their *residues* modulo a set of *coprime* moduli. Thus, for example, the 10 integers in the range 0 to 9 can be reconstructed from their two residues modulo 2 and modulo 5 (the coprime factors of 10). Say the known residues of a decimal digit are $r_2 = 0$ and $r_5 = 3$; then the unknown digit is 8 (uniquely!).

It turns out that representing numbers in a notation based on the CRT has very useful applications, to wit: fast digital computations (convolutions and Fourier transforms), superfast *optical* transforms and, the solution of quadratic congruences, which are discussed in Chap. 18.

16.1 Simultaneous Congruences

So far, we have only talked about one congruence at a time. We shall now explore the inherent attractions of two or more *simultaneous* congruences:

$$x \equiv a_i \pmod{m_i} \; ; \; i = 1,2, \dots , k \; . \tag{16.1}$$

Provided the moduli m_i are pairwise coprime:

$$(m_i, m_j) = 1 \quad \text{for } i \neq j \; , \tag{16.2}$$

a solution exists mod $(m_1 m_2 \dots m_k)$ and is unique.

Next we shall construct the solution. Define:

$$M := \prod_{i=1}^{k} m_i \quad \text{and} \quad M_i = \frac{M}{m_i} \; . \tag{16.3}$$

Note that $(m_i, M_i) = 1$. Thus, there *are* solutions N_i of

$$N_i M_i \equiv 1 \pmod{m_i} \; . \tag{16.4}$$

With these N_i, the solution x to the simultaneous congruences is [16.1]:

$$x \equiv a_1 N_1 M_1 + \ldots + a_k N_k M_k \pmod{M} \; . \tag{16.5}$$

To see this, we introduce as a convenient notation *acute brackets* with a subscript:

$$\langle x \rangle_m \tag{16.6}$$

by which we mean the least positive (or, occasionally, the least nonnegative) residue of x modulo m. With this notation we have

$$\langle x \rangle_{m_i} \equiv a_i N_i M_i \equiv a_i \pmod{m_i}, \tag{16.7}$$

because all other terms in the sum (16.5) making up x contain the factor m_i, and therefore do not contribute to the residue modulo m_i. Because $N_i M_i \equiv 1 \pmod{m_i}$, the solution is also unique modulo M. This is the *Chinese Remainder Theorem*.

Example: $m_1 = 3$, $m_2 = 5$; $M = 15$, $M_1 = 5$, $M_2 = 3$.

$$\langle 5N_1 \rangle_3 = 1 \text{ yields } N_1 = 2 \qquad \text{and}$$

$$\langle 3N_2 \rangle_5 = 1 \text{ yields } N_2 = 2 \; .$$

Thus, the solution of the simultaneous congruences

$$x \equiv a_1 \pmod{m_1} \; , \quad x \equiv a_2 \pmod{m_2} \qquad \text{is}$$

$$x \equiv 10a_1 + 6a_2 \pmod{15} \; .$$

Example: $a_1 = 2$ and $a_2 = 4$.

$$x \equiv 20 + 24 = 44 \equiv 14 \pmod{15} \; .$$

Check: $\langle 14 \rangle_3 = 2$, $\langle 14 \rangle_5 = 4$. Check!

16.2 The Sino-Representation: A Chinese Number System

Based on the Chinese Remainder Theorem (CRT), any positive integer x not exceeding M, where M can be written as a product of coprime integers:

$$0 < x \leqslant M = m_1 \ldots m_k \; , \qquad \text{with} \tag{16.8}$$

$$(m_i, m_j) = 1 \quad \text{for } i \neq j \; , \tag{16.9}$$

can be represented uniquely by its least positive residues modulo the m_i. Thus, x is represented by a k-tuple of residues. In "vector" notation

$$x = (r_1, r_2, ..., r_k) , \quad \text{with} \tag{16.10}$$

$$r_i = \langle x \rangle_{m_i} , \quad m_1 < m_2 < ... < m_k . \tag{16.11}$$

This is the so-called *Sino-representation*. For only two moduli, the Sino-representation admits of an illuminating graphic representation, in which r_1 and r_2 form the ordinate and abscissa, respectively. For $m_1 = 3$ and $m_2 = 5$, and using least nonnegative residues, we have

$r_2 =$	0	1	2	3	4
$r_1 = 0$	0	6	12	3	9
$r_1 = 1$	10	1	7	13	4
$r_1 = 2$	5	11	2	8	14

$$\tag{16.12}$$

Here the position $(1,1)$ is occupied by 1 and the successive integers follow a $45°$-degree downward diagonal jumping back to $r_1 = 0$ after $r_1 = 2$ and back to $r_2 = 0$ after $r_2 = 4$.

As can be seen from the above table, the integer 14, for example, has the representation $14 = (2,4)$. To go from the residues back to the integer, we use (16.5):

$$2 \cdot 2 \cdot 5 + 4 \cdot 2 \cdot 3 = 44 \equiv 14 \pmod{15} .$$

The Sino-representation is much more useful, in our digital age, than other ancient systems, for example the Roman number system (although *Shannon* [16.2] at Bell Laboratories did build a computer that truly calculated with Roman numerals — a masochistic numerical exercise if ever there was one).

Addition in the Sino-representation is "column by column": with

$$a = (a_1, a_2) \quad \text{and} \quad b = (b_1, b_2) ,$$

$$a + b = \left[\langle a_1 + b_1 \rangle_{m_1} , \langle a_2 + b_2 \rangle_{m_2} \right] , \tag{16.13}$$

which is easy enough to show.

Similarly, multiplication proceeds columnwise:

$$a \cdot b = \left[\langle a_1 b_1 \rangle_{m_1} , \langle a_2 b_2 \rangle_{m_2} \right] . \tag{16.14}$$

These rules generalize to any finite number of moduli.

16.3   Applications of the Sino-Representation

Consider the two sequences, each of length M:

$$a(k) \text{ and } b(k); \quad k = 0,1,..., M-1 . \tag{16.15}$$

In numerous applications, the *convolution sequence* $c(k)$ of two such sequences is required [16.3]; it is defined by

$$\sum_{l=1}^{M} a(l) \cdot b\left[\langle k-l\rangle_M\right], \, k = 0,1, ..., M-1 . \tag{16.16}$$

Here arguments are taken modulo M, which makes this a *circular* convolution.

A *two*-dimensional circular convolution is defined as follows:

$$c(k_1,k_2) := \sum_{l_1=1}^{m_1} \sum_{l_2=1}^{m_2} a(l_1,l_2) \cdot b(\langle k_1-l_1\rangle_{m_1}, \, \langle k_2-l_2\rangle_{m_2}) . \tag{16.17}$$

If we introduce the Sino-representation for the index l in the *one*-dimensional convolution (16.17), we get

$$\tilde{c}(k_1,k_2) = \sum_{(l_1,l_2)} \tilde{a}(l_1,l_2) \cdot \tilde{b}\left[\langle k_1-l_1\rangle_{m_1}, \, \langle k_2-l_2\rangle_{m_2}\right] , \tag{16.18}$$

which is identical with the two-dimensional circular convolution (16.17). Thus, a one-dimensional circular convolution in Sino-representation is equivalent to a two- (or higher) dimensional circular convolution!

This has important consequences for the efficacy of calculating circular convolutions if a and b can be factored. In the one-dimensional case, we need M multiplications for each of the M values of the result $c(k)$, i.e., a total of M^2 multiplications. By contrast, in the two-dimensional Sino-representation, we need only $m_1 + m_2$ multiplications for each value of $c(k_1, k_2)$, or a total of $(m_1 + m_2)M$ multiplications. The reduction factor η_s in the number of multiplications is thus $(m_1 + m_2)/M$ or, in the general case of decomposing M into r coprime factors,

$$\eta_s = \frac{m_1 + m_2 + ... + m_r}{M} . \tag{16.19}$$

Example: say $M \approx 10^6$. Make $M = 13 \cdot 15 \cdot 16 \cdot 17 \cdot 19 = 1,007,760$ (by appending 0's to the data if necessary). Then a five-dimensional Sino-representation yields a reduction in the number of multiplications by a factor $\eta_s^{-1} = 12597$, a very substantial saving!

In fact, the saving realized by the Sino-representation is exactly the same as that realizable by the famous Fast Fourier Transform (FFT) method [16.4]. In the FFT, too, the reduction factor equals

$$\eta_F = \frac{m_1 + m_2 + ... + m_r}{M} , \tag{16.20}$$

except that the m_i are no longer required to be coprime (at the price of some other complication, such as "twiddle" factors).

The greatest saving is of course realized if M is decomposed into as many small factors as possible. Thus, for $M \approx 10^6$, we can choose $M = 2^{20} = 1,048,576$ and get a reduction factor of $2 \cdot 20/2^{20} = 3.81 \cdot 10^{-5}$, which is, however, only twice as good as the Sino-decomposition.

In general for the FFT, choosing all $m_i = 2$, the optimum reduction factor is

$$\eta_F (m_i = 2) = \frac{2\log_2 M}{M} , \tag{16.21}$$

a well-known result [16.4].

Of course, if we wanted to apply FFT methods to circular convolution, 2 FFT's would be required [16.5] and the advantage of FFT over Sino-decomposition would be lost. And, of course, many more integers can be decomposed into, say, 5 coprime factors than factors of 2.

16.4 Discrete Fourier Transformation in Sino

The Discrete Fourier Transformation (DFT) of a sequence $a(n)$, $n = 0,1,..., M-1$, is defined by

$$A(k) := \sum_{n=1}^{M} a(n) W^{nk}, k = 0,1,..., M-1 . \tag{16.22}$$

Here W is a (primitive) Mth root of unity, for example

$$W = e^{-2\pi i/M} . \tag{16.23}$$

Ordinarily, to compute all M values of $A(k)$ takes M^2 (complex) multiplications.

Now if M can be decomposed into coprime factors (we use only two factors in this exposition), $M = m_1 \cdot m_2$, and if we represent indices k and n in Sino:

$$k = (k_1 = \langle k \rangle_{m_1}, \quad k_2 = \langle k \rangle_{m_2}) \qquad \text{and} \qquad (16.24)$$

$$n = (n_1 = \langle n \rangle_{m_1}, \quad n_1 = \langle n \rangle_{m_2}) ,$$

then the above DFT (16.22) can be rewritten

$$\tilde{A}(k_1,k_2) = \sum_{n_1=1}^{m_1} \sum_{n_2=1}^{m_2} \tilde{a}(n_1,n_2) \, W^{\langle n_1 k_1 \rangle_{m_1} N_1 M_1 + \langle n_2 k_2 \rangle_{m_2} N_2 M_2} , \qquad (16.25)$$

where as before, $M_i = M/m_i$, and N_i is the (unique) solution of $N_i M_i \equiv 1 \pmod{m_i}$.

Because of the sum in the exponent in (16.25), W can be represented by the *product* of two factors, each one depending only on one n_i. Thus, the total number of multiplications is only $(m_1 + m_2)M$ instead of M^2. Again the reduction in the number of multiplications due to the Sino-representation is $\eta_s = (m_1 + m_2)/M$ or, in the general case, where M is decomposed into r coprime factors, $M = m_1 \cdot m_2 \cdots m_r$,

$$\eta_s = \frac{m_1 + m_2 + \cdots m_r}{M} .$$

Here, in the case of the DFT, too, the Sino-representation has converted the one-dimensional operation into a multidimensional one. In fact, a two-dimensional DFT is defined as follows:

$$A(k_1',k_2') := \sum_{n_1=1}^{m_1} \sum_{n_2=1}^{m_2} a(n_1,n_2) \, W^{M_1 n_1 k_1' + M_2 n_2 k_2'} , \qquad (16.26)$$

where the factors $M_i = M/m_i$ in the exponent convert the W from an Mth root of unity to an m_ith root of unity, as required.

Comparing the two-dimensional DFT (16.26) with the Sino-representation, we see that they are identical if we set

$$k_i' = \langle N_i k_i \rangle_{m_i} . \qquad (16.27)$$

In other words the Fourier array, indexed by k_i', differs from the Sino-array only by the indexing (k_i instead of k_i').

16.5 A Sino-Optical Fourier Transformer

The difference in indexing noted in the previous section becomes important in the following application of the Sino-representation to the DFT of one-dimensional data. Using Sino-representations, we can lay down the data in a

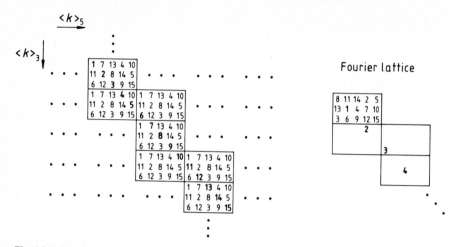

Fig. 16.1. One-dimensional data of period 15, put into two-dimensional array according to Sino-representation of index k modulo 3 and modulo 5 (the coprime factors of 15). On the right, the corresponding Fourier (DFT) array

two-dimensional array; see Fig. 16.1 for the case $M = 15$, $m_1 = 3$, $m_2 = 5$. As shown in this illustration, the "funny"-looking way of doing this becomes highly regular if we repeat the array periodically: consecutive numbers simply follow a $-45°$ diagonal straight line.

If we Fourier-transform this data (one basic 3×5 array of it), we obtain the Fourier array[1] shown on the right of Fig. 16.1. Now, if we want to reconvert to one-dimensional data (the original data never was two-dimensional!), we have to know the sequence in which to "read" the data from the array. The proper "reading instruction" is embodied in the above index conversion equation $k_i' = \langle N_i k_i \rangle_{m_i}$. With $N_1 = N_2 = 2$, as we saw above, the proper reading "jumps" *two* units down and to the right to go from one sample to the next. These samples, of course, also fall on a straight line if the Fourier array is thought of as periodically repeated, as hinted in Fig. 16.1.

16.6 Generalized Sino-Representation

Instead of representing an integer k in the range 1 to M by its residues modulo the coprime factors m_i of M:

---

[1] If we do this optically by means of *Fourier optics* methods [16.6] using coherent light from a laser, the Fourier transform will be "calculated" (by interfering light waves) literally at the speed of light!

$$k = (k_1, k_2, ..., k_r) \qquad \text{where}$$

$$k_i = \langle k \rangle_{m_i}, \tag{16.28}$$

we can, instead, choose the "components" k_i as follows:

$$k_i = \langle \alpha_i k \rangle_{m_i}, \tag{16.29}$$

where each α_i has to be chosen coprime to its modulus m_i : $(\alpha_i, m_i) = 1$. The inverse relation, i.e., the calculation of k from its components k_i, is the same as before:

$$k = \left\langle \sum_i k_i N_i M_i \right\rangle_M, \tag{16.30}$$

except that the congruences that determine the N_i look a little different:

$$\langle \alpha_i N_i M_i \rangle_{m_i} = 1 . \tag{16.31}$$

These congruences have unique solutions, because $(\alpha_i M_i, m_i) = 1$.
    To check the above solution:

$$\langle \alpha_j k \rangle_{m_j} = k_j \langle \alpha_j N_j M_j \rangle_{m_j} = k_j . \qquad \text{Check!}$$

Figure 16.2 illustrates the Sino-array for $\alpha_1 = 1$ and $\alpha_2 = 2$; in other words, the data is laid down in a "knight's-move" manner, as defined in chess.

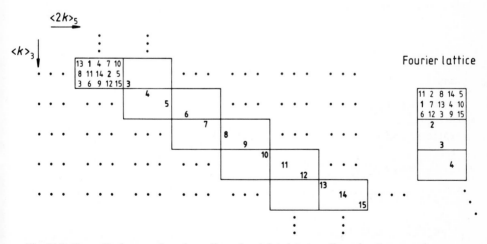

Fig. 16.2. Generalized conversion of one-dimensional data into two-dimensional array and corresponding Fourier array

The solutions of $\langle \alpha_i N_i M_i \rangle_{m_i} = 1$ are $N_1 = 2$ and $N_2 = 1$. (Check: $\langle 1 \cdot 2 \cdot 5 \rangle_3 = 1$ and $\langle 2 \cdot 1 \cdot 3 \rangle_5 = 1$. Check!). Thus, the Fourier array also has to be read in a knight's-move fashion:

$$k_1' = \langle 2k_1 \rangle_3 \quad \text{and} \quad k_2' = \langle k_2 \rangle_5$$

as illustrated on the right of Fig. 16.2.

16.7 Fast Prime-Length Fourier Transform

In the preceding sections we have learned how to exploit compositeness of the number M of data samples to realize fast convolution and Fourier transform algorithms. But suppose M is *prime* and we cannot "pad" the data with zero-value samples to make the number of samples highly composite; suppose we are stuck with a prime M. What then?

Fortunately, *Rader* [16.7] comes to our rescue with a beautiful idea that bridges the prime-M gap. In fact, for Rader's algorithm to work, M *has* to be prime.

With (16.22) and (16.23) for the DFT, and assuming for the moment that $a(M) = 0$, we have

$$A(k) = \sum_{n=1}^{M-1} a(n) \exp\left[-\frac{2\pi i}{M} nk \right] . \tag{16.32}$$

Since M is prime, it has a primitive root g, which we put to work to *permute* our indices k and n as follows:

$$k =: \langle g^k \rangle_M ,$$

and

$$n =: \langle g^n \rangle_M , \tag{16.33}$$

where k' and n' are the permuted indices; they cover the same range as k and n.

Defining further

$$a(g^{n'}) =: a'(n') ,$$

$$A(g^{k'}) =: A'(k') ,$$

we can write (16.32) as

$$A'(k') = \sum_{n'=1}^{M-1} a'(n') \exp\left[- \frac{2\pi i}{M} g^{n'+k'}\right] \quad k' = 1,2, ..., M-1 , \qquad (16.34)$$

which we recognize as a circular convolution. Thus, the availability of a primitive root has allowed us to convert the *product* nk into the *sum* of the permuted indices $n'+k'$: we have turned a discrete Fourier transform into a circular convolution for which fast algorithms are available, even if $M-1$ is not highly composite [16.4]. (Note: the number of data samples is no longer the prime M, but the composite $M-1$.)

Since $k \equiv 0 \pmod{M}$ is not included in our permutation (16.33), we have to calculate $A(0)$ separately, which causes no sweat:

$$A(0) = \sum_{n=1}^{M-1} a(n) . \qquad (16.35)$$

If $a(M)$ [or $a(0)$] does not equal 0 as initially assumed, then all $A(k)$ have to be augmented by $a(0) = a(M)$.

But these are small technical details. The important thing to remember is that the existence of a primitive root has allowed us to convert a product (Fourier transform) into a sum (convolution). At the same time, the number of data samples has been reduced by 1, thereby offering an escape from the unpleasant (for fast algorithms) primeness of M.

Converting products into sums smacks, of course, of logarithms, and indeed, although we never made it explicit, logarithms are implied by our permutation formulae (16.33). Instead of (16.33) we could have introduced the permuted indices as follows:

$$k' := \text{ind}_g k \qquad \text{and}$$

$$n' := \text{ind}_g n , \qquad (16.36)$$

where, as we recall (see Chap. 13), the index function is the number-theoretic equivalent of the usual logarithm.

17. Fast Transformations and Kronecker Products

While on the subject of fast computational algorithms based on the Chinese Remainder Theorem and primitive roots (discussed in the preceding chapter), we will now take time out for a glance at another basic principle of fast computation: decomposition into *direct* or *Kronecker* products. We illustrate this by showing how to factor Hadamard and Fourier matrices — leading to a Fast Hadamard Transform (FHT) and the well-known Fast Fourier Transform (FFT).

The Hadamard transformation is related to certain "Galois sequences" introduced in Chap. 24, and the FHT discussed here permits the computationally efficient utilization of these sequences in measurement processes of astounding precision.

17.1 A Fast Hadamard Transform

A Hadamard matrix is an $n \times n$ matrix of elements that equal $+1$ or -1 and whose rows and columns are mutually orthogonal [17.1]. Since the "norm" (sum of squared elements) of each row or column equals n, we have

$$H_n H_n^T = H_n^T H_n = nI_n \; ,$$

(17.1)

where the raised T stands for "transpose" and I_n is the $n \times n$ unit matrix, having 1's in the main diagonal and 0's elsewhere.

The simplest nontrivial Hadamard matrix H_n is

$$H_2 = \begin{bmatrix} 1 & 1 \\ 1 & -1 \end{bmatrix} \; .$$

(17.2)

All other Hadamard matrices necessarily have $n = 4k$.

Of special interest are the Hadamard matrices of Sylvester type with $n = 2^m$, which are obtained from H_2 by a "direct" or Kronecker product:

$$H_4 = H_2 \otimes H_2 \; ,$$

(17.3)

where $\otimes$ indicates the direct product, defined by replacing each element in the

$$V \otimes W = \begin{bmatrix} v_{11} W & v_{12} W & \cdots & v_{1m} W \\ \cdots & \cdots & \cdots & \cdots \\ v_{m1} W & v_{m2} W & \cdots & v_{mm} W \end{bmatrix}$$

Fig. 17.1. Construction of direct product of two matrices

first factor by the product of that element with the second factor (see Fig. 17.1). Thus,

$$H_4 = \begin{bmatrix} 1 & 1 & 1 & 1 \\ 1 & -1 & 1 & -1 \\ 1 & 1 & -1 & -1 \\ 1 & -1 & -1 & 1 \end{bmatrix} . \tag{17.4}$$

All Sylvester-type Hadamard matrices are formed in the same manner:

$$H_{2^m} = \underbrace{H_2 \otimes H_2 \otimes \cdots \otimes H_2}_{m \text{ factors}} . \tag{17.5}$$

In a shorter notation, this is also expressed as a *Kronecker power* or exponentiation, indicated by a bracket in the exponent:

$$H_{2^m} =: H_2^{[m]} . \tag{17.6}$$

As an example H_8 is shown in Fig. 17.2, where a horizontal line stands for -1.

$$H_8 = \begin{bmatrix} 1 & 1 & 1 & 1 & 1 & 1 & 1 & 1 \\ 1 & - & 1 & - & 1 & - & 1 & - \\ 1 & 1 & - & - & 1 & 1 & - & - \\ 1 & - & - & 1 & 1 & - & - & 1 \\ 1 & 1 & 1 & 1 & - & - & - & - \\ 1 & - & 1 & - & - & 1 & - & 1 \\ 1 & 1 & - & - & - & - & 1 & 1 \\ 1 & - & - & 1 & - & 1 & 1 & - \end{bmatrix}$$

Fig. 17.2. The Sylvester-Hadamard matrix $H_8 = H_2^{[3]}$

Various properties of the direct product are the following:

$$(X+Y) \otimes W = X \otimes W + Y \otimes W$$

$$(X \otimes Y)^T = X^T \otimes Y^T$$

$$(X \otimes Y)(V \otimes W) = (XV) \otimes (YW)$$

$$(X \otimes Y)^{-1} = X^{-1} \otimes Y^{-1} \tag{17.7}$$

$$\text{Trace}(X \otimes Y) = \text{Trace}(X)\,\text{Trace}(Y)$$

$$\det(X \otimes Y) = (\det X)^n (\det Y)^m ,$$

where X is $n \times n$ and Y is $m \times m$.

The eigenvalues of $X \otimes Y$ are the nm products of the n eigenvalues of X and the m eigenvalues of Y. This last property is particularly revealing, because it shows that if we have a $k \times k$ matrix A, with $k = nm$, and if we can decompose A into two Kronecker factors

$$A = X \otimes Y , \tag{17.8}$$

then instead of having to solve an nm-dimensional eigenvalue problem, we only have one n-dimensional and one m-dimensional problem to solve. The product nm is replaced by the sum $n + m$, just as in the Sino attack on convolutions and Fourier transforms, with a potentially enormous saving — especially if further Kronecker factors can be found, as in the case of the Sylvester-Hadamard matrices which completely decompose into the direct product of the 2×2 Hadamard matrices.

We shall demonstrate the attendant saving due to such decomposition for matrix-vector multiplications [17.2]. Let

$$z = H_4 \begin{pmatrix} a \\ b \\ c \\ d \end{pmatrix} = \begin{pmatrix} a + b + c + d \\ a - b + c - d \\ a + b - c - d \\ a - b - c + d \end{pmatrix} . \tag{17.9}$$

To calculate z, 12 additions/subtractions are required. In general, the number of operations is $n^2 - n$. [In (17.9) $n = 4$.]

However, by calculating intermediate results

$$a' = a + b$$

$$b' = a - b$$

$$c' = c + d \tag{17.10}$$

$$d' = c - d ,$$

we get the end result by only four more additions/subtractions:

$$z = \begin{pmatrix} a' + c' \\ b' + d' \\ a' - c' \\ b' - d' \end{pmatrix} , \tag{17.11}$$

a total of 8 operations. A saving of $12 - 8 = 4$ operations does not seem very impressive until we see the general formula: multiplication by a Sylvester-Hadamard matrix in the manner illustrated above for H_4 (possible because $H_4 = H_2 \otimes H_2$) requires

$$n \log_2 n \quad \text{instead of} \quad n^2 - n \tag{17.12}$$

additions/subtractions. For $n = 2^{16}$, the reduction factor equals $2.44 \cdot 10^{-4}$, a savings that may spell the difference between a problem being affordable (or even possible) to run on a given facility and impossible to run — perhaps on any machine. For further details and applications to spectroscopy see the excellent book by *Harwit* and *Sloane* [17.2].

Hadamard matrix operations owe their importance in part to the great technical significance of certain "maximum-length" sequences that are generated by primitive polynomials over finite number (Galois) fields, $GF(2^m)$. We shall return to this fascinating topic when we discuss cyclotomic polynomials over Galois fields (Chaps. 24, 25).

17.2 The Basic Principle of the Fast Fourier Transforms

As we saw in Chap. 16, the Discrete Fourier Transform (DFT) is defined by [17.3]:

$$A_k := \sum_{n=1}^{M} a_n e^{-2\pi i nk/M} , \tag{17.13}$$

which can also be written as a matrix-vector product

$$A = F_M a , \tag{17.14}$$

where F_M is the $M \times M$ DFT matrix, the simplest nontrivial one being

$$F_2 = \begin{bmatrix} 1 & 1 \\ 1 & -1 \end{bmatrix} , \tag{17.15}$$

which is exactly like the Hadamard matrix H_2. Is F_4 perhaps the direct product of F_2 with itself? Not quite:

$$F_4 = \begin{bmatrix} 1 & 1 & 1 & 1 \\ 1 & -i & -1 & i \\ 1 & -1 & 1 & -1 \\ 1 & i & -1 & -i \end{bmatrix} . \tag{17.16}$$

However, if we permute rows 2 and 3 and multiply all imaginary terms by a "twiddle factor" i (not to be confused with the "fudge factor," so beloved among order-loving scientists), we obtain

$$\tilde{F}_4 = \begin{bmatrix} 1 & 1 & 1 & 1 \\ 1 & -1 & 1 & -1 \\ 1 & 1 & -1 & -1 \\ 1 & -1 & -1 & 1 \end{bmatrix} = F_2 \otimes F_2 \quad (!) \tag{17.17}$$

Thus, with a little twiddling and some permuting, we have succeeded in decomposing the 4 × 4 DFT matrix into the direct product of two (identical) 2 × 2 matrices, which — as an extra reward — are made up exclusively of ±1's, requiring only additions or subtractions.

More generally, a properly twiddled and permuted $2^m \times 2^m$ DFT matrix can be written as a Kronecker power of F_2:

$$\tilde{F}_{2^m} = F_2^{[m]} , \tag{17.18}$$

which results in a fast algorithm as demonstrated for the Hadamard transform. This is the basic principle of all Fast Fourier Transforms (FFT). The savings in the number of operations is again governed by the ratio $\log_2 n / n$, where n is the number of data samples.

The rest is exhausting technical detail that is covered by a vast body of literature commensurate with the importance of the FFT (see [17.3] and the references therein).

18. Quadratic Congruences

Here we shall see how the Chinese Remainder Theorem allows us to solve *quadratic congruences* for composite moduli. Quadratic congruences play a role in such digital communication tasks as certified receipts, remote signing of contracts, and coin tossing — or playing poker over the telephone (discussed in Chap. 19). Finally, quadratic congruences are needed in the definition of pseudoprimes, which were once almost as important as actual primes in digital encryption (see Chap. 19).

18.1 Application of the Chinese Remainder Theorem (CRT)

First consider quadratic congruences [18.1] modulo an odd prime p:

$$x^2 \equiv b \pmod{p} , \qquad\qquad (18.1)$$

where p does not divide b : $p \nmid b$. This congruence has precisely two solutions: x and $p-x$. The same is true for moduli that are powers of an odd prime: $p_i^{e_i}$.

Quadratic congruences modulo a general modulus

$$m = p_1^{e_1} \cdots p_k^{e_k} \qquad\qquad (18.2)$$

are solved by solving k *simultaneous* congruences modulo the coprime factors of m [18.2]:

$$x^2 \equiv b \pmod{p_i^{e_i}} ,$$

because owing to the CRT, solutions x to these simultaneous congruences will also solve the congruence

$$x^2 \equiv b \pmod{m = p_1^{e_1} \cdots p_k^{e_k}} . \qquad\qquad (18.3)$$

Example: $x^2 \equiv 9 \pmod{28}$. The coprime factors of 28 are 4 and 7. Thus, we solve

$$x^2 \equiv 9 \pmod{4} \quad \text{and} \quad x^2 \equiv 9 \pmod{7}$$

with the solutions

$$\langle x \rangle_4 = 3 \qquad\qquad \langle x \rangle_7 = 3 \qquad \text{and}$$

$$\langle x \rangle_4 = 4 - 3 = 1 \qquad \langle x \rangle_7 = 7 - 3 = 4 \; .$$

With these four values of x we can now construct four incongruent solutions modulo 28:

$$x = \langle \langle x \rangle_4 \cdot M_1 \cdot N_1 + \langle x \rangle_7 \cdot M_2 \cdot N_2 \rangle_{28} \; ,$$

where the N_i are given by

$$\langle N_i M_i \rangle_{m_i} = 1 \; ,$$

or with $M_1 = 7$ and $M_2 = 4$: $N_1 = 3$ and $N_2 = 2$ and

$$x = \langle \langle x \rangle_4 \cdot 21 + \langle x \rangle_7 \cdot 8 \rangle_{28} \; .$$

Substituting the above values for $\langle x \rangle_{m_i}$, we get the following four solutions

$$x_{++} = \langle 3 \cdot 21 + 3 \cdot 8 \rangle_{28} = 3$$

$$x_{+-} = \langle 3 \cdot 21 + 4 \cdot 8 \rangle_{28} = 11$$

$$x_{-+} = \langle 1 \cdot 21 + 3 \cdot 8 \rangle_{28} = 17$$

$$x_{--} = \langle 1 \cdot 21 + 4 \cdot 8 \rangle_{28} = 25 \; .$$

Check: $3^2 \equiv 11^2 \equiv 17^2 \equiv 25^2 \equiv 9 \pmod{28}$.   Check!
    Of course, these solutions still come in pairs:

$$x_{++} \equiv -x_{--} \pmod{28} \qquad \text{and}$$

$$x_{+-} \equiv -x_{-+} \pmod{28} \; .$$

If we know one solution from each pair, we can find factors of the modulus by computing the greatest common divisor of their sum and the modulus. For example $(3+11,28) = 14$ or $(3+17,28) = 4$. We shall return to this point in Chap. 19 when we discuss "coin tossing" by telephone.

19. Pseudoprimes, Poker and Remote Coin Tossing

In this chapter we take a closer look at numbers that are not primes, but are tantalizingly close to primes in some respects. Of course, a given number $n > 1$ is either prime or composite — in other words, n is either "pregnant" with factors or not; there is no third alternative. But nevertheless, it makes sense to define and, as we do in this chapter, discuss such odd entities as pseudoprimes, absolute (or universal) pseudoprimes and strong pseudoprimes. When talking about extremely large numbers, pseudoprimality is sometimes the only evidence we can go by.

19.1 Pulling Roots to Ferret Out Composites

Fermat says

$$b^{n-1} \equiv 1 \pmod{n} , \tag{19.1}$$

if b and n are coprimes and n is prime (cf. Chap. 8). However, as is well known, the above congruence can also hold for composite n.

Example:

$$2^{340} \equiv 1 \pmod{341} , \tag{19.2}$$

in spite of the fact that $341 = 11 \cdot 31$ is composite. Because of this primelike property, 341 is called a *pseudoprime to the base* 2: $341 = \text{psp}(2)$.

More generally, a psp(a) is defined as an *odd composite* n for which

$$a^{n-1} \equiv 1 \pmod{n}$$

holds.

There are 245 pseudoprimes to the base 2 below one million, and the 6000th psp(2) is 1,187,235,193 [19.1]. The number of actual primes, by comparison, is about ten thousand times larger. Thus, numbers that are psp(2) are rather rare. Nevertheless, there are infinitely many[1] [19.2] and, in

---

[1] Indeed, every Fermat number, prime or not, obeys (19.1) (see Sect. 19.5).

fact, infinitely many in every prime residue class [19.3]. Numerically, the number of pseudoprimes less than n, $\Psi(n)$, seems asymptotically related [19.1] to the prime-counting function $\pi(n)$ by

$$\Psi(n) = k\ \pi(\sqrt{n})\ ,$$

with $k \approx 1.62$ (the Golden ratio??).

The exponent 340 in (19.2) is even, and now that we have learned to pull square roots, we might give it a try to see what happens. Of course, if 341 were prime, the only possible results would be $+1$ or -1.

$$2^{170} = (2^{10})^{17} \equiv 1^{17} = 1 \pmod{341}\ . \tag{19.3}$$

Again, 341 does not reveal its compositeness.

In general, composite numbers n for which

$$a^{\frac{n-1}{2}} \equiv \left|\frac{a}{n}\right| \pmod{n}\ ,$$

are called *Euler* pseudoprimes, abbreviated epsp(a) [19.4]. Here (a/n) is the Legendre symbol. Thus, because of (19.3), 341 is an epsp(2). Another epsp(2) is 1905. Check: $2^{952} \equiv (2/1905) = 1 \pmod{1905}$. Check!

Since 170 is also even, we can execute another square-root operation:

$$2^{85} = 32(2^{10})^8 \equiv 32 \pmod{341}\ , \tag{19.4}$$

since $2^{10} \equiv 1 \pmod{341}$.

Here we have trapped 341 and discovered its compositeness, because 32 is neither congruent 1 nor $-1 \pmod{341}$.

By repeated square rooting, we have, incidentally, also discovered the four solutions to the quadratic congruence

$$x^2 \equiv 1 \pmod{341}\ , \tag{19.5}$$

without inconveniencing the Chinese Remainder Theorem. Two of the solutions are obviously ± 1, and the other two are ± 32. In our repeated square rooting of 2^{340}, the solution $+1$ occurred twice and the solution $+32$ once.

Again (see Chap. 18), two absolutely different solutions tell us "something" about the factors of 341. Taking the greatest common divisors of the sum and difference of the two solutions with the modulus, we obtain

$$(32+1,341) = 11$$

and

$$(32-1,341) = 31 \, , \qquad\qquad (19.6)$$

the factors of 341.

19.2 Factors from a Square Root

This looks too good to be true: we have unearthed both factors of 341. Is this an accident? Or have we discovered a new approach to factoring the product of two primes? If so, we had better know more about it, even if the algorithm relies on the knowledge of two absolutely different square roots. To answer this question, we write down the solution to

$$x^2 \equiv 1 \pmod{341}$$

in terms of simultaneous congruences. Since both $\langle x \rangle_{11}$ and $\langle x \rangle_{31}$ equal ± 1, we have

$$x = \left\langle \pm \frac{341}{11} \left\langle \frac{341}{11} \right\rangle_{11}^{-1} \pm \frac{341}{31} \left\langle \frac{341}{31} \right\rangle_{31}^{-1} \right\rangle_{341} . \qquad (19.7)$$

Here we have made use of a new but obvious symbol: if

$$N_i M_i \equiv 1 \pmod{m_i} \qquad\qquad (19.8)$$

or, equivalently,

$$\langle N_i M_i \rangle_{m_i} = 1 \, , \qquad\qquad (19.9)$$

we write

$$N_i =: \langle M_i \rangle_{m_i}^{-1} \, , \qquad\qquad (19.10)$$

and say N_i is the inverse of M_i modulo m_i. In the above numerical example, $N_{11} = 5$ and $N_{31} = 17$, and we have

$$x = \langle \pm 155 \pm 187 \rangle_{341} . \qquad\qquad (19.11)$$

In terms of least residues, we have $x = \pm 1$ or $x = \pm 32$, as we saw before. Now, looking at the general solutions of

$$x^2 \equiv 1 \pmod{pq}, \ p \text{ and } q \text{ odd primes} \, , \qquad (19.12)$$

we have

$$x = \pm \langle q \langle q \rangle_p^{-1} + p \langle p \rangle_q^{-1} \rangle_{pq} \qquad (19.13)$$

and, using a different symbol, y, for the absolutely different solutions

$$y = \pm \langle q \langle q \rangle_p^{-1} - p \langle p \rangle_q^{-1} \rangle_{pq} . \qquad (19.14)$$

Taking the upper signs, we see immediately that the sum equals

$$\langle x + y \rangle_{pq} = 2 \langle q \langle q \rangle_p^{-1} \rangle_{pq} \qquad (19.15)$$

and the difference

$$\langle x - y \rangle_{pq} = 2 \langle p \langle p \rangle_q^{-1} \rangle_{pq} . \qquad (19.16)$$

Consequently, since we have assumed that p and q are odd,

$$(x + y, pq) = q \qquad \text{and} \qquad (19.17)$$

$$(x - y, pq) = p . \qquad (19.18)$$

Thus, calculating greatest common divisors of n and the sum and differences of quadratic congruences modulo n can indeed reveal the two factors of n.

19.3 Coin Tossing by Telephone

Alice and Bruno are divorcing (each other). Who gets the old jalopy, once the couple's precious chromium-embroidered self-propelled couch (rhymes with ouch)? Alice telephones Bruno and says: "Let's toss coins." Bruno: "Fine, I'll toss." But Alice, tossing her head (invisible to Bruno, because they are not connected by Picturephone®), objects: "Let *me* do the tossing — you know, you always liked my salads, if nothing else." So Bruno gives in: "O.K. I choose TAILS." Alice tosses (the coin) and up comes HEADS. Alice to Bruno: "Sorry, you lost — you'd better look around for another car."

How can we make this situation immune to cheating without resorting to video or witnesses? By extracting square roots! And here is how [19.5].

PROTOCOL

1)   Alice selects two very large odd primes p and q and calculates $n = pq$.

2)   Alice sends Bruno n.

3) Bruno picks a random number x smaller than n and calculates the least remainder of x^2 modulo n: $\langle x^2 \rangle_n$.

4) Bruno sends Alice $\langle x^2 \rangle_n$.

5) Alice, knowing p and q (and remembering the Chinese Remainder Theorem), calculates the four square roots, $\pm x$ and $\pm y$.

6) Alice sends Bruno one of these four values, not knowing which one Bruno had originally selected under (3).

7) Bruno squares the received number to check that Alice has not made a mistake (or cheated). If Alice sent $+x$ or $-x$, Bruno has lost, admits it, and Alice gets the remains of their car. However, if Alice sent $+y$ or $-y$, Bruno tells Alice that he has won and would she please have the car delivered to his new pad.

8) Alice does not believe that Bruno has won and challenges him to prove it.

9) Bruno adds his x to the y Alice sent him and, using Euclid's time-honored algorithm, determines the greatest common divisor with n, which is one of the two factors of n:

$$(x + y, n) = p \quad \text{or} \quad q .$$

10) Bruno sends Alice $(x + y, n)$ and $n/(x + y, n)$, i.e., the two factors p and q of n.

11) Alice, knowing that Bruno has succeeded in factoring n, throws in the towel and hopes for better luck next time. Of course, Alice had picked p and q *so* large that Bruno (or anyone else, except perhaps the Government) could not factor n without knowledge of an absolutely different square root.

END OF PROTOCOL.

Example: (too small, except perhaps to wager on a set of seven-year-old snow tires): $p = 101$, $q = 113$, $n = 11413$.

$$x = 1234, \qquad \langle x^2 \rangle_{11413} = 4827 .$$

$$\langle x \rangle_{101} = 22 , \qquad \langle x \rangle_{113} = 104$$

$$\langle 113 \rangle_{101}^{-1} = 59 \qquad \langle 101 \rangle_{113}^{-1} = 47$$

$$x, y = \pm \langle 22 \cdot 113 \cdot 59 \pm 104 \cdot 101 \cdot 47 \rangle_{11413}$$

$$x, y = \pm 1234, \pm 4624; \quad \langle 4624^2 \rangle_{11413} = 4827 . \quad \text{Check!}$$

$$x + y = 5858$$

$$(5858,11413) = 101 . \text{ Check!}$$

$$y - x = 3390$$

$$(3390,11413) = 113 . \text{ Check!}$$

Suppose Bruno does not *want* to win. Then, even when Alice sends him $\pm y$ to his $\pm x$, he can pretend that what Alice sent him was the square root he already knew, and that therefore he has lost. If Alice *wants* Bruno to win, and if he did win the toss, then under the present protocol she has no way of telling whether Bruno lied when he said "I lost."

To take care of this peculiar situation, a different protocol has been advanced by *Silvio Micali* [19.6] of the University of California at Berkeley. It is called "tossing coins into wells." We will not describe Micali's protocol, but simply mention that the "well" into which Alice tosses the coin is so deep that Bruno cannot reach in and turn the coin over. But he can *see* whether "heads" or "tails" is up, and if Alice does not believe him, he can ask her to look at the "well" herself.

It is noteworthy, and perhaps surprising, that number theory, of all disciplines, provides the solution to such mundane problems.

It is obvious that the general idea exploited here, namely the exchange of partial information, can be exploited in other kinds of transactions such as

"I will tell you my secret only
if you tell me yours."

The problem, of course, arises when the two "clams" are remote from each other. If clam A releases her secret, clam B may say "thank you" and remain shut up. To avoid this kind of cheating, the secrets are represented by strings of bits, and one bit from each clam is exchanged at any one time following a procedure similar to the coin-tossing protocol. If either partner clams up in the middle of the exchange, then both have acquired the same number of bits (give or take one bit). Of course, the early bits of B (or A) could be relatively worthless. I shall leave it to the reader to find a way out of this semantic quandary.

Another potential application of the coin-tossing protocol is to play poker [19.7] over the wire. The author, not being a poker player himself, invites the interested reader to devise a safe protocol. But then, poker might lose much of its charm without the faces in a smoke-filled room.

19.4 Absolute and Strong Pseudoprimes

We showed above that 341 was a pseudoprime to the base 2 because it is composite and $341 | (2^{341} - 2)$. There are some numbers N called *Carmichael numbers* [19.8] or *absolute pseudoprimes*, composites defined by $N | (b^{N-1} - 1)$ for all $(b,N) = 1$. The smallest member of this dangerous species (dangerous for primality tests) is $561 = 3 \cdot 11 \cdot 17$.

How can it be possible for b^{N-1} to be congruent to 1 modulo N for *all* b coprime to N? For $N = 561$ there are $\phi(561) = 320$ such values b. Credulity is strained! Yet the mystery behind the Carmichael numbers is not very deep. All we need are three (or more) odd primes p_i such that each $p_i - 1$ divides $N - 1 = \Pi p_i - 1$. This is obviously the case for $p_1 = 3$, $p_2 = 11$ and $p_3 = 17$ because 2, 10 and 16 all divide $3 \cdot 11 \cdot 17 - 1 = 560$.

Why is this sufficient? First note that according to Fermat,

$$b^{p_i - 1} \equiv 1 \pmod{p_i}$$

for all $(b,p_i) = 1$. These are simultaneous congruences (see Chap. 16), and by writing them

$$x \equiv 1 - b^{p_i - 1} \pmod{p_i}$$

we have brought them into the form of (16.1). These congruences have a *unique* solution modulo $N = \Pi p_i$. (In Chap. 16 N is called M, and the coprime moduli m_i in (16.2) are in fact primes here, the primes p_i.) This unique solution is, of course, $x = 0$. Thus, we have

$$0 \equiv 1 - b^{p_i - 1} \pmod{N} , \quad \text{or}$$

$$b^{p_i - 1} \equiv 1 \pmod{N} .$$

Since $p_i - 1$ divides $N - 1$:

$$N - 1 = k_i (p_i - 1) ,$$

for some k_i, we have

$$b^{N-1} \equiv 1^{k_i} \equiv 1 \pmod{N} .$$

This congruence holds for all $(b,p_i) = 1$, $i = 1,2, \dots$, which implies $(b,N) = 1$. In other words, N is an *absolute* pseudoprime.

Another example is $N = 2821 = 7 \cdot 13 \cdot 31$, because 6, 12 and 30 all divide $N - 1 = 2820$. With $\phi(2820) = 2160$, there are 2160(!) bases b for which 2821 divides $b^{2820} - 1$.

The reader may wish to find other n-tuples of primes which give Carmichael numbers.

To get a better grip on pseudoprimes, the concept of a *strong pseudoprime* [19.4] has been introduced. Definition: N is a strong pseudoprime to the base b if it is composite and the following holds. With $N-1 = K2^t$, K odd, consider

$$x_t = \langle b^{K2^t} \rangle_N \, ,$$

$$x_{t-1} = \langle b^{K2^{t-1}} \rangle_N \tag{19.19}$$

$$\vdots$$

$$x_0 = \langle b^K \rangle_N \, ,$$

where $\langle \; \rangle_N$ means least absolute remainder modulo N. Now let x_i be the remainder with the largest index i that differs from $+1$. If $x_i = -1$, then N is called a *strong* pseudoprime to the base b, abbreviated spsp(b).

The strong pseudoprimality test is capable of unmasking even absolute pseudoprimes such as 561: While

$$2^{560} \equiv 2^{280} \equiv 1 \pmod{561} \, ,$$

the next halving of the exponent reveals 561 to be composite:

$$2^{140} \equiv 67 \not\equiv -1 \pmod{561} \, .$$

And 2821 is exposed as nonprime by the first halving of the exponent:

$$2^{1410} \equiv 1520 \not\equiv -1 \pmod{2821} \, .$$

According to *Rabin* [19.9], the probability is less than 0.25 that an odd N is in fact composite when N passes the above test. Thus, if the test for strong pseudoprimality is repeated, say, 50 times with 50 different bases, the probability that a composite number would have "slipped through the net" would be less than $(0.25)^{50} = 2^{-100} \approx 10^{-30}$.

As a consequence, if since the beginning of the universe (10^{17} seconds ago), one *million* primes had been tested every *second,* then the probability that a single composite number would be among the strong pseudoprimes so found would be is less than 10^{-7}. Of course, 10^{-7} is not zero, and with many extremely large numbers we shall probably *never* know with *certainty* whether they are composite or prime: they will keep their mystery forever.

Even so, strong pseudoprimality is a much more reliable indicator of primality than some "unambiguous" deterministic computations that take, say, one hundred (or one trillion) years. Who will guarantee the fail-safety of the computer over such long running times?

And while we are talking of complexity and absolute provability, what about a proof that is more than 5000 pages long (or ten good-sized book volumes), like the recent proof that all finite simple groups are now known and that the so-called *monster group* is the largest sporadic simple group (i.e., has the largest order)?

Nevertheless, recent progress (1980-1982) in *deterministic* primality testing promises reasonable speeds for numbers with fewer than 600 digits (Sect. 19.6).

19.5 Fermat and Strong Pseudoprimes

It is interesting to note that *all* Fermat numbers

$$F_n := 2^{2^n} + 1 \tag{19.20}$$

are either primes or strong pseudoprimes to the base 2. To see this, we write:

$$2^{F_n-1} = 2^{2^{2^n}} = (2^{2^n})^{2^{2^n-n}} = (F_n - 1)^{2^{2^n-n}} . \tag{19.21}$$

Taking the remainder modulo F_n yields

$$\langle 2^{F_n-1} \rangle_{F_n} = (-1)^{2^{2^n-n}} = 1 . \tag{19.22}$$

Now, taking square roots will always give 1's until eventually, after 2^n-n square-rootings, a -1 will appear. Thus, all Fermat numbers are primes or strong pseudoprimes to the base 2 — a weak partial vindication of Fermat's erroneous conjecture ("quasi persuadé," to use Fermat's own words) that all F_n are in fact prime.

On the other hand

$$\langle 3^{F_5-1} \rangle_{F_5} = 3,816,461,520 \neq 1 , \tag{19.23}$$

so that F_5 is not even an ordinary pseudoprime to the base 3. The horrendous expression on the left side of (19.23) was calculated with the calculator program listed in the Appendix.

19.6 Deterministic Primality Testing

Here we briefly review some background and the latest published advances in *deterministic* tests for primality [19.2]. Mersenne once remarked (in translation) that "to tell if a given number of 15 or 20 digits is prime or not, all time would not suffice for the test." How astonished he would be if he learned that the primality of 100-digit numbers can now, albeit 340 years later, be *ascertained* in a matter of minutes! And Gauss would be delighted too. In *Disquisitiones Arithmeticae* [19.10] he wrote (in Latin): "The dignity of the science itself [number theory, of course] seems to require that every possible means be explored for the solution of a problem so celebrated" (distinguishing prime numbers from composite numbers and resolving the latter into their prime factors).

However, the progress reported here is only one part of Gauss's "problem," namely *primality testing*. No comparable progress has been made on the second part of his problem, which is *factoring* [19.11]. We must always bear in mind the distinction between these two tasks. Primality testing has become very efficient in recent years, and that is good for public-key cryptography, because large primes are at a premium. Factoring, on the other hand, is still very difficult, but that is also good, nay vital, for the security of public-key cryptosystems.

The first deterministic primality test that we encountered was based on Wilson's theorem (Chap. 8): n is prime *iff* $n \mid (n-1)! + 1$. Unfortunately, the factorial $(n-1)!$ destroys any attraction that Wilson's theorem might have evoked: the slowest sieve method will be lightning-fast compared to checking the divisibility of $(n-1)! + 1$ for large n. If n has 100 digits, then $(n-1)!$ has roughly 10^{102} *digits* — not to be confused with the "relatively" small number 10^{102}; the utter futility is obvious.

Along came *Edouard A. Lucas* [19.12], who showed in 1876 that n is prime *iff* there is a b, coprime to n, such that

$$b^{n-1} \equiv 1 \pmod{n} \tag{19.24}$$

and

$$b^{\frac{n-1}{p}} \not\equiv 1 \pmod{n}$$

for *each* prime factor p of $(n-1)$.

Lucas's test is especially expeditious for $n = 2^m + 1$, because then $n-1$ has only one prime factor: 2.

Example: $n = 17$, $n-1 = 2^4 = 16$, $2^{16} \equiv 1 \pmod{17}$ and $2^8 = 256 \equiv 1 \pmod{17}$. Thus $b = 2$ does not tell us anything. But for $b = 3$, $3^{16} \equiv 1 \pmod{17}$ and $3^8 = 6561 \equiv 16 \not\equiv 1 \pmod{17}$. Thus, 17 *is* a prime.

The problem, of course, is that for large n not all the prime factors of $n-1$ may be known. Thus it came as a relief when in 1945, a primality test for n was formulated requiring knowledge of only *some* of the prime factors of $n-1$ and $n+1$. Later, a primality test was devised [19.13] for which partial factorization of $n^2 + 1$ or $n^2 - n + 1$ or $n^2 + n + 1$ is required. However, if none of these numbers factor easily, the test does not fly.

Finally, in 1980, *Adleman* and *Rumely* [19.14] constructed a test that has drastically improved the efficacy of checking the primality of large numbers without restriction. With some improvements by H. Cohen and H. W. Lenstra, Jr., the test can now check a 100-digit number in less than a minute! And even a 1000-digit number can now be tested in about a week.

More generally, it is conjectured [19.2] that to test the primality of n requires a time not exceeding

$$c \, (\ln \, n)^{b \, \ln(\ln(\ln \, n))} \,,$$

where b and c are constants *independent of* n, and b is of order 1 and c depends, of course, on the calculating speed of the computer. With $c = 5 \cdot 10^{-5}$ (1982) and $b \approx 1.5$ (empirical), the maximal testing time for a 2000-digit number is less than a year — only moderately unreasonable.

If we take present life expectancy at birth to be about 80 years, then a teenager can know the primality of any 5000-digit number during his expected lifetime — and sooner, if there is further progress in primality testing during the next decades, or perhaps the next few weeks. However, there will always be a "primality horizon" beyond which numbers will keep their prime secret "forever" — the half-life of a proton (ca. 10^{31} years), say.

19.7 A Very Simple Factoring Algorithm

The simplest method for factoring N is to divide N by all primes up to $N^{1/2}$ and to watch for zero remainders. Divisibility by 2, 3, 5 and 11 is of course easily tested (see Sect. 6.1). But the main problem with this approach is the necessity to store many primes (more than 400 million primes for $N \leqslant 10^{20}$).

The need to store any primes evaporates if we divide by *all* integers up to $N^{1/2}$. This method is of course very time consuming. A neat and simple algorithm, suitable for *pocket calculators,* follows from the following observation.

Suppose we *trial divide* N first by 2. If 2 is not a divisor of N, then trial dividing by any other even number is redundant. Thus, it suffices to trial divide only by 2 and odd numbers.

Similarly, after trial division with 3, trial division by any multiple of 3 is unnecessary. Thus, only trial divisors congruent 1 or 5 modulo 6 are still in the running.

Finally, after also trial dividing by 5, which is just as easy, only trial divisors congruent to 1, 7, 11, 13, 17, 19, 23 and 29 modulo 30 need be considered, i.e., 8 out of each 30 integers. I encourage the reader to program her or his calculator for this simple algorithm, which requires storing fewer than 10 integers.

Other factoring algorithms for calculators, home computers and large machines are described by H. C. Williams in "Factoring on a Computer" in a recent issue of *The Mathematical Intelligencer* (Volume 6, Number 3, 1984, pp. 29-36).

What are the factors (other than 2) of $11^{64} + 1$, a 67-digit number? Is R_{1031} a "repunit" consisting of 1031 1's (see Sect. 3.6) prime? These are difficult questions when this was written (22 February, 1985)!

20. The Möbius Function and the Möbius Transform

After Euler's totient function, the Möbius function (named after the Möbius of strip fame) is one of the most important tools of number theory. It allows us to *invert* certain number-theoretic relations. In a sense, if we liken summation over divisors to integration, then taking the Möbius function is like differentiating. It is defined as follows:

$$\mu(n) := \begin{cases} 1 & \text{for } n = 1 \\ 0 & \text{if } n \text{ is divisible by a square} \\ (-1)^k & \text{if } n \text{ is the product of } k \text{ distinct } primes \, . \end{cases} \quad (20.1)$$

Thus, $\mu(n) \neq 0$ only for 1 and squarefree integers. The Möbius function is *multiplicative:*

$$\mu(mn) = \mu(m)\mu(n) \quad \text{if } (m,n) = 1 \, .$$

Also,

$$\mu(mn) = 0 \qquad \text{if } (m,n) > 1 \, , \quad (20.2)$$

because then mn is not squarefree.

The most important property of the Möbius function is that its summatory function (Sect. 10.4) equals zero, except for $n = 1$:

$$\sum_{d|n} \mu(d) = \delta_{n1} \, . \quad (20.3)$$

Here δ_{nm} is the *Kronecker symbol* defined by

$$\delta_{nm} := \begin{cases} 1 & \text{for} \quad n = m \\ 0 & \text{for} \quad n \neq m \end{cases} \, . \quad (20.4)$$

20.1 The Möbius Transform and Its Inverse

Let F be the summatory function of f:

$$F(n) = \sum_{d|n} f(d) \, . \quad (20.5)$$

Then

$$f(n) = \sum_{d|n} \mu(\frac{n}{d}) \, F(d) \, , \tag{20.6}$$

and conversely, (20.6) implies (20.5). Thus, the Möbius function emerges as the one that permits us to determine a function from its summatory function.

$F(n)$ is also called the *Möbius transform* of $f(n)$ and $f(n)$ is called the *inverse* Möbius transform of $F(n)$ [20.1].

Example: As we have seen (and repeatedly used) in previous chapters, the following identity holds for Euler's function:

$$n = \sum_{d|n} \phi(d) \, . \tag{20.7}$$

Thus, n is the Möbius transform of $\phi(n)$.

By applying the above inversion formula, we obtain the following expression for Euler's function:

$$\phi(n) = \sum_{d|n} \mu(\frac{n}{d}) \, d = n \sum_{d|n} \mu(d) \, \frac{1}{d} \, , \tag{20.8}$$

a relation between Euler and Möbius that is good to know. How can we evaluate the sum on the right? We only need to consider squarefree divisors d^* of n because $\mu(d) = 0$ for nonsquarefree d. Writing

$$n = \prod_i p_i^{e_i} \, , \tag{20.9}$$

such divisors d^* are of the form p_i or $p_i p_j (i \neq j)$, etc. Thus, with (20.1),

$$\sum_{d|n} \frac{1}{d} \mu(d) = 1 - \sum_i \frac{1}{p_i} + \sum_{i \neq j} \frac{1}{p_i p_j} - \dots$$

$$\tag{20.10}$$

$$= \prod_i \left[1 - \frac{1}{p_i} \right] \, .$$

In this manner, we have found another proof of

$$\frac{\phi(n)}{n} = \prod_i \left[1 - \frac{1}{p_i} \right] \, , \tag{20.11}$$

the average of which, as we know, tends to $6/\pi^2$ (see Sect. 4.4).

The above proof is, of course, not complete without showing that (20.3) holds. Defining

$$\epsilon(n) := \sum_{d|n} \mu(d) , \qquad\qquad (20.12)$$

we notice that trivially, $\epsilon(1) = 1$. Next, we consider n to be a prime power p^α ($\alpha > 0$). In that case, the divisors of n are $p, p^2, ..., p^\alpha$. Thus,

$$\epsilon(p^\alpha) = 1 + \mu(p) + \mu(p^2) + ... + \mu(p^\alpha) . \qquad\qquad (20.13)$$

Here all terms vanish except the first two, which cancel. Hence

$$\epsilon(p^\alpha) = 0 , \qquad (\alpha > 0) . \qquad\qquad (20.14)$$

The result for general n now follows from the fact that $\mu(n)$ is multiplicative, making its summatory function also multiplicative (which we do not prove here, but is easy to show).

20.2 Proof of the Inversion Formula

We want to show that for

$$F(n) = \sum_{d|n} f(d) ,$$

the inverse Möbius transform is

$$f(n) = \sum_{d|n} \mu(d) F(\frac{n}{d}) ,$$

and vice versa. We note that

$$F(\frac{n}{d}) = \sum_{d'|\frac{n}{d}} f(d') \qquad\qquad (20.15)$$

and introduce this in the above equation, yielding

$$f(n) = \sum_{d|n} \mu(d) \sum_{d'|\frac{n}{d}} f(d') . \qquad\qquad (20.16)$$

By inverting the order of summation over d and d', we obtain

$$f(n) = \sum_{d'|n} f(d') \sum_{d|\frac{n}{d'}} \mu(d) , \qquad (20.17)$$

where the last sum equals zero, except for $n/d' = 1$, i.e., $d' = n$ — which proves (20.6). The converse is proved similarly.

According to the basic relation (20.3) the Möbius function $\mu(n)$. is the Möbius transform of the *Kronecker deltafunction* $\delta(n) = \delta_{n1}$ as defined in (20.4). The Möbius transform of $\delta(n)$ is the constant 1.

The Möbius transform of n^k is the generalized divisor function $\sigma_k(n)$ as defined in Chap. 10.

The *inverse* Möbius transform of n^k could be called a generalized Euler function $\phi_k(n)$, in view of the fact (20.8) that the inverse Möbius transform of n is $\phi(n) = \phi_1(n)$. The so defined generalized Euler function is given by

$$\phi_k(n) = n^k \prod_{p|n} \left[1 - \frac{1}{p^k} \right], \quad k = 1,2, \dots .$$

We mention one more function with interesting Möbius transform properties, the von Mangoldt function, which plays an important role in analytic number theory [20.2]:

$$\Lambda(n) = \begin{cases} \ln p & \text{if } p \text{ is the only prime factor of } n \\ 0 & \text{if } n \neq p^m . \end{cases}$$

Its Möbius transform is the logarithm $\ln n$. The inverse Möbius transform of the von Mangoldt function is $-\mu(n) \ln n$ [20.3].

20.3 Second Inversion Formula

Let

$$G(n) = \prod_{d|n} g(d) ; \quad \text{then} \qquad (20.18)$$

$$g(n) = \prod_{d|n} G(d)^{\mu(n/d)} \qquad (20.19)$$

and vice versa. The proof follows from the one just given by setting $g = e^f$ and $G = e^F$.

Applying this formula to the one previously derived (Chap. 10)

$$\prod_{d|n} d = n^{\frac{1}{2}d(n)} ,$$

where $d(n)$ is the number of divisors of n, we obtain the following curious (and probably useless) representation of n^2:

$$n^2 = \prod_{d|n} d^{d(d)\mu(n/d)} .$$ (20.20)

20.4 Third Inversion Formula

For $x > 0$, let

$$H(x) = \sum_{n=1}^{\infty} h(nx) . \quad \text{Then}$$ (20.21)

$$h(x) = \sum_{n=1}^{\infty} \mu(n) H(nx)$$ (20.22)

and vice versa, provided

$$\sum_{m,n} |f(mnx)|$$

converges.

20.5 Fourth Inversion Formula

For $x > 0$, let

$$H(x) = \sum_{n=1}^{\lfloor x \rfloor} h(\frac{x}{n}) . \quad \text{Then}$$ (20.23)

$$h(x) = \sum_{n=1}^{\lfloor x \rfloor} \mu(n) H(\frac{x}{n}) .$$ (20.24)

20.6 Riemann's Hypothesis and the Disproof of the Mertens Conjecture

The Möbius function is involved in the following noteworthy sums:

$$\sum_{n=1}^{\infty} \frac{\mu(n)}{n} = 0 ,$$ (20.25)

$$\sum_{n=1}^{\infty} \frac{\mu(n)}{n} \ln n = -1 \; , \tag{20.26}$$

the former one being the result of E. Landau's dissertation.
The partial sums over the Möbius function

$$M(N) := \sum_{n=1}^{N} \mu(n)$$

were conjectured by Mertens in 1897 to obey the bound $|M(N)| < N^{\frac{1}{2}}$ for all $x > 1$. The truth of this conjecture, which has attracted much attention over the last 87 years, would imply the Rieman hypothesis (see Sect. 4.2). But, alas, the Mertens conjecture was disproved in 1984 by A. Odlyzko and H. J. J. te Riele, although no counterexamples had been found for N up to 10^{10}. In fact, there may be no counterexamples below 10^{20} or even 10^{30}, but Odlyzko and te Riele proved that, for some N, $M(N)N^{-\frac{1}{2}}$ exceeds 1.06 (and may, in fact, be unbounded for $N \rightarrow \infty$). They did not actually give a counterexample but expect one to occur "near" $10^{10^{65}}$. Here, once more, *numerical* evidence was — and would continue to be! — very misleading. But, ironically, the disproof involved heavy numerical computation (of the zeroes of the zetafunction).

20.7 Dirichlet Series and the Möbius Function

A Dirichlet series is of the form [20.1]:

$$F(s) = \sum_{n=1}^{\infty} \frac{f(n)}{n^s} \; , \tag{20.27}$$

where $F(s)$ is called the *generating function* of the function $f(n)$, which appears as the coefficient of n^{-s} in a Dirichlet series. In the following we shall ignore questions of convergence.
If $f(n)$ is multiplicative, then

$$F(s) = \prod_{p} \left[1 + \frac{f(p)}{p^s} + \frac{f(p^2)}{p^{2s}} + \ldots \right] \; , \tag{20.28}$$

and if $f(n)$ is *completely* multiplicative then

$$F(s) = \prod_{p} \left[1 - \frac{f(p)}{p^s} \right]^{-1} \; . \tag{20.29}$$

For $f(n) = 1$ in (20.27) or (20.29), $F(s)$ equals the Riemann zetafunction $\zeta(s)$.

If we have a second Dirichlet series:

$$G(s) = \sum_{n-1}^{\infty} \frac{g(n)}{n^s} ,\tag{20.30}$$

then the product of $F(s)$ and $G(s)$ is given by

$$F(s) \, G(s) = \sum_{n-1}^{\infty} \frac{h(n)}{n^s} , \qquad \text{where} \tag{20.31}$$

$$h(n) = \sum_{d|n} f(d) g\left(\frac{n}{d}\right) , \tag{20.32}$$

which is called a *number-theoretic convolution*. In other words, if two arithmetic functions are "convolved" as in (20.32), then the generating function is the product of their Dirichlet series.

Because, by definition (20.1):

$$\mu(p^k) = 0 \quad \text{for } k > 1 ,$$

we have

$$1 - \frac{1}{p^s} = 1 + \frac{\mu(p)}{p^s} + \frac{\mu(p^2)}{p^{2s}} + \cdots , \tag{20.33}$$

and therefore

$$\frac{1}{\zeta(s)} = \prod_p \left[1 - \frac{1}{p^s} \right] = \prod_p \left[1 + \frac{\mu(p)}{p^s} + \frac{\mu(p^2)}{p^{2s}} + \cdots \right] , \tag{20.34}$$

which, with (20.28) and (20.27), can be written:

$$\sum_{n-1}^{\infty} \frac{\mu(n)}{n^s} = \zeta^{-1}(s) . \tag{20.35}$$

For $s = 2$, we obtain the interesting result:

$$\sum_{n-1}^{\infty} \frac{\mu(n)}{n^2} = \frac{6}{\pi^2} . \tag{20.36}$$

In parting, we cite a few similarly derived Dirichlet series results, which the reader may wish to prove:

$$\sum_{n=1}^{\infty} \frac{|\mu(n)|}{n^s} = \frac{\zeta(s)}{\zeta(2s)} \, , \tag{20.37}$$

$$\sum_{n=1}^{\infty} \frac{d(n)}{n^s} = \zeta^2(s) \, , \tag{20.38}$$

$$\sum_{n=2}^{\infty} \frac{\ln n}{n^s} = -\zeta'(s) \, , \tag{20.39}$$

$$\sum_{n=1}^{\infty} \frac{(\ln n)^2}{n^s} = \zeta''(s) \, , \tag{20.40}$$

$$\sum_{u=2}^{\infty} \frac{\Lambda(n)}{n^s} = - \frac{\zeta'(s)}{\zeta(s)} \, . \tag{20.41}$$

If $g(n)$ is the Möbius transform of $f(n)$, then their generating functions, $G(s)$ and $F(s)$, are related by

$$G(s) = \zeta(s) \, F(s) \, . \tag{20.42}$$

Setting $f(n) = \phi(n)$, so that $g(n) = n$ and therefore

$$G(s) = \sum_{n=1}^{\infty} \frac{n}{n^s} = \zeta(s-1) \, ,$$

we obtain the relationship:

$$\sum_{n=1}^{\infty} \frac{\phi(n)}{n^s} = \frac{\zeta(s-1)}{\zeta(s)} \tag{20.43}$$

and, in a similar manner,

$$\sum_{n=1}^{\infty} \frac{\phi_k(n)}{n^s} = \zeta(s)^{-1} \, \zeta(s-k) \, . \tag{20.44}$$

Relations between *prime* divisor functions and the zetafunction were already mentioned in Sect. 11.3 (11.40-43), although the relation with $\zeta(s)$ was not made explicit there, it is given by the following two identities:

$$\frac{\zeta(2s)}{\zeta(s)} = \sum_{n=1}^{\infty} \frac{(-1)^{\Omega(n)}}{n^s} \quad \text{and} \tag{20.45}$$

$$\frac{\zeta^2(s)}{\zeta(2s)} = \sum_{n=1}^{\infty} \frac{2^{\omega(n)}}{n^s} \, . \tag{20.46}$$

21. Generating Functions and Partitions

The generating functions introduced in Chap. 20 and defined by Dirichlet series are not the only kind of generating functions. Here we shall briefly get to know another type of generating function with many useful properties that are applicable in numerous fields of mathematics and other sciences. As an illustration of that utility, we shall acquaint ourselves with various partition problems such as the partitions of our main subject: the positive integers. For example, the integer 4 has 5 unrestricted partitions into integers:

$$4 = 3 + 1 = 2 + 2 = 2 + 1 + 1 = 1 + 1 + 1 + 1 .$$

The subject of partitions is closely connected with Elliptic Modular functions [21.1], one of the more fertile fields of mathematics, which, however, is beyond the scope of this book.

Applications of partition functions abound throughout the natural sciences — we mention only (quantum) statistical physics, where questions such as how to partition a given number of energy quanta or, perhaps, a fluctuating number of photons among various "bins" or *eigenstates* are at the very root of our understanding of what radiation and matter are about.

21.1 Generating Functions

A *generating function* $F(s)$ for the function $a(n)$ can be defined quite generally by:

$$F(s) = \sum a(n) \, u(n,s) , \tag{21.1}$$

where, again, we ignore questions of convergence [21.2].

With $u(n,s) = n^{-s}$, (21.1) is a Dirichlet series as discussed in Sect. 20.7. Another choice of $u(n,s)$ is:

$$u(n,s) = e^{-sn} ,$$

or, with

$$z = e^{-n} ,$$

$$F(z) = \sum a(n) \, z^{-n} \, . \tag{21.2}$$

$F(z)$ is also known as the *z transform* of $a(n)$, especially in engineering applications — to time-discrete electrical circuits, for example [21.3].

While Dirichlet series, as a result of (20.31) and (20.32), play an important role in *multiplicative* problems of number theory (in particular the theory of primes and divisibility), the z transform is dominant in *additive* problems, because of the following all-important convolution property: If $F(z)$ is the z transform of $a(n)$ as defined in (21.2), and $G(z)$ is the z transform of $b(n)$, then the z transform of the *convolution* (ubiquitous in linear systems)

$$c(n) = \sum_k a(k) \, b(n-k) \tag{21.3}$$

equals the *product* of the z transforms:

$$\sum c(n) z^{-n} = F(z) G(z) \, . \tag{21.4}$$

The decisive difference between Dirichlet series and z transforms is, of course, that in the former the variable (s) appears in the *exponent*, whereas in the latter the variable (z) appears as a base.

If we take z as a complex variable and consider it on the unit circle:

$$z = e^{i\theta} \, ,$$

then the z transform turns into a *Fourier series* — another important property of the power-series type generating function as defined by (21.2).

Generating functions of this type also play a great role in *probability theory*, where they are known as *characteristic functions*. They generate statistical *moments* or *cumulants* of probability distributions and permit their convenient manipulation [21.4]. Since summing two or more independent random variables results in a convolution of their distributions, the corresponding characteristic functions are simply multiplied to obtain the characteristic function of the summed random variables. (Thus, for example, the sum of several Poisson variables is immediately seen to be another Poisson variable.)

As a simple example of a generating function, we consider

$$F(z) = (1 + z)^n \, ,$$

which, of course, generates the binomial coefficients $\binom{N}{n}$:

$$(1 + z)^N = \sum_{n=0}^{N} \binom{N}{n} z^n \, , \tag{21.5}$$

which tell us how many ways n objects can be selected from a set of N distinct objects without regard to order. (Often power series as in (21.5) are in z rather than z^{-1}, but that is a matter of choice depending on the field of application.)

We may also mention here the simple generating function of the Fibonacci numbers (Chap. 5):

$$\sum_{n=0}^{\infty} F_n z^n = \frac{z}{1-z-z^2} , \qquad (21.6)$$

which is very useful in obtaining closed-form expressions for F_n as in (5.30).

21.2 Partitions of Integers

Let us ask ourselves how many ways $p(n)$ a positive integer n can be partitioned so that the sum of the parts equals n — without regard to order and with no restrictions (other than that the parts shall also be positive integers).

Example: $p(5) = 7$, because:

$$5 = 4 + 1 = 3 + 2 = 3 + 1 + 1 = 2 + 2 + 1$$
$$= 2 + 1 + 1 + 1 = 1 + 1 + 1 + 1 + 1 ,$$

i.e., there are 7 ways to partition 5 in the prescribed manner.

Partitions obey an endearing symmetry. For example, consider the 5 dots:

This geometrical representation of a partition, called a Ferrer graph, illustrates *two* partitions of the integer 5. Reading column-wise (vertically) we see that $5 = 2 + 2 + 1$ (which we already knew), and reading row-wise (horizontally) we discover that $5 = 3 + 2$. Thus, each partition has a (not necessarily distinct) *conjugate.*[1] Note that the first partition ($5 = 2 + 2 + 1$) is into 3 integers of which the largest are 2, and the second partition

---

[1] In quantum mechanics such geometric representations of partitions are known as Young tableaux, after the inventor who introduced them to the study of symmetric groups. They are important in the analysis of the symmetries of many-electron systems.

$(5 = 3 + 2)$ is into 2 integers of which 3 is the largest. Some partitions are self-conjugate, as for example $3 + 1 + 1$.

The following two theorems are an interesting consequence of conjugacy [21.2]:

1) *The number of partitions of n into m integers is equal to the number of partitions of n into integers the largest of which is m.*

For example, there are 2 partitions of 5 into 2 integers $(4 + 1$ and $3 + 2)$ and there are equally 2 partitions of which 2 is the largest integer $(2 + 2 + 1$ and $2 + 1 + 1 + 1)$.

2) *The number of partitions of n into at most m integers is equal to the number of partitions of n into integers which do not exceed m.*

For example, there are 3 partitions of 5 into *at most* 2 integers $(5, 4 + 1$ and $3 + 2)$ and there are equally 3 partitions whose integers do not exceed 2, namely $2 + 2 + 1$, $2 + 1 + 1 + 1$ and $1 + 1 + 1 + 1 + 1$. Astounding consequences, considering the simplicity of the underlying symmetry!

21.3 Generating Functions of Partitions

The useful generating functions for *partitions* are of the power-series (or z-transform) type, because when we talk about partitioning, we are talking *additively*.

The generating function of $p(n)$ was found by Euler. Defining $p(0) = 1$, it is

$$\sum_{n=0}^{\infty} p(n)\, z^n = \frac{1}{(1-z)(1-z^2)(1-z^3)\,\ldots}\,. \tag{21.7}$$

To see that Euler was right, we expand each factor into a geometric series:

$$
\begin{aligned}
&\textit{integer } 1: \quad (\underline{z^0} + \underline{z^1} + z^2 + \overset{\equiv}{z^3} + \ldots) \\[4pt]
&\textit{integer } 2: \quad (\overset{\equiv}{z^0} + \underline{z^2} + z^4 + z^6 + \ldots) \\[4pt]
&\textit{integer } 3: \quad (\overset{\equiv}{z^0} + \underline{z^3} + z^6 + z^9 + \ldots)
\end{aligned}
\tag{21.8}
$$

$$\textit{taken} \quad 0 \quad 1 \quad 2 \quad 3 \quad \ldots \textit{times}\,, \text{ etc.}$$

In multiplying this out, we have to take one term from each factor. For example, what is the coefficient of z^3, i.e., how many ways are there to get z^3?

First, we take the term z^3 in the third factor (1-column) and the two 1's in the first two factors. (These 3 terms, one from each factor, are identified by underlining with a single horizontal stroke.) Next we take the term z^1 in the first factor, the term z^2 in the second factor, and the 1 in the third factor, indicated by underlining with a double stroke. This gives us another z^3 term in the product. Third, we take the factor z^3 from the first factor (3-column) and the 1's from the other two factors, as shown by overlining with a triple stroke. This gives us a third z^3 term. Since there are no more ways of making z^3 terms (the next factor — not shown in (21.8) — is $1 + x^4 + x^8 + \dots$), we conclude, tentatively, that

$$p(3) = 3 .$$

And indeed, there are precisely 3 partitions of 3: $3 = 2 + 1 = 1 + 1 + 1$.

Why is the Euler formula (21.7) correct? In what way did our way of picking z^3 terms correspond to the 3 partitions of 3? Let us identify the *first* factor in (21.8) with the partitioning integer 1. Having picked the term z^3 (3-column) from it (triple overlined) we say we have used the integer 1 exactly 3 times. In other words, we identify this z^3 term with the partition $1 + 1 + 1$: the integer 1 (first row) taken 3 times (3-column). Similarly, we identify the second and third factors in (21.8) with the partitioning integers 2 and 3, respectively. Thus, the double underlined contribution to the z^3 term in the product corresponds to the partition $1 + 2$. The single underlined terms correspond to the partition 3; in other words, the integer 3 is taken once (1-column) and the integers 1 and 2 are both taken 0 times (0-column). *This is how generating functions function.*

A more formal way to define partitions is to consider the expression

$$k_1 + 2k_2 + 3k_3 + \dots = n , \tag{21.9}$$

and to ask how many sets of nonnegative integers k_m there are that satisfy (21.9). Here k_m is the number of times the integer m appears in the partition of n.

21.4 Restricted Partitions

The above discussion of the generating function (21.7) for unrestricted partitions immediately allows us to write down the generating functions $F_m(z)$ for the partition of n into integers the largest of which is m:

$$F_m(z) = \frac{1}{(1-z)(1-z^2) \dots (1-z^m)} .$$

(21.10)

Similarly, the generating function for partitions into *even* integers exceeding 4, for example, is

$$_6F_{even}(z) = \frac{1}{(1-z^6)(1-z^8)(1-z^{10}) \dots} .$$

If we are interested in partitions of n into *distinct* integers, a moment's thought will show that the generating function is

$$F_{dist}(z) = (1+z)(1+z^2)(1+z^3) \dots .$$

(21.11)

Since each factor in (21.11) has only one term other than 1, we can take each integer at *most* once, i.e., the integers in any partition are unequal or *distinct*. (Remember that the individual factors in these generating functions stand for the different integers.)

We shall illustrate the supreme usefulness of generating functions by the following example. $F_{dist}(z)$ according to (21.11) can be rewritten

$$F_{dist}(z) = \frac{1-z^2}{1-z} \frac{1-z^4}{1-z^2} \frac{1-z^6}{1-z^3} \dots$$

$$= \frac{1}{(1-z)(1-z^3)(1-z^5) \dots} .$$

(21.12)

But the latter generating function is $F_o(z)$, corresponding to partitions into *odd* integers.

Thus, we have found the remarkable result that *distinct* partitions (that take each integer *at most once,* without repetition) and partitions into *odd* integers are equally numerous:

$$p_{dist}(n) = p_o(n) .$$

(21.13)

Example: There are three partitions of 5 into distinct integers, namely 5, 4 + 1 and 3 + 2. And there are also three partitions of 5 into *odd* integers: 5, 3 + 1 + 1 and 1 + 1 + 1 + 1 + 1. Check!

Suppose a hypothetical physical situation depends critically on the partition of n elementary particles or quanta over a number of bins or eigenstates. Then the purely arithmetic relation (21.13) would tell us that we could never distinguish between two different interpretations or "statistics," namely "distinct" *versus* "odd." Distinctly odd!

Could the reader have discovered the "parity" law (21.13) without generating functions? He may wish to prove his prowess with generating functions by proving the following simple generalization of (21.13): The number of partitions of n with no part repeated more than r times is equal to the number of partitions of n with no part divisible by $r + 1$.

For $r = 1$, we obtain the above result (2.13). But for $1 < r < n-3$, interesting new results emerge. For example, for $n = 6$ and $r = 2$, there are 7 (out of a total of 11) partitions that do not use the integers 3 and 6: $5 + 1 = 4 + 2 = 4 + 1 + 1 = 2 + 2 + 2 = 2 + 2 + 1 + 1 = 2 + 1 + 1 + 1 + 1 + 1 = 1 + 1 + 1 + 1 + 1 + 1$. And there are also 7 partitions not using any integer more than twice $6 = 5 + 1 = 4 + 2 = 3 + 3 = 4 + 1 + 1 = 3 + 2 + 1 = 2 + 2 + 1 + 1$.

Let us turn the argument around and write down certain polynomial identities and ask what they mean in terms of partitions. Consider the well-known (?) identity:

$$\frac{1}{1-z} = (1+z)(1+z^2)(1+z^4)(1+z^8) \dots . \tag{21.14}$$

Its correctness can be established by multiplying it with $(1-z)$, which yields

$$1 = (1-z^2)(1+z^2)(1+z^4)(1+z^8) \dots .$$

And now a mathematical chain reaction sets in: Combining the first two factors yields

$$(1-z^4)(1+z^4)(1+z^8) \dots ,$$

which in turn yields

$$(1-z^8)(1+z^8) \dots , \text{ etc. },$$

which converges to 1 for $|z| < 1$.

But what does (21.14) imply in terms of partitions? The left side is the generating function of the integers themselves; i.e., it generates each integer exactly once. The right side of (21.14) is the partition of the integers into powers of 2 without repetition. Thus (21.14) tells us that each integer can be represented uniquely as a sum of powers of 2 (including $2^0 = 1$). How comforting to know!

Manipulations of generating functions, as illustrated by the identity (21.14), lead to several fascinating interpretations as partitions. For example, Euler proved that

$$(1-z)(1-z^2)(1-z^3) \cdots = \sum_{n=-\infty}^{\infty} (-1)^n z^{n(3n+1)/2}$$

$$= 1-z-z^2+z^5+z^7-z^{12}-z^{15}+ \dots , \tag{21.15}$$

which, by multiplying with $F_{\text{dist}}(z)$ from (21.12), can be converted into a more easily recognizable form:

$$\frac{(1-z^2)(1-z^4)(1-z^6)\cdots}{(1-z)(1-z^3)(1-z^5)\cdots} = 1 + z + z^3 + z^6 + z^{10} + \ldots, \tag{21.16}$$

where the exponents are the familiar triangular numbers $n(n+1)/2$ (see Sect. 7.4). (In other words, the left side of (21.16) is the generating function of the triangular numbers.) The exponents in (21.15) are the pentagonal numbers $g_5(n)$ and $g_5(-n)$, see Sect. 7.4.

The identity (21.15) has a striking combinatorial interpretation. The coefficient of z^n on the left side of (21.15) is

$$\sum (-1)^k ,$$

where the summation is over all partitions into distinct integers and k is the number of integers in such a partition. Thus, the left side of (21.15) represents the difference between partitions (into distinct integers) with an even and an odd number of integers, respectively.

Calling the number of partitions of n into an *even* number of *distinct* integers $p_{\text{even}}(n)$, and the number of partitions into an odd number of distinct integers $p_{\text{odd}}(n)$, we get

$$p_{\text{even}}(n) - p_{\text{odd}}(n) = \begin{cases} (-1)^k & \text{for } n = k(3k \pm 1)/2 \\ 0 & \text{otherwise ,} \end{cases} \tag{21.17}$$

where $n = k(3k \pm 1)/2$ corresponds to the terms occurring on the right side of (21.15).

Example: $n = 8$, which does not appear as an exponent on the right side of (21.15). Thus, $p_{\text{even}}(8)$ should equal $p_{\text{odd}}(8)$. And indeed, there are three partitions into distinct integers of each kind:

$$7 + 1 = 6 + 2 = 5 + 3 \qquad \text{(even number of distinct integers) ,}$$
$$8 = 5 + 2 + 1 = 4 + 3 + 1 \quad \text{(odd number of distinct integers) .}$$

The identity (21.15) also leads to a recurrence formula [21.5] for $p(n)$:

$$p(n) = \frac{1}{n} \sum_{k=1}^{n} \sigma(k)\, p(n-k) , \quad p(0) = 1 , \tag{21.18}$$

where $\sigma(k)$ is the sum of the divisors of k (see Chap. 10).

With the asymptotic behavior of $\sigma(k)$ (Sect. 10.4), one obtains an asymptotic law [21.5] for $p(n)$ for large n:

$$p(n) \approx \exp[\pi(2n/3)^{\frac{1}{2}}]/4n3^{\frac{1}{2}} . \tag{21.19}$$

Example: for $n = 243$ (21.19) gives $p(n) \approx 1.38 \cdot 10^{14}$. The exact result,

$$p(243) = 133978259344888 ,$$

contradicts a famous conjecture by S. Ramanujan [21.6,7] (see also [21.1], p. 289) that *if*

$$d = 5^a 7^b 11^c$$

and

$$24n \equiv 1 \pmod{d} ,$$

then

$$p(n) \equiv 0 \pmod{d} .$$

But $p(243)$ is *not* divisible by $d = 7^3 = 343$, although $24 \cdot 243 \equiv 1 \pmod{343}$.

However, Ramanujan found a number of striking congruence relations for $p(n)$ that stood the test of time (and proof). One of them involves the modulus 5:

$$p(5m + 4) \equiv 0 \pmod{5} . \tag{21.20}$$

In the meantime, many more congruence relations for partitions (and other combinatorial identities) have been discovered and proved, stimulating a lot of good mathematics [21.8].

We conclude this chapter with another striking discovery by Ramanujan involving a (generalized) continued fraction (Chap. 5):

$$1 + \frac{e^{-2\pi}}{1+} \frac{e^{-4\pi}}{1+} \frac{e^{-6\pi}}{1+} = e^{-2\pi/5} (5^{1/4}g^{1/2} - g)^{-1}, \tag{21.21}$$

where g is the Golden ratio:

$$g = \tfrac{1}{2}(5^{1/2} + 1) = 1.618 \dots .$$

22. Cyclotomic Polynomials

Cyclotomy, the art of dividing a circle into equal parts, was a Greek specialty, and the only tools allowed were a straightedge and a compass. The subject is deeply related to number theory, as we saw in our discussion of Fermat primes in Sect. 3.8. In addition, *cyclotomic polynomials* play an important role in modern digital processes and fast computation (Sect. 24.3).

We begin this chapter with an informal discussion on dividing a circle into equal parts and constructing regular polygons. After that we explore some of the important properties of cyclotomic polynomials and their applications in physics, engineering and, in Chap. 28, artistic design.

22.1 How to Divide a Circle into Equal Parts

A circle in the complex number plane $z = x + iy$, centered at the origin $z = 0$, is described by the equation

$$z = re^{i\alpha} , \tag{22.1}$$

where r is the radius of the circle and α the angle between the direction of z and the real axis. As α goes from 0 to 2π, the circle is traversed exactly once.

If we only care about angles α, and not about the radius r of the circle, we might as well set $r = 1$ and consider the unit circle

$$z = e^{i\alpha} . \tag{22.2}$$

Dividing the circle into, say, n equal parts means finding an angle $\alpha_{1/n}$ such that

$$\alpha_{1/n} = \frac{2\pi}{n} . \tag{22.3}$$

Then the incongruent (mod 2π) multiples of $\alpha_{1/n}$ divide the circle into n equal parts.

Of course, $\alpha_{1/n}$ may not be unique. In fact, for $(k,n) = 1$,

$$\alpha_{k/n} = k \frac{2\pi}{n} \tag{22.4}$$

will also give the desired "cyclotomy" — the learned expression for circle division. Since there are $\phi(n)$ incongruent coprime values of k, there are $\phi(n)$ primitive, or generating, angles $\alpha_{k/n}$, but of course, knowledge of one of these suffices to do the job.

The n equidistant points on the unit circle

$$z_m = e^{i2\pi m/n} , \quad m = 0,1,...,n-1 , \tag{22.5}$$

not only divide the circle into equal parts but are also the corners of a regular n-gon inscribed in the circle. Thus, we see that the tasks of equal division of the circle and construction of a regular n-gon are related.

Mathematically, finding the z_m of (22.5) is the same as finding the roots of the polynomial

$$z^n - 1 = 0 , \tag{22.6}$$

i.e., we have converted our geometrical problem into an arithmetic one. Upon setting z in (22.6) equal to z_m from (22.5), we obtain

$$(e^{i2\pi m/n})^n - 1 = e^{i2\pi m} - 1 = 1 - 1 = 0 ,$$

for *all* integer values of m, including the n incongruent ones:

$$m = 0,1, ..., n-1. \tag{22.7}$$

But how do we find the z_m, or at least *one* primitive root?

The Case $n = 2$.

Dividing the circle into two parts $(n = 2)$ is simple: the equation

$$z^2 - 1 = 0 \tag{22.8}$$

has the well-known solutions $z = 1$ and $z = -1$, corresponding to the angles $\alpha = 0$ and $\alpha = \pi$ (180°), which indeed divides the circle into two equal parts. In other words, a straight line through the (known!) center of the circle divides it into two equal parts.

The Case $n = 3$.

To divide the circle into three equal parts, the equation we have to solve is

$$z^3 - 1 = 0 , \tag{22.9}$$

with the solutions

$$z_{m/3} = e^{i2\pi m/3} , \quad m = 0,1,2, ... , \tag{22.10}$$

corresponding to the angles 0, 120° and 240°.

Now comes a crucial question. The reader may recall that we started out with a geometrical problem: the equal division of a circle or the construction of a regular polygon. It is therefore legitimate — and quite natural — to ask whether we can, in fact, solve $z^n - 1 = 0$ by "geometric means," i.e., with a *straightedge* and a *compass* but no other tools.

Let us look more closely at the problem for $n = 3$, the regular trigon — otherwise known as an equilateral triangle. First we notice that $z^3 - 1$ can be factored

$$z^3 - 1 = (z-1)(z^2+z+1) . \tag{22.11}$$

Here the first factor gives the universal solution (valid for all n) $z = 1$. To find the two remaining solutions, we have to solve a *quadratic* equation, namely

$$z^2 + z + 1 = 0 , \tag{22.12}$$

which is easy:

$$z = -\frac{1}{2} \pm \frac{1}{2}\sqrt{-3} = e^{\pm i2\pi/3} , \tag{22.13}$$

as expected.

But the important point now is that we only had to solve a *quadratic* equation — and this is something that can be accomplished purely by straightedge and compass, as desired by the purist Greeks. In fact, (22.13) tells us that we should bisect the negative real axis between $z = 0$ and $z = -1$ (the second intersection of the straight line through $z = 0$ and $z = 1$, both of which can be selected arbitrarily). Erecting the normal at $z = -1/2$ (also possible with the means at hand) we find that it intersects the circle at $x = -1/2$ and $y = \pm\sqrt{3}/2$.

Unwittingly, by this construction we have solved the quadratic equation

$$y = \sqrt{1^2-\left(\frac{1}{2}\right)^2} = \pm\frac{1}{2}\sqrt{3} , \tag{22.14}$$

because by erecting the normal at $z = -1/2$ and locating its intersection with the circle $|z| = 1$, we have constructed a Pythagorean triangle with a hypotenuse of length 1 and one side of length $1/2$. The length of the other side is then of course given by (22.14).

The Case $n = 4$.

How about the regular tetragon, widely known (in mathematics and other disciplines) as a square? Since we have already solved the case $n = 2$, a bisection of the angle between the two solutions $z = 1$ and $z = -1$ will give the two additional solutions for $n = 4$: $z = i$ $(\alpha = 90°)$ and $z = -i$ $(\alpha = 270°)$.

In arithmetic terms, we can express our ability to solve the case $n = 4$ geometrically as follows. The equation whose roots we need is $z^4 - 1$, which factors as follows:

$$z^4 - 1 = (z^2-1)(z^2+1) . \tag{22.15}$$

Thus, instead of an equation of the fourth degree, we only have to deal with second-degree equations which, as we already know, pose no problems to geometrical purists employing only straightedge and compass.

This process of bisection can be repeated arbitrarily often. Thus, if we have a solution for the case n, we "automatically" have solutions for $n \cdot 2^k$ with $k = 1,2,3,...$. So far, we have therefore shown how to construct regular polygons with $n = 2,3,4,6,8,12,16,24,32$, etc. edges. In some of these cases there are particularly simple solutions, as for $n = 6$, the regular hexagon, whose edge length equals the radius of the circle.

The Case $n = 5$.

The lowest-degree case so far unsolved is that of the regular pentagon. The equation $z^5 - 1$ factors into

$$z^5 - 1 = (z-1)(z^4+z^3+z^2+z+1) . \tag{22.16}$$

At first blush, this case looks impossible because of the fourth-degree polynomial on the right of (22.16). But this polynomial must be factorizable into polynomials of no higher degree than quadratic, or the Greeks couldn't have constructed the regular pentagon as they did.

Now instead of giving the Greek solution, we will use the case $n = 5$ to present the gist of Gauss's sensational solution of the case $n = 17$ [22.1]. This makes the case $n = 5$ look slightly more difficult than necessary, but we can then see how Gauss went about solving $n = 17$ and the other basic cases still possible, namely when n equals a Fermat prime, i.e., a prime of the form $n = 2^{2^m} + 1$. For $m = 0$ we get the triangle, for $m = 1$ the pentagon and for $m = 2$ the heptadecagon $(n = 17)$. The only other *known* basic cases are $m = 3$ $(n = 257)$ and $m = 4$ $(n = 65537)$.

22.2 Gauss's Great Insight

Gauss noticed that if n is a *Fermat prime,* then $n - 1$ is a power of 2 (in fact, it must be a *power* of a power of 2). Furthermore, Fermat primes, like all primes, have *primitive roots*. And Gauss proceeded to order the roots of $z^n - 1$ according to powers of a primitive root of n, say w. We shall illustrate this, as we said, for $n = 5$ because it is easier to write down than the case $n = 17$ — but the principle is the same.

First, as always, we factor out the universal factor $z - 1$ (giving the solution $z = 1$):

$$z^5 - 1 = (z-1)(z^4+z^3+z^2+z+1) .$$

Now we are looking for 4 values of z such that

$$z + z^2 + z^3 + z^4 = -1 . \tag{22.17}$$

A primitive root of 5 is $w = 2$. (Check: $w^k \equiv 2,4,3,1 \pmod 5$ for $k = 1,2,3,4$. Check!) We now combine exponents in (22.17) according to alternating terms in the series $w^k = 2,4,3,1$, i.e., we write

$$z + z^2 + z^3 + z^4 = r_1 + r_2 , \quad \text{where} \tag{22.18}$$

$$r_1 = z^2 + z^3 \quad \text{and} \quad r_2 = z^4 + z^1 . \tag{22.19}$$

Can we determine r_1 and r_2? Yes, we can! With (22.17) we have

$$r_1 + r_2 = -1 , \tag{22.20}$$

and with (22.19),

$$r_1 \cdot r_2 = z^6 + z^3 + z^7 + z^4 , \tag{22.21}$$

which, because of $z^5 = z^0$, can be written

$$r_1 \cdot r_2 = z^1 + z^3 + z^2 + z^4 .$$

Here the sum on the right is the sum of all different fifth roots of 1, except 1 itself. Since the sum of *all* roots equals zero, we have

$$r_1 \cdot r_2 = -1 . \tag{22.22}$$

Now (22.20) and (22.22) are of the form of the two solutions r_1 and r_2 of a *quadratic* equation in r:

$$r^2 + ar + b = 0 , \tag{22.23}$$

where, according to Vieta's rule [22.2],

$$a = -(r_1+r_2) = 1 \qquad \text{and} \tag{22.24}$$

$$b = r_1 \cdot r_2 = -1 . \tag{22.25}$$

Hence our quadratic equation (22.23) for r is

$$r^2 + r - 1 = 0 , \tag{22.26}$$

with the solution

$$r_1 = -\frac{\sqrt{5}+1}{2} \quad \text{and} \quad r_2 = \frac{\sqrt{5}-1}{2} , \tag{22.27}$$

where, incidentally, r_2 equals the reciprocal of the Golden ratio, $g = 1.618...$, and r_1 equals the negative of the Golden ratio (Sect. 5.5 and Fig. 5.4).

With these intermediate solutions it is easy to obtain the final ones. Setting $z^4 = z^{-1}$ in (22.19), we may write

$$z^{-1} + z = r_2 = \frac{1}{g} , \tag{22.28}$$

another *quadratic* equation! Its solutions are

$$z_{1,2} = \frac{1}{2g} \pm i \sqrt{1 - \frac{1}{4g^2}} , \tag{22.29}$$

where, of course, $|z_{1,2}| = 1$ and the real part is

$$x_{1,2} = \frac{1}{2g} = \frac{\sqrt{5}-1}{4} . \tag{22.30}$$

Thus, two of the roots are obtained by constructions of $\sqrt{5}$ (as the hypotenuse of a right triangle with sides of lengths 1 and 2) and by proceeding according to (22.30). Since all roots of the polynomial $z^4 + z^3 + z^2 + z + 1$ are *primitive* roots, the remaining roots can be found geometrically by taking integral multiples of the angle associated with, say, $z_{1,2}$:

$$\alpha_{1,2} = \arccos \frac{1}{2g} = \pm \frac{2\pi}{5} = \pm 72° . \tag{22.31}$$

This equation tells us, incidentally, that $72°$ is an angle whose cosine is not transcendental but is a quadratic irrational.

Thus, we have succeeded in factoring $z^4 + z^3 + z^2 + z + 1$ using only quadratic irrationals:

$$z^4 + z^3 + z^2 + z + 1 = (z-z_1)(z-z_2)(z-z_3)(z-z_4) \tag{22.32}$$

with $z_{1,2}$ as given by (22.29) and $z_{3,4}$ given as the solutions of

$$z + z^{-1} = r_1 = -g \; ,$$

namely

$$z_{3,4} = \frac{g}{2} \pm i \sqrt{1 - \frac{g^2}{4}} \; , \tag{22.33}$$

with associated angles

$$\alpha_{3,4} = \arccos\left(\frac{-g}{2}\right) = \pm \frac{4\pi}{5} = \pm 144° \; . \tag{22.34}$$

If our solutions are correct, they must lie on the unit circle, which they do, and their sum must equal -1. And indeed,

$$z_1 + z_2 + z_3 + z_4 = \frac{1}{g} - g = -1 \; . \quad \text{Check!}$$

The case $n = 17$ proceeds quite analogously but requires one extra step of the kind used in (22.19), namely, assigning powers of z to two different sets. *Rademacher* [22.3] wrote a lucid exposition of the case $n = 17$.

The Case $n = 6$.

The polynomial $z^6 - 1$ can be written as follows:

$$z^6 - 1 = (z^2)^3 - 1 \; , \tag{22.35}$$

which resembles the case for $n = 3$, except that z in $z^3 - 1$ has been replaced by z^2. Thus, $n = 6$ can be solved geometrically by obtaining the angle for $n = 3$, as already described, and halving it. Further halvings crack the cases $n = 12, 24, 48$, etc.

The Case $n = 7$.

The polynomial $z^7 - 1$ factors over the rationals into

$$z^7 - 1 = (z-1)(z^6 + z^5 + z^4 + z^3 + z^2 + z + 1) \; . \tag{22.36}$$

Since 7 is a prime, the second factor can be shown to be irreducible over the rationals. But 7 is not a *Fermat prime,* and 6 is not a power of 2. Thus, the grouping exemplified by (22.19), which works for $n = 5, 17, 257$ and 65537, leading to nested *quadratic* equations, does not work for $n = 7$, and the regular heptagon can therefore not be constructed by straightedge and compass. This insight by Gauss should have ended a vain search that had lasted over 2000 years, but some people are still trying!

The Case $n = 8$.

The case $n = 8$ follows from the case $n = 4$ by angle halving, as do the cases $n = 16, 32, 64$, etc.

The Case $n = 9$.

While $9 - 1 = 8$ *is* a power of 2, the number 9 unfortunately is composite. The polynomial $z^9 - 1$ can be written as

$$z^9 - 1 = (z^3)^3 - 1 , \tag{22.37}$$

which requires solving the case $n = 3$ and then trisecting the resulting angle (120°), which unfortunately is still not possible. Alternatively, we could say that solving $n = 9$ requires taking a third root, which is impossible geometrically. Thus, $n = 9$ also fails, but for a different reason than $n = 7$: $n = 7$ *has* a primitive root but $7 - 1$ is not a power of 2, while $n = 9$ exceeds a power of 2 by 1, as required, but 9 is composite. *Both* conditions must be fulfilled, and they are fulfilled only by the Fermat primes.

The Case $n = 10$.

The case $n = 10$ reduces to the case $n = 5$ and angle halving, which also yields geometrical solutions for $n = 20, 40, 80$, etc.

The Cases $n = 11$ and $n = 13$ are like $n = 7$: impossible.

The Case $n = 15$.

The case $n = 15$ is interesting. The composite $15 = 3 \cdot 5$ is the product of precisely *two* Fermat primes. As we know, we can construct geometrically the angle for the equilateral triangle (120°) and the regular pentagon (72°), and half the difference angle, $(120° - 72°)/2 = 24°$, is the angle for the regular 15-gon. In like manner, we can solve any n that is the product of *different* Fermat primes, such as 51, 85, 355, etc.

The Case $n = 17$.

The case $n = 17$ is the famous case first solved by *Gauss* [22.1, 3] using a primitive root of 17 and "noting" that $17 - 1 = 16$ is a power of 2. (See the discussion under $n = 5$.)

The Case $n = 257$.

The cases $n = 257$ and 65537 are similar to $n = 5$ and $n = 17$ but require one and two more decomposing steps, respectively, than the case $n = 17$.

The next *new and interesting* case would be given by a prime $n = 2^m + 1$ which requires m to be a power of 2. The next *candidate* is $m = 2^{32}$, i.e., $n = 4294967297$, but unfortunately this is composite. This is the end of the road for geometrical cyclotomy until another Fermat prime is found. (But see Sect. 3.8 for the odds of this being successful.)

After this brief excursion into the realm of cyclotomy, we shall study, in a somewhat more formal manner, some general properties of cyclotomic polynomials that give rise to a host of enticing applications (fast algorithms, periodic sequences with unique and very useful spectral properties, etc.).

22.3 Factoring in Different Fields

In numerous and diverse applications it will be important to factor the polynomial

$$P_N(z) := z^N - 1 . \tag{22.38}$$

In fact, the applications we shall discuss here are so diverse that we seek factors whose coefficients are from three different number fields:

1)   The complex number field C

2)   The rational number field Q

3)   Finite (Galois) number fields $GF(p^m)$

In the following we will continue our discussion of cyclotomy and related factoring on a somewhat more formal level, and in view of the importance of the subject, we will even permit ourselves a bit of redundancy.

22.4 Cyclotomy in the Complex Plane

Factorization of $z^N - 1$ is easy if we admit complex coefficients. Setting

$$P_N(z) = z^N - 1 = 0 , \tag{22.39}$$

we see immediately that the zeroes of $P_N(z)$ are the N distinct Nth roots of unity

$$W_N^r , \quad r = 0,1, ..., N-1 , \tag{22.40}$$

where W_N is a *primitive* Nth root of unity, for example:

$$W_N = e^{2\pi i/N} . \tag{22.41}$$

Insisting on a primitive root here insures that W_N^r will indeed run through *all* N distinct Nth roots as r goes from 0 to $N-1$. For example, for $N = 4$,

$$W_4 = e^{2\pi i/4} = i \tag{22.42}$$

is a primitive 4th root, and so is

$$W_4^3 = e^{6\pi i/4} = -i , \tag{22.43}$$

because each will generate the remaining two roots of $z^4 - 1 = 0$, namely $z = 1$ and $z = -1$.

All the primitive Nth roots are given by

$$W_N^r , \quad \text{where } (r,N) = 1 . \tag{22.44}$$

Thus, we see that there are $\phi(N)$ primitive Nth roots.

Since if $(r,N) = 1$, so is $(N - r,N)$, and because r and $N - r$ are distinct for $N > 2$, the primitive roots come in "conjugate" pairs, which are seen to be conjugate *complex:*

$$W_N^r \quad \text{and} \quad W_N^{N-r} = W_N^{-r} = W_N^{r*} , \tag{22.45}$$

where * stands for conjugate complex in this book. (Physicists and engineers need the horizontal bar for averages.) It is interesting to note that the term "conjugate complex" comes from the more general algebraic concept of "conjugate" and not vice versa, as one might be led to think if one is familiar only with *complex* conjugates.

Thus, our factorization of $P_N(z)$ is

$$z^N - 1 = \prod_{n=0}^{N-1} (z - W_N^n) , \tag{22.46}$$

which is as far as one can go, because the individual factors are all linear in z.

Replacing the primitive root W_N above by another primitive root, W_N^r, with $(r,N) = 1$, changes only the order of the factors. Thus, the above factorization is essentially unique.

22.5 How to Divide a Circle with Compass and Straightedge

How to divide a circle into N equal parts by compass and straightedge, i.e., by purely geometric means, was one of the classical problems of geometry first brought into prominence by the ancient Greeks. They discovered that a circle could be so divided if $N = 3$ or 5. Since angles can also be halved geometrically, additional possibilities are $N = 6, 12, 24, \ldots$ and $N = 5, 10, 20, \ldots$. Finally, it is not difficult to see that by dividing a circle both into 3 *and* 5 equal parts, the angle $2\pi/3 \cdot 5$ can also be obtained, thereby permitting division of the circle into 15 equal parts.

In general, according to Euclid, a circle can be divided into N equal parts by compass and straightedge if

$$N = 2^k \cdot 3^m \cdot 5^n , \tag{22.47}$$

where k is any nonnegative integer and n and m are either 0 or 1.

Since dividing a circle into N equal parts is equivalent to constructing a regular N-gon, i.e., a polygon with N vertices (or edges), the lowest-order regular N-gon that could not be constructed was the 7-gon or *heptagon*.

Since the time of Euclid, for almost 2000 years, mathematicians and amateurs alike had been trying to smash the boundary at $N = 7$, but in vain. Then, on March 30, 1796, an 18-year-old Brunswick (Germany) youth scribbled in his brand-new notebook, at the top of page 1, (see Fig. 5.1) freely translated from the Latin that he was using: "How to divide the circle by geometric means into 17 equal parts." In other words Gauss (the name of the young man) had just discovered that the numbers 3 and 5 of the ancient Greeks had to be supplemented by 17 and, in general, by primes of the form

$$2^{2^n} + 1 , \tag{22.48}$$

i.e., the Fermat primes F_n of which then, and to this day, only 5 are known: $F_0 = 3$, $F_1 = 5$, $F_2 = 17$, $F_3 = 257$ and $F_4 = 65537$.

Of equal significance, Gauss proved that the *only* regular N-gons that can be constructed by geometric means are of the form

$$N = 2^k \prod_n F_n , \tag{22.49}$$

where the product is over *distinct* Fermat primes. Thus, apart from the factor 2^k, there are at present $2^5 - 1 = 31$ different regular odd N-gons that can be constructed geometrically: from the triangle and the pentagon to the 17-gon and the $3 \cdot 5 \cdot 17 \cdot 257 \cdot 65537 = 4\ 294\ 967\ 295$-gon. (The author strongly advises against attempting the latter case. Gauss said it is possible and that should suffice. On the other hand, there is a *suitcase* at the Mathematics Institute of the University of Göttingen which is jam-packed with the details of constructing the regular 65537-gon.)

How is Gauss's great discovery related to our factorization? One way to state it is to say that the primitive Nth root W_N involves only *square* roots, and square roots can be constructed with a compass and a straightedge (following old Pythagoras and his right triangle).

Example: $N = 5$:

$$W_5 = e^{2\pi i/5} = \cos(2\pi/5) + i \sin(2\pi/5) , \qquad \text{where} \qquad (22.50)$$

$$\cos(2\pi/5) = (\sqrt{5} - 1)/4 ,$$

$$\sin^2(2\pi/5) = (\sqrt{5} + 5)/8 . \qquad (22.51)$$

22.5.1 Rational Factors of $z^N - 1$

One of our main aims is to factor

$$P_N(z) = z^N - 1 \qquad (22.52)$$

into polynomials with rational coefficients. This is trivial for

$$P_1(z) = z - 1 \qquad (22.53)$$

(which leaves nothing to factor) and easy for $P_2(z)$:

$$P_2(z) = (z - 1)(z + 1) . \qquad (22.54)$$

Furthermore, remembering geometric series, we write

$$P_3(z) = (z - 1)(z^2 + z + 1) . \qquad (22.55)$$

But now we have reached the end of Easy Street — unless, that is, we recall Möbius and the second inversion formula involving his function (Chap. 20).

To be able to apply Möbius, we factor $P_N(z)$ as follows (perhaps as a long shot, not knowing what else to do):

$$z^N - 1 = \prod_{n|N} C_n(z) , \qquad (22.56)$$

where according to the inversion formula (20.19), the factors $C_n(z)$ are given by

$$C_n(z) = \prod_{d|n} (z^d - 1)^{\mu(n/d)} . \qquad (22.57)$$

Since $\mu = 0, \pm 1$, these factors of $z^N - 1$ are indeed rational. Also, they are not further reducible over the rationals. The degree of $C_n(z)$ is

$$\deg[C_n(z)] = \sum_{d|n} d\mu\left(\frac{n}{d}\right) = \phi(n) . \qquad (22.58)$$

Hence the degree of the product is

$$\deg\left[\prod_{n|N} C_n(z)\right] = \sum_{n|N} \phi(n) = N . \quad \text{Check!} \qquad (22.59)$$

Example: $N = 6$. With (22.56):

$$P_6(z) = z^6 - 1 = C_1(z) \cdot C_2(z) \cdot C_3(z) \cdot C_6(z)$$

and with (22.57):

$$C_1(z) = z - 1$$

$$C_2(z) = \frac{z^2-1}{z-1} = z + 1$$

$$C_3(z) = \frac{z^3-1}{z-1} = z^2 + z + 1$$

$$C_6(z) = \frac{(z-1)(z^6-1)}{(z^2-1)(z^3-1)} = z^2 - z + 1 .$$

Check: $z^6-1 = (z-1)(z+1)(z^2+z+1)(z^2-z+1)$. Check! Here the first factor, $z - 1$, "captures" the common root of all $P_n(z)$: $z_1 = 1$. The second factor, $C_2(z) = z + 1$, captures the remaining root of $z^2 - 1 = 0$, namely the primitive root $z_2 = e^{2\pi i/2} = -1$. The third factor, $z^2 + z + 1$, captures the two primitive roots of $z^3 - 1$, namely $z_3 = e^{2\pi i/3}$ and $z_4 = e^{4\pi i/3}$. Finally, the fourth factor, $z^2 - z + 1$, captures the two primitive roots of $z^6 - 1$, namely $z_5 = e^{2\pi i/6}$ and $z_6 = e^{10\pi i/6}$. The angles in the complex plane subtended between the real axis and these roots are, in the above order, $0°$, $180°$, $120°$, $240°$, $60°$ and $300°$, i.e., all distinct multiples of $60°$ — as expected.

22.6 An Alternative Rational Factorization

Another way of representing the factors of $C_n(z)$ is as follows. We first replace d by n/d, which only changes the order of the factors:

$$C_n(z) = \prod_{d|n} (z^{n/d} - 1)^{\mu(d)} . \qquad (22.60)$$

Now it is apparent that we need only those divisors of n which are squarefree, because otherwise $\mu = 0$. These squarefree divisors can be put into classes depending on the number of distinct prime factors. We introduce the product

$$\Pi_k := \prod_{d_k} (z^{n/d_k} - 1) , \tag{22.61}$$

where the d_k are all the squarefree divisors that have exactly k distinct prime factors. Thus, for example,

$$\Pi_o = z^n - 1 ,$$

$$\Pi_1 = \prod_{p_i} (z^{n/p_i} - 1) , \tag{22.62}$$

$$\Pi_2 = \prod_{p_i \neq p_j} (z^{n/p_i p_j} - 1) , \quad \text{etc.} ,$$

where the p_i are the prime factors of n.

With this notation, and because by definition $\mu(d_k) = (-1)^k$, we get the factorization

$$C_n(z) = \frac{\Pi_0 \, \Pi_2 \, \Pi_4 \, \cdots}{\Pi_1 \, \Pi_3 \, \Pi_5 \, \cdots} , \tag{22.63}$$

which has several interesting theoretical applications. Specifically, it can be shown that $C_n(z)$ has only *integral* coefficients. Furthermore, for $n < 105$, the only coefficients that appear in $C_n(z)$ other than 0 are ± 1.[1]

22.7 Relation Between Rational Factors and Complex Roots

We have found two seemingly independent factorizations of $z^N - 1$ so far, one involving the complex Nth roots of 1 but with factors linear in z, and the other involving higher powers of z but with rational (in fact, integer) coefficients. How are these two factorizations related? Can we express the cyclotomic polynomials $C_n(z)$ in terms of the nth roots of 1, W_n? Yes, we can:

$$C_n(z) = \prod_{\substack{0 < r < n \\ (r,n)=1}} (z - W_n^r) . \tag{22.64}$$

---

[1] $C_{105}(z)$ is the lowest-order cyclotomic polynomial in which other coefficients (-2 in two places) appear. The reason is that 105 is the smallest integer that is the product of three distinct odd primes [22.4].

Here we recall that W_n is the "first" nth root of 1:

$$W_n = e^{2\pi i/n} , \tag{22.65}$$

which is a *primitive* root, and the roots appearing in the above factorization are *all* the primitive nth roots of 1. There are precisely $\phi(n)$ such roots, so that the degree of $C_n(z)$ becomes $\phi(n)$. Check!

But that does not suffice to establish the correctness of the factorization. Instead of proving the above factorization of $C_n(z)$ in terms of the $(z - W_n^r)$, we will first illuminate it for a special case, $n = 6$ (see also the above example). The Möbius inversion gives us

$$C_6(z) = \frac{(z-1)(z^6-1)}{(z^2-1)(z^3-1)} . \tag{22.66}$$

Here the factor z^3-1, for example, can be written

$$z^3 - 1 = (z-1)(z-W_3^1)(z-W_3^2) . \tag{22.67}$$

However, we shall express W_3 by W_6:

$$W_3 = W_6^2 . \tag{22.68}$$

Thus,

$$z^3 - 1 = (z-1)(z-W_6^2)(z-W_6^4) , \tag{22.69}$$

or, more generally, we write

$$z^d - 1 = (z-1) \prod_{(k,n)=\frac{n}{d}} (z-W_n^k) . \tag{22.70}$$

In the special case of $C_6(z)$, this factorization is as follows:[2]

$$C_6(z) = \frac{(z-W_6^1)(z-W_6^2)\cdots(z-W_6^5)}{(z-W_6^3)(z-W_6^2)(z-W_6^4)} . \tag{22.71}$$

Here, in the numerator the roots W_6^k, $k = 1,2,3,4,5$ appear, i.e., *all* the 6th roots of 1 except 1 itself. In the denominator all those W_6^k appear, where

---

[2]   Leaving out the four factors $(z-1)$ which always cancel, except for $C_1(z) = z-1$, because, as we derived earlier,

$$\sum_{d|n} \mu(d) = 0 \quad \text{for } n > 1 .$$

$(k,6) > 1$. Thus, $C_6(z)$ is the product of precisely all factors W_6^r with $0 < r < 6$ and $(r,6) = 1$:

$$C_6(z) = \prod_{\substack{0 < r < n \\ (r,6)=1}} (z - W_6^r) , \tag{22.72}$$

i.e., $C_6(z)$ is the product of linear factors whose roots are precisely all the *primitive* roots of $z^6 - 1$.

The demonstration would have been essentially the same for other squarefree orders. For non-squarefree orders, some of the factors are missing. We leave it to the reader to generalize this result to arbitrary orders n.

22.8 How to Calculate with Cyclotomic Polynomials

For prime order $n = p$, one has

$$C_p(z) = \prod_{r=1}^{p-1} (z - W_p^r) = \frac{z^p - 1}{z - 1} , \tag{22.73}$$

or,

RULE I: $$C_p(z) = z^{p-1} + z^{p-2} + \cdots + z + 1 . \tag{22.74}$$

Without proof,

RULE II: $$C_{mp^k}(z) = C_{mp}(z^{p^{k-1}}) . \tag{22.75}$$

For $(m,p) = 1$,

RULE III: $$C_{mp} = \frac{C_m(z^p)}{C_m(z)} . \tag{22.76}$$

For odd $n > 2$,

RULE IV: $$C_{2n}(z) = C_n(-z) , \quad \text{for odd } n \geqslant 3 . \tag{22.77}$$

For $z = 1$, one has

RULE V: $$C_n(1) = \begin{cases} 0 & \text{for } n = 1 \\ p & \text{for } n = p^k \\ 1 & \text{otherwise} . \end{cases} \tag{22.78}$$

These rules expedite the calculation of cyclotomic polynomials:

$$C_1(z) \;=\; z - 1$$

$$C_2(z) \;=\; z + 1 \qquad\qquad \text{(by RULE I)}$$

$$C_3(z) \;=\; z^2 + z + 1 \qquad\qquad \text{(by RULE I)}$$

$$C_4(z) \;=\; C_2(z^2) = z^2 + 1 \qquad\qquad \text{(by RULE II)}$$

$$C_6(z) \;=\; C_2(z^3)/C_2(z) = z^2 - z + 1 \qquad \text{(by RULE III)}$$

Check: $C_6(z) \;=\; C_3(-z)$. Check! $\qquad\qquad$ (by RULE IV)

Check: $C_6(1) \;=\;$ 1. Check! $\qquad\qquad$ (by RULE V)

Check: $C_4(1) \;=\;$ 2. Check! $\qquad\qquad$ (by RULE V)

And, incidentally,

$$C_{17}(z) = z^{16} + z^{15} + \cdots + z + 1 , \qquad\qquad (22.79)$$

so that

$$z^{17} - 1 = (z-1)C_{17}(z) . \qquad\qquad (22.80)$$

Now, while $C_{17}(z)$ cannot be further factored into polynomials with rational coefficients, Gauss observed that $C_{17}(z)$ can be decomposed into a set of nested *quadratic* equations, each solvable by "geometric means," leading to the constructibility of the regular 17-gon with compass and straightedge (Sect. 22.2).

23. Linear Systems and Polynomials

One of the main applications of polynomial theory occurs in the analysis of linear electrical circuits and the many other physical situations that are customarily represented by linear-circuit analogs. With the advent of computers and digital signal processing, *time-discrete* systems have taken on a special significance, and these are effectively represented by polynomials called z transforms that are akin to *generating functions* in other branches of mathematics. The application of cyclotomic polynomials, in particular, leads to fast computational algorithms, excellent error-correcting codes, and special signals for precision measurement (Chap. 26).

23.1 Impulse Responses

If the capacitor C in Fig. 23.1 is charged up to voltage y_0 at time $t = 0$ and thereafter allowed to discharge via the resistor R, its voltage as a function of time will decay exponentially:

$$y(t) = y_0 e^{-t/\tau} \quad \text{for } t > 0. \tag{23.1}$$

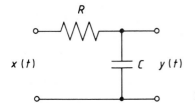

Fig. 23.1. A simple linear passive system

Here the "time constant" (in seconds) is $\tau = RC$, where R is the resistance (in ohms) and C is the capacitance (in farads). (The symbols R and C do double duty here.)

If the charging impulse was of very short duration and the capacitor was "empty" for negative times, then

$$y(t) = 0 \quad \text{for } t < 0. \tag{23.2}$$

If the charging voltage is a Dirac-delta function $\delta(t)$, an idealized pulse which is 0 everywhere except for $t = 0$ and represents a unit charge, then the output of the linear system is called its *impulse response* $h(t)$. For the circuit in Fig. 23.1,

$$
h(t) = \begin{cases} 0 & \text{for} \quad t < 0 \\ \dfrac{1}{C}\, e^{-t/\tau} & \text{for} \quad t > 0 \ . \end{cases} \tag{23.3}
$$

The term "linear" in linear system means that its parameters (here the values R and C) do not change with the relevant variables (here electrical current, voltage and charge). Roughly: the resistor does not heat up and the capacitor does not break down.

If an arbitrary voltage as a function of time, $x(t)$, is applied to the input of the linear system of Fig. 23.1, the output voltage is a *convolution integral* [23.1]:

$$
y(t) = \int x(t')h(t-t')dt' \ . \tag{23.4}
$$

This expression follows directly from the assumed linearity of the system (and its time invariance and passivity; i.e., its parameters do not change with time and do not *generate* energy).

23.2 Time-Discrete Systems and the z Transform

If we want to simulate linear systems on digital computers, we have to "discretize," i.e., represent by rational numbers, all constants and variables, including time. The impulse response of a time-discrete system is a sequence h_n, which is its output if at "time" $n = 0$ a unit impulse was applied to the quiescent system.

The impulse response h_n is often represented by its z transform (Chap. 21):

$$
H(z) := \sum_{n=-\infty}^{\infty} h_n z^{-n} \ , \tag{23.5}
$$

which (except for the sign in the exponent of z) is nothing but the familiar *generating function* for the sequence h_n. If we set $z = e^{i\omega T}$, then $H(z)$ becomes the Fourier transform, where ω is the radian frequency and T is the "sampling interval," i.e., the time interval corresponding to a difference of 1 in the "time" index n.

23.3 Discrete Convolution

If we apply an input sequence x_n to a linear system with impulse response h_n, the output sequence (because of linear superposition) becomes

$$y_n = \sum_k x_k h_{n-k} \; . \tag{23.6}$$

Such a *discrete convolution* is often abbreviated by the convolution star:

$$y_n =: x_n * h_n \; , \tag{23.7}$$

a convenient notation.

But if we express the above sequences y_n, x_n and h_n by their z transforms, then the (somewhat messy) discrete convolution becomes a simple product:

$$Y(z) = X(z) \cdot H(z) \; , \tag{23.8}$$

which is verified by comparing equal powers of z on the two sides of the equation.

23.4 Cyclotomic Polynomials and z Transform

For many applications it is either appropriate or convenient to consider the data and the impulse responses to be periodic or *circular* [23.2]:

$$x_n = x_{n+N} \; , \tag{23.9}$$

where N is the (fundamental) period. More generally, we may write

$$\left. \begin{array}{l} x_n = x_m \\ h_n = h_m \end{array} \right\} \quad \text{for} \quad n \equiv m \pmod{N} \; , \tag{23.10}$$

and consequently also

$$y_n = y_m \quad \text{for} \quad n \equiv m \pmod{N}.$$

Circular convolution looks like this:

$$y_n = \sum_{k=0}^{N-1} x_k h_{<n-k>_N} \; , \tag{23.11}$$

where we have used the acute brackets $\langle\ \rangle_N$ to indicate a least nonnegative residue modulo N.

For circular data, the z transform contains the factor $z^N - 1$, and the possibility of factorizing $z^N - 1$ leads to a reduction in the number of multiplications. We will not pursue the details here, but simply cite a theorem by Winograd which states that the number of multiplications can be reduced by as much as m, where m is the number of irreducible factors of $z^N - 1$, i.e., $m = N$ if we factor over the complex number field, and $m = d(N)$ if we factor over the rationals [23.3].

24. Polynomial Theory

In this chapter we further develop our acquaintance with polynomials, especially those with integer coefficients and discover that in many ways, their arithmetic is akin to that of integers: there are coprime polynomials, divisor polynomials, and Diophantine equations in polynomials, and there is even a version of Euclid's algorithm and a Chinese Remainder Theorem for polynomials.

The most important application of integer polynomials is in the construction of finite number fields, also called Galois fields (Chap. 25), which play a dominant role in today's digital world.

24.1 Some Basic Facts of Polynomial Life

Many of the things we have learned about numbers (residue systems, primitive elements, etc.) can be generalized to polynomials — with virtually unbounded applications. Since we will subsequently need at least the elements of *polynomial theory*, we will familiarize ourselves with this fascinating extension now. A polynomial $p(z)$ of *degree n over a field F* looks like this [24.1]:

$$p(z) = \sum_{k=0}^{n} a_k z^k , \qquad n \geqslant 0 , \qquad a_n \neq 0 , \tag{24.1}$$

where the coefficients a_k are elements of the number field F, for example the complex number field C, the rational number field Q, or some *finite* number field such as $GF(2)$, the Galois field of order 2 consisting typically of the elements 0 and 1.

If the leading term $a_n = 1$, the polynomial is said to be *monic*. The degree of $p(z)$ is denoted by $\deg[p(z)]$. With $a_n \neq 0$, $\deg[p(z)] = n$.

A polynomial $d(z)$ *divides* another polynomial $p(z)$ if there exists a polynomial (over the same field!) such that $p(z) = q(z)d(z)$ holds. The polynomial $d(z)$ is then called a divisor of $p(z)$, and we write $d(z)|p(z)$. A polynomial whose only divisors are of degree 0 or $\deg[p(z)]$ is called *irreducible* (over the chosen field) [24.2].

For example, $z^2 + 1$ *is* reducible over C: $z^2 + 1 = (z + i)(z - i)$ but irreducible over Q. Interestingly, $z^2 + 1$ *is* reducible over the finite field

$GF(2)$: $z^2 + 1 = (z + 1)^2$, because $1 + 1 = 0$ in $GF(2)$. Thus, possible factorizations depend on the field chosen, and moreover they do so in a seemingly haphazard manner.

Just as integers can be uniquely factored into primes, so polynomials $p(z)$ can be uniquely factored into irreducible polynomials $p_i(z)$:

$$p(z) = K \prod_{i=1}^{s} [p_i(z)]^{r_i} ,\qquad(24.2)$$

where K is a constant that allows us to restrict the $p_i(z)$ to *monic* polynomials. The degree of $p(z)$ is given by

$$\deg[p(z)] = \sum_{i=1}^{s} r_i \deg[p_i(z)] .\qquad(24.3)$$

The "ultimate" factorization is, of course, into polynomials $p_i(z)$ of the first degree — if that is possible in the chosen field. As a result of the Fundamental Theorem of Algebra[1] the number of such first-degree factors, including multiplicity, is precisely equal to $\deg[p(z)]$. Polynomials that have no common factors are called mutually prime or *coprime*.

24.2 Polynomial Residues

It is always possible to write, for two polynomials $p(z)$ and $d(z)$:

$$p(z) = q(z)d(z) + r(z) ,\qquad(24.4)$$

and this representation is *unique*, if

$$\deg[r(z)] < \deg[d(z)] .\qquad(24.5)$$

Here, for obvious reasons, $r(z)$ is called the *remainder polynomial*.

In a well-wearing fashion of notation we will also express the above relationship by [24.3]:

$$r(z) = \langle p(z) \rangle_{d(z)} .\qquad(24.6)$$

These latter notations, of course, recommend themselves if we care not about the *quotient polynomial* $q(z)$ but only about the remainder. (It is peculiar that in this field of human endeavor — as in few others — *remainders* should be the main pickings.)

---

[1] First complete proof by Gauss in his Ph.D. thesis at Helmstedt, (barely West) Germany.

The operation of obtaining $r(z)$ from $p(z)$ and $d(z)$ is called *polynomial residue reduction*. Two polynomials $p_1(z)$ and $p_2(z)$ are said to be congruent modulo $d(z)$ if they leave the same remainder $r(z)$ upon residue reduction. Equivalently, they are congruent modulo $d(z)$ if their difference is divisible by $d(z)$: We write

$$p_1(z) \equiv p_2(z) \pmod{d(z)} , \tag{24.7}$$

or equivalently,

$$d(z) | (p_1(z) - p_2(z)) . \tag{24.8}$$

Example:

$$\langle z^2 \rangle_{z+1} = 1 , \tag{24.9}$$

or

$$z^2 \equiv 1 \pmod{(z + 1)} , \tag{24.10}$$

because

$$(z + 1) | (z^2 - 1) , \tag{24.11}$$

namely:

$$\frac{z^2 - 1}{z + 1} = z - 1 . \tag{24.12}$$

More generally,

$$\langle p(z) \rangle_{z-a} = p(a) , \tag{24.13}$$

and even more generally, for monic divisor polynomials

$$d(z) = z^n + \sum_{k=0}^{n-1} d_k z^k , \tag{24.14}$$

we get:

$$\langle z^n \rangle_{d(z)} = - \sum_{k=0}^{n-1} d_k z^k , \tag{24.15}$$

i.e., the highest power can always be reduced to the negative of the "tail" of the divisor polynomial.

Example:

$$\langle x^4 + x^6 \rangle_{x^4+x+1} = \langle x^4 \rangle_{x^4+x+1} (1 + x^2) = -(x + 1)(1 + x^2) .$$

Of course, if $d(z)$ is not monic, $d(z)/d_n$ will be, and we can proceed as above.

24.3 Chinese Remainders for Polynomials

Whenever it comes to *simultaneous* congruences (and a few other delicacies) we turn to the Chinese, and sure enough, there is a Chinese Remainder Theorem for polynomials also. To wit:

There exists a unique polynomial satisfying

$$y(z) \equiv y_i(z) \pmod{P_i(z)} , \quad i = 1, 2, \ldots, m , \tag{24.16}$$

such that

$$0 \leqslant \deg[y(z)] < \sum_{i=1}^{m} \deg[P_i(z)] , \tag{24.17}$$

provided the monic polynomials $P_i(z)$ are pairwise coprime.

For a constructive proof, let us first consider a simpler problem: find m "inverse" polynomials $R_i(z)$ such that each satisfies the following m congruences:

$$R_i(z) \equiv \delta_{ij} \pmod{P_j(z)} , \quad j = 1, 2, \ldots, m . \tag{24.18}$$

With these, the solution is obviously

$$y(z) = \sum_{i=1}^{m} R_i(z) y_i(z) \pmod{P(z)} , \tag{24.19}$$

where

$$P(z) = \prod_{i=1}^{m} P_i(z) . \tag{24.20}$$

Now, the inverse polynomials must be of the form

$$R_i(z) = S_i(z) P(z) / P_i(z) , \tag{24.21}$$

where $S_i(z)$ is obtained from the *single* congruence

$$S_i(z)P(z)/P_i(z) \equiv 1 \pmod{P_i(z)} .$$ (24.22)

This will automatically satisfy the m congruences for $R_i(z)$.
    The above congruence can be written as

$$S_i(z)P(z)/P_i(z) + T_i(z)P_i(z) = 1 ,$$ (24.23)

which is the generalization to polynomials of the Diophantine equation that we studied earlier and solved by Euclid's algorithm.

24.4 Euclid's Algorithm for Polynomials

To solve the above Diophantine equation, following Euclid, we expand $P_i^2(z)/P(z)$ into a continued fraction:

$$\frac{P_i^2(z)}{P(z)} = C_0(z) + \cfrac{1}{C_1(z) + \cfrac{1}{C_2(z) + \cdots + \cfrac{1}{C_k(z)}}} ,$$ (24.24)

where the $C_r(z)$ are determined successively by long division. Truncating the continued fraction yields the polynomial convergents $A_r(z)$ and $B_r(z)$, which obey the recursions

$$A_r(z) = C_r(z)A_{r-1}(z) + A_{r-2}(z)$$ (24.25)

and

$$B_r(z) = C_r(z)B_{r-1}(z) + B_{r-2}(z) .$$ (24.26)

The solution is then given by

$$S_i(z) = (-1)^k K A_{k-1}(z)$$ (24.27)

and

$$T_i(z) = (-1)^k K B_{k-1}(z) ,$$ (24.28)

where K is the leading coefficient of $A_k(z)$ and $B_k(z)$.
    In addition, the Euclidean algorithm gives the greatest common divisor of the two polynomials: $B_{k-1}(z)$.

Example:

$$R(z) \equiv \begin{cases} 1 & (\text{mod} \quad (z^2 + z + 1)) \\ 0 & (\text{mod} \quad (z - 1)) \end{cases} \tag{24.29}$$

i.e., with (24.23) (and dropping the index $i = 1$),

$$S(z)(z-1) + T(z)(z^2 + z + 1) = 1 . \tag{24.30}$$

We thus need the continued-fraction expansion

$$\frac{z^2 + z + 1}{z - 1} = z + 2 + \cfrac{1}{\frac{1}{3}(z - 1)} , \tag{24.31}$$

yielding the approximants

$$\frac{A_0(z)}{B_0(z)} = z + 2 \qquad \text{and} \tag{24.32}$$

$$\frac{A_1(z)}{B_1(z)} = \frac{\frac{1}{3}(z^2 + z + 1)}{\frac{1}{3}(z - 1)} , \tag{24.33}$$

and consequently, since $k = 1$,

$$S(z) = -\frac{1}{3}(z + 2) . \tag{24.34}$$

Thus, the solution is

$$R(z) = S(z)(z - 1) = -\frac{1}{3}(z^2 + z + 2) . \tag{24.35}$$

Check:

$$\langle R(z) \rangle_{z-1} = -\frac{1}{3}(1^2 + 1 - 2) = 0 . \quad \text{Check!}$$

$$\langle R(z) \rangle_{z^2+z+1} = -\frac{1}{3}(-1 - 2) = 1 . \quad \text{Check!}$$

The greatest common divider of $z^2 + z + 1$ and $z - 1$ equals $B_{k-1}(z) = B_0(z) = 1$. Check!

These rules and algorithms are about all we need to operate successfully with polynomials.

25. Galois Fields

In Galois fields, full of flowers,
Primitive elements dance for hours ...[1]

A residue system modulo a prime p forms a finite number field of order p. For many applications, we need number fields of order p^m. Here, with the knowledge acquired in Chaps. 23 and 24, we learn how to construct and represent them, and how to calculate in them.

25.1 Prime Order

We have already encountered finite number fields of prime order p. We now designate them by $GF(p)$, where GF stands for *Galois field* [25.1]. They consist, for example, of the elements $0,1,2, ... , p-1$, for which addition, subtraction, multiplication and division (except by 0) are defined, obeying the usual commutative, distributive and associative laws. Thus, in $GF(3)$, for example, consisting of the elements 0, 1 and 2, we have, by way of illustration, $1 + 2 = 0$, $1 - 2 = 2$, $2 \cdot 2 = 1$, $1/2 = 2$, etc.

25.2 Prime Power Order

We shall now construct finite number fields of order equal to a prime power p^m, designated $GF(p^m)$. These Galois fields have virtually unlimited application in such diverse fields as physics (diffraction, precision measurements), communications (error-correcting codes, cryptography) and artistic design (necklaces, etc.).

All realizations of $GF(p^m)$ are isomorphic. We will choose as field elements either m-tuples ("vectors"), $m \times m$ matrices, or polynomials of degree $m - 1$, all with components and coefficients, respectively, from $GF(p)$.

Since $GF^*(p^m)$, i.e., $GF(p^m)$ without the 0 element, forms a cyclic group with multiplication as the group operation, we can also represent $GF^*(p^m)$,

---

[1]   S. B. Weinstein, found in [25.1]

which has order $p^m - 1$, by a primitive element α and its $p^m - 1$ distinct powers:

$$\alpha, \alpha^2, \alpha^3, \ldots, \alpha^{p^m-1} = 1 . \tag{25.1}$$

We shall illustrate this for $p = 2$, i.e., $GF(2^m)$, which has order 2^m. We begin with the representation by m-tuples with components from $GF(2)$, i.e., 0 or 1. There are exactly 2^m such m-tuples in agreement with the order 2^m. For $m = 4$, $GF(2^4)$ then has the following 16 elements:

$$
\begin{array}{cccc}
0 & 0 & 0 & 0 \\
1 & 0 & 0 & 0 \\
0 & 1 & 0 & 0 \\
1 & 1 & 0 & 0 \\
0 & 0 & 1 & 0 \\
\end{array}
\tag{25.2}
$$

$$\vdots$$

$$
\begin{array}{cccc}
1 & 1 & 1 & 1
\end{array}
$$

The 4-tuples are written down here in binary sequence from 0 to 15 (with the least significant digit on the left, for a change).

Addition and subtraction are defined column-wise modulo 2. Thus, for example, $1100 + 1111 = 0011$ (which does *not* correspond to binary addition, because there are no "carries").

Multiplication is defined as follows. First, the m-tuples are represented by polynomials of degree up to $m - 1$. There are exactly p^m such polynomials, i.e., as many as the order of $GF(p^m)$. Each polynomial corresponds to one element of the field. For example, the element 1100 corresponds to

$$x^0 + x^1 + 0 + 0 = 1 + x , \tag{25.3}$$

and the element 0101 corresponds to

$$0 + x^1 + 0 + x^3 = x + x^3 . \tag{25.4}$$

Now, multiplication in $GF(p^m)$ is defined as multiplication of polynomials *modulo a given irreducible polynomial* $\pi(x)$ over $GF(p)$ of degree m. For $p = 2$ and $m = 4$, there are precisely three such polynomials: $1 + x + x^4$, (its "reciprocal") $1 + x^3 + x^4$ and $1 + x + x^2 + x^3 + x^4$. We choose

$$\pi(x) = 1 + x + x^4 , \tag{25.5}$$

as our modulus and define the product of two elements of $GF(2^4)$, represented by two polynomials $g(x)$ and $h(x)$, as

$$g(x) \cdot h(x) = \langle g(x) \cdot h(x) \rangle_{\pi(x)} , \tag{25.6}$$

where the acute brackets indicate a least remainder, i.e. a polynomial of degree smaller than the degree of $\pi(x)$.

Thus, for example,

$$1101 \cdot 1001 \stackrel{\wedge}{=} (1+x+x^3)(1+x^3) = 1 + x + x^4 + x^6 . \tag{25.7}$$

As we saw before, calculating the remainder modulo $\pi(x)$ is the same as setting $\pi(x) = 0$, or replacing x^4 by $1 + x$. Thus, we may write (remember the coefficients are from $GF(2)$, i.e., $1 + 1 = 0$, etc.):

$$1 + x + x^4 + x^6 = 1 + x + 1 + x + x^2(1+x) = x^2 + x^3 . \tag{25.8}$$

Thus, we get

$$1101 \cdot 1001 = 0011 . \tag{25.9}$$

For the multiplication to have an inverse in $GF^*(p^m)$, $\pi(x)$ must be irreducible — just as we require that for the multiplication of the integers $1,2,...,n-1$ to have an inverse modulo n, n must be "irreducible," i.e., prime.

To show that $1 + x + x^4$ is, in fact, irreducible, we have to show that it is not divisible by any polynomial over $GF(2)$ up to degree $4/2 = 2$.

Obviously, $1 + x + x^4$ is not divisible by x or $1 + x$. It is also not divisible by x^2 or $1 + x^2$ [because $(1+x^2)^2 = 1 + x^4$], which leaves only $1 + x + x^2$ as a potential divisor. By long division we get

$$x^4 + x + 1 : x^2 + x + 1 = x^2 + x$$

$$x^4 + x^3 + x^2$$

$$x^3 + x^2 + x + 1 \tag{25.10}$$

$$x^3 + x^2 + x$$

$$\text{remainder: } 1 \neq 0 .$$

Thus, $1 + x + x^4$ is indeed irreducible.

Another polynomial of degree 4 over $GF(2)$ is the *reciprocal* defined by

$$\hat{\pi}(x) := x^{\deg[\pi(x)]} \pi(x^{-1}) , \tag{25.11}$$

or, for $\pi(x) = 1 + x + x^4$:

$$\hat{\pi}(x) = 1 + x^3 + x^4 . \tag{25.12}$$

In general, in a reciprocal polynomial, the exponents are "flipped," i.e., the exponent k becomes the exponent $\deg[\pi(x)] - k$.

25.3 Generation of $GF(2^4)$

We shall now try to generate the finite number field $GF(2^4)$ of order 16 by making use of the irreducible polynomial $\pi(x) = 1 + x + x^4$ and the primitive element

$$\alpha = 0100 \triangleq x .\tag{25.13}$$

We start with the 0 element $(\alpha^{-\infty})$ and the 1 element (α^0) and proceed by multiplying with α, i.e., multiplying by the polynomial x (corresponding to a right shift in the m-tuple) and residue reduction modulo $1 + x + x^4$, which corresponds to adding 1's (modulo 2) to the two left-most places of the m-tuple when a 1 "disappears" on the right:

α power	polynomial	4-tuple	
$-\infty$	0	0000	
0	1	1000	
1	x	0100	
2	x^2	0010	
3	x^3	0001	
4	$1 + x$	1100	
5	$x + x^2$	0110	
6	$x^2 + x^3$	0011	(25.14)
7	$1 + x + x^3$	1101	
8	$1 + x^2$	1010	
9	$x + x^3$	0101	
10	$1 + x + x^2$	1110	
11	$x + x^2 + x^3$	0111	
12	$1 + x + x^2 + x^3$	1111	
13	$1 + x^2 + x^3$	1011	
14	$1 + x^3$	1001	
15	1	1000	

Thus, we see that $\alpha^{15} = \alpha^0 = 1$ and $\alpha^n \neq 1$ for $0 < n < 15$, as is expected of a primitive element. And we also see that the process has generated all $15 + 1 = 16$ binary-valued 4-tuples and all polynomials over $GF(2)$ of maximal degree 3.

It is interesting to note that the above listing also contains the representation of $GF(2^4)$ by 4×4 matrices with elements 0 or 1. Consider the matrix consisting of the third to sixth 4-tuples:

$$M = \begin{bmatrix} 0 & 1 & 0 & 0 \\ 0 & 0 & 1 & 0 \\ 0 & 0 & 0 & 1 \\ 1 & 1 & 0 & 0 \end{bmatrix}. \qquad (25.15)$$

M is a primitive element of this representation. The matrix representation is redundant, but its advantage is that the other nonzero elements are generated by ordinary matrix multiplication! For example,

$$M^2 = \begin{bmatrix} 0 & 0 & 1 & 0 \\ 0 & 0 & 0 & 1 \\ 1 & 1 & 0 & 0 \\ 0 & 1 & 1 & 0 \end{bmatrix}, \qquad (25.16)$$

which corresponds to the fourth to seventh 4-tuple in the listing above. In general, M^{n+1} is obtained from M^n by deleting the upper row of M^n and appending the next row from the (cyclical) list (25.14) to the bottom of M^n. In fact, M^k is the 4×4 matrix whose top row is α^k.

Note that we have a set of matrices here whose multiplication is commutative.

25.4 How Many Primitive Elements?

In generating $GF(2^4)$, we used α as the primitive element, generating all 15 nonzero elements by α^k, $k = 1,2,...,15$. How many such distinct primitive elements are there? Obviously $\phi(15) = 8$, because with α, α^n is also a primitive element *iff* $(n,15) = 1$. Thus, α^2, α^4, α^7, α^8, α^{11}, α^{13}, and α^{14} are the other primitive elements.

In general, the order of an element α^n of $GF(p^m)$ equals $(p^m-1)/(n,p^m-1)$. Thus, for $GF(2^4)$, α^3, α^6, α^9 and α^{12} have order 5, while α^5 and α^{10} have order 3. The number of elements having order T, where T must be a divisor of p^m-1, is $\phi(T)$, as we saw in Chap. 13. As always, there is one element of order $T = 1$, namely $\alpha^0 = 1$.

25.5 Recursive Relations

Still another way of looking at $GF(p^m)$ is to focus on the first column of the m-tuple representation. Skipping the 0 element, we obtain a binary-valued periodic sequence $\{a_k\}$, one period of which for $GF(2^4)$ looks like this:

$$\{a_k\} = 1\ 0\ 0\ 0\ 1\ 0\ 0\ 1\ 1\ 0\ 1\ 0\ 1\ 1\ 1 . \tag{25.17}$$

It is easy to see that this sequence is generated by the recursion

$$a_{k+4} = a_{k+1} + a_k \tag{25.18}$$

with the initial condition $a_1 = 1$, $a_2 = a_3 = a_4 = 0$.

The recursion is a direct consequence of having generated the m-tuples by means of polynomial residue reduction. For our example of $GF(2^4)$ we had chosen for the residue reduction the polynomial modulus $\pi(x) = x^4 + x + 1$ which, as we saw, corresponds to setting

$$x^4 = x + 1 = x^1 + x^0 . \tag{25.19}$$

Now, multiplication by x engenders a right shift in the m-tuple. This polynomial equation therefore corresponds directly to the recursion

$$a_{k+4} = a_{k+1} + a_{k+0} ,$$

as already stated.

The above recursion has a "memory" of four binary digits. Once four consecutive a_k are specified, the sequence is uniquely determined by the recursion. One way to implement a physical generator for the above sequence is by a *finite-state* machine, a so-called shift register as shown in Fig. 25.1.

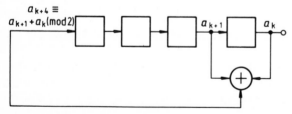

Fig. 25.1. Four-stage linear shift register generating maximum-length Galois sequence corresponding to the finite field $GF(2^4)$. The contents of each of the four registers are transferred to the next neighbor to the right upon application of a clock pulse (not shown). The sequence of pulses repeats after $2^4 - 1 = 15$ clock pulses

At the time of each clock pulse, the content of a register (0 or 1) is transferred to its next neighbor to the right. The recursion is implemented by the linear feedback connection to the input of the first register, also shown in Fig. 25.1.

Since this finite-state machine has $m = 4$ registers which can hold either a 0 or a 1, there are $2^4 = 16$ possible states. However, the state 0 0 0 0 never generates anything but more 0's and is therefore of no interest to us here. Thus, there remain $p^m - 1 = 2^4 - 1 = 15$ distinct nonzero states and the above sequence *must* repeat after at *most* 15 steps — which it does. In this connection it is instructive to check that the consecutive 4-tuples of $\{a_k\}$ cover all 15 possibilities except the all-zero 4-tuple 0 0 0 0.

It can be shown that such *maximum-length* sequences exist for all periods of length $p^m - 1$ [25.1]. What is needed is an irreducible polynomial of degree m, of which there are, in fact, see also [27.2]:

$$\frac{1}{m} \sum_{d|m} \mu\left(\frac{m}{d}\right) p^d \ .$$

For $p = 2$ and $m = 4$, there are three irreducible polynomials, as already noted (Sect. 25.2).

If the sequence is to be generated by a shift register, as illustrated in Fig. 25.1, then, in addition, the root x of the irreducible polynomial must be a primitive element. Polynomials with this property are called *primitive polynomials*. There are exactly

$$\frac{\phi(p^m - 1)}{m} \tag{25.20}$$

such polynomials, each leading to a distinct maximum-length sequence. For $p^m > 4$, these sequences come in pairs, generated by pairs of reciprocal polynomials, that are mirror images of each other (reflected in "time").

For $p = 2$ and $m = 4$, (25.20) tell us there are two primitive polynomials. Hence one of the three irreducible polynomials of degree 4 cannot be primitive. Indeed, the polynomial $1 + x + x^2 + x^3 + x^4$, while irreducible, is not primitive because it has no roots that are primitive elements. For example, the root x generates only 5 different elements, i.e. it only has order 5. To wit: $x^5 = x \, x^4 = x + x^2 + x^3 + x^4 = 1$. The corresponding shift register connection or recursion

$$a_{n+4} = a_{n+3} + a_{n+2} + a_{n+1} + a_n \tag{25.21}$$

produces sequences with period 5. In fact, starting with 1 0 0 0, we get

1 0 0 0 1 1 0 0 0 1 1 0 0 0 1 etc.

Irreducible polynomials that are not primitive are also "distinguished" by the fact that they are factors of $1 + x^n$ with $n < p^m - 1$. Thus, for $p = 2$ and $m = 4$, $1 + x + x^2 + x^3 + x^4$ is indeed a factor not only of $1 + x^{15}$ but also of $1 + x^5$:

$$(1 + x + x^2 + x^3 + x^4)(1 + x) = 1 + x^5 .$$

This fact can be used to show, in a different way, why the corresponding recursion has only period 5. First one shows (easily) that the recursion produces a sequence that corresponds to the *reciprocal* of the corresponding polynomial. This, in turn, can be written as

$$(1 + x + x^2 + x^3 + x^4)^{-1} = \frac{(1+x)}{1+x^5} = (1 + x)(1 + x^5 + x^{10} + ...) ,$$

which has obviously period 5. As a binary sequence, the right side corresponds to 1 1 0 0 0 1 1 0 0 0, etc. which agrees, within a shift, with the above sequence found directly by the recursion (25.21).

By the same token, a *primitive* polynomial is a factor of $1 + x^n$ with $n = 2^m - 1$ *and no smaller* n. Thus, its reciprocal repeats after $2^m - 1$ terms and not before!

Interestingly, sequences generated by linear-recursion that do not have maximum length depend on the initial condition (even when lateral shifts are discounted). For example, the initial condition 1 1 1 1, with the above recursion (25.21), produces a different sequence:

1 1 1 1 0 1 1 1 1 0 1 1 1 0 etc.

By contrast, different initial conditions (excepting 0 0 ... 0), for a recursion based on a *primitive* polynomial, always give the same sequence (starting of course with the initial condition). This must be so because a *maximum-length* sequence runs through all m-tuples (except 0 0 ... 0), and each possible initial condition therefore occurs in the sequence; there is no variety to be obtained by different choices of initial conditions other than a shift.

25.6 How to Calculate in $GF(p^m)$

For adding and subtracting elements of $GF(p^m)$, the m-tuple representation is recommended. The m-tuples are added column-wise modulo p.

For multiplying, dividing, and exponentiating, the α-power representation is preferable.

Examples from GF (2^4):

$$0111 \cdot 1111 \overset{\wedge}{=} \alpha^{11}\alpha^{12} = \alpha^{23} = \alpha^8 \overset{\wedge}{=} 1010 .$$

$$(1010)^{-1} \overset{\wedge}{=} (\alpha^8)^{-1} = \alpha^7 \overset{\wedge}{=} 1101$$

$$(0110)^{\frac{1}{2}} \overset{\wedge}{=} (\alpha^5)^{\frac{1}{2}} = (\alpha^{20})^{\frac{1}{2}} = \alpha^{10} \overset{\wedge}{=} 1110 .$$

The Zech logarithm $Z(n)$, defined by

$$\alpha^{Z(n)} := 1 + \alpha^n , \tag{25.22}$$

permits *both* adding and multiplying using the α-power notation without anti-logging!

Example:

$$(\alpha^9 + \alpha^{10})^2 = \alpha^3 + \alpha^5 = \alpha^3(1 + \alpha^2) ,$$

and, from a table of $Z(n)$, $Z(2) = 8$. Thus, $1 + \alpha^2 = \alpha^8$ and

$$(\alpha^9 + \alpha^{10})^2 = \alpha^{11} .$$

Check: $(\alpha^9 + \alpha^{10})^2 \overset{\wedge}{=} (0101 + 1110)^2 = (1011)^2 \overset{\wedge}{=} (\alpha^{13})^2 = \alpha^{11}$. Check! But the latter (checking) operation takes three table-lookups (instead of one) and requires going from α powers to m-tuples and back again to α powers. So *Zech* does something quite sobering for us.

25.7 Zech Logarithm, Doppler Radar and Optimum Ambiguity Functions

The Zech logarithm has a noteworthy property that leads directly to several interesting applications in some rather unlikely fields of human endeavor. Imagine, if you can, the year 1943 — not a prime year. Radar had been developed to a fine tool for locating U-boats and air planes. But then a new trick was first tried that foiled conventional radars: metal foil strips, called chaff, tuned to the searching radar's wavelength. When they were released by the target, they so overwhelmed the radar scope with phoney reflections that the real target was all but invisible and often escaped unharmed.

Of course, for every weapon there is a counter weapon, and for the chaff (code-named, somewhat unimaginatively, *Düppel-Streifen* in German) it was the Doppler effect, named after the Austrian physicist Johann Christian Doppler (1803-1853) who discovered it. The Doppler effect says that a wave emitted or reflected from a moving target suffers a change in its wavelength or frequency. (Interestingly, but perhaps not surprisingly, considering that Doppler was born in Salzburg, the city of Mozart, his effect was first verified in 1845 by using *trumpeters* on a moving train.)

To separate dangerous targets from harmless chaff, engineers (electrical) soon hit upon the idea to measure the frequency shift of the radar echo; only the frequency-shifted echoes were harvested for further action. But early Doppler radars, as they were called, using frequency-modulated ("chirp") pulses had a disturbing trait: they confused range (target distance) with range rate (speed of approach). The same echo could signal either stationary chaff at 10 km, say, or a fast approaching plane at 9 km — a potentially fatal ambiguity. To lasso this electronic double entendre into a neat mathematical form, theoretically inclined engineers defined an *ambiguity function*. After that, everyone in the Doppler arena began hunting for the *ideal* ambiguity function, dubbed *thumb-tack* function: a function of range and range rate that vanishs everywhere except at one sharp point, thereby defining both distance and speed unambiguously.

What does an ideally unambiguous radar pulse look like? First, radar pulses are periodic, with period T, so that target range and speed can be periodically updated, say every $T = 10$ milliseconds, corresponding to a maximal range of 15 km. Radars must also have a certain range accuracy $\Delta r = 2c \cdot \Delta t$, where c is the speed of light and Δt the corresponding timing accuracy (typically a fraction of a microsecond). Without imposing any severe constraints on the design of the radar, we shall assume an integer relationship between T and Δt:

$$T = (p^m - 1)\Delta t , \qquad (25.23)$$

where p is a prime and m is an integer.

To achieve a timing accuracy Δt, the frequency bandwidth of the radar, according to the uncertainty principle, must exceed $1/\Delta t$. If all the frequencies contained in a sufficiently wide frequency band were transmitted simultaneously in one time slot Δt, the peak power limitation of the radar transmitter would be ill used. Instead we assume that during each time slot, the nth one beginning at

$$t_n = n\Delta t , \quad n = 1, ... , p^m - 2 , \qquad (25.24)$$

only *one* discrete frequency is transmitted; every Δt seconds the radar frequency hops to a new value and then remains constant at that value for a time interval Δt. Calling these discrete frequencies

$$f(n) = f_0 + z(n)\Delta f , \qquad (25.25)$$

the remaining problem in our radar design is to determine in *which order* they are to be radiated. We want to make sure that for each delay difference k any frequency difference occurs at most once. Mathematically speaking, we are looking for a permutation $z(n)$ of the integers from $n = 1$ to $n = p^m - 2$ such that

$$z(n+k) - z(n) \not\equiv z(n'+k) - z(n') \tag{25.26}$$

for $k \not\equiv 0$ and *all* $n \not\equiv n' \bmod p^m - 1$.

Such "perfect" permutations are not so easy to generate. For example, the "random" permutation $z(n) = 3,1,4,5,2,6$ of the integers from 1 to 6 does not have property (25.26): for $k = 2$, $n = 1$ and $n' = 4$, we have $4 - 3 = 6 - 5$. In fact, perfect permutations, which obey (25.26), form a small minority among all possible permutations.

This is where our Zech logarithm comes in: more than 30 years after the original Doppler radar excitement it was suddenly discovered that primitive roots and Zech logarithms can generate the required permutations with property (25.26) and thus provide a new answer to the ambiguity problem [25.2].

Let us define the Zech logarithm $z(n)$ by

$$\alpha^{z(n)} := 1 - \alpha^n , \tag{25.27}$$

where α is a primitive element of the Galois field $GF(p^m)$. (Note that, for $p = 2$, $z(n) = Z(n)$ as defined in (25.22). The minus sign in (25.27) was chosen for subsequent notational convenience.)

To illustrate the use of the Zech logarithm, let us construct a permutation of the integers from 1 to 6 using the Galois field $GF(2^3)$ with the primitive element $\alpha = x \triangleq 0\ 1\ 0$ and the modulus polynomial $x^3 = 1 + x$:

$$
\begin{array}{ll}
\alpha^{-\infty} & 0\ 0\ 0 \\
\alpha^0 & 1\ 0\ 0 \\
\alpha^1 & 0\ 1\ 0 \\
\alpha^2 & 0\ 0\ 1 \\
\alpha^3 & 1\ 1\ 0 \\
\alpha^4 & 0\ 1\ 1 \\
\alpha^5 & 1\ 1\ 1 \\
\alpha^6 & 1\ 0\ 1
\end{array}
\tag{25.28}
$$

Now with (25.27) we have, for $n = 1,2, \ldots 6$, $z(n) = 3,6,1,5,4,2$, which is indeed a perfect permutation having property (25.26). This is easily checked, noting that both $z(n)$ and n are taken modulo $p^m - 1 = 7$.

A general (indirect) proof that this is so proceeds as follows. Suppose that (25.26) is violated for some $k \not\equiv 0$ and some pair $n \not\equiv n'$ mod $p^m - 1$. Then, with (25.27),

$$(1-\alpha^{n+k})/(1-\alpha^n) \equiv (1-\alpha^{n'+k})/(1-\alpha^{n'}) \mod p^m - 1 , \quad \text{or}$$

$$\alpha^n(1-\alpha^k) \equiv \alpha^{n'}(1-\alpha^k) \mod p^m - 1 .$$

Since $k \not\equiv 0$, and therefore $1 - \alpha^k \not\equiv 0$, it follows that

$$\alpha^n \equiv \alpha^{n'} \mod p^m - 1 ,$$

which, for primitive α, is true if and only if $n \equiv n'$ mod $p^m - 1$, contradicting our assumption.

To construct a radar pulse of period T and time resolution Δt according to this principle, a prime power p^m near $1 + T/\Delta t$ is first selected. A primitive element α in $GF(p^m)$ is then used to generate the $p^m - 2$ successive frequencies $f(n)$ in the radar pulse according to (25.25) where $z(n)$ is the Zech logarithm of n defined by (25.27). To give the pulse the proper period $T = (p^m - 1)\Delta t$, one empty interval, during which no energy is radiated, is inserted into each period at $n \equiv 0$ mod $p^m - 1$.

In addition to the permutation $z(n)$, generated directly by the Zech logarithm, the product $rz(n)$, where r is coprime to $p^m - 1$, yields more permutations with property (25.26). There are thus a total of $\phi(p^m-1)$ "Zech-like" permutations, anyone of which leads to the desired ambiguity function and can therefore be used in the Doppler radar frequency hopping scheme (25.25). However, while the original permutation $z(n)$ is symmetric, in that $z(z(n)) = n$, this is no longer true for $r \not\equiv 1$ mod $p^m - 1$. This absence of symmetry may be an advantage in certain applications where a potential adversary might exploit symmetry for his advantage. The additional degree of freedom offered by different choices of the parameter r could also be helpful in foiling attempts at jamming.

For $m = 1$, another method of generating perfect permutations of the integers $1,2, \ldots ,p-1$ is by means of successive powers of a primitive root g of the prime p: $g^1, g^2, \ldots ,g^{p-1}$. This method is, however, limited to primes p, while the Zech logarithm is applicable to all prime *powers* p^m, a denser set. (How dense?)

Quadratic residues modulo an odd prime can also generate integers with the property (25.26). However, for the periodic sequence n^2 mod p, half the integers in $1,2, \ldots ,p-1$ do not occur and the other half occurs twice, violating our requirement of fully utilizing the available frequency space.

25.8 A Unique Phase-Array Based on the Zech Logarithm

Consider the periodic sequence with period $p^m - 1$:

$$a_0 = 0 \tag{25.29}$$

$$a_n = \exp[2\pi i z(n)/(p^m - 1)], \quad \text{for} \quad n = 1, 2, \dots, p^m - 2,$$

where $z(n)$ is a Zech logarithm in $GF(p^m)$ defined according to (25.27). All a_n, $n \not\equiv 0 \mod p^m - 1$, have magnitude 1. The periodic correlation sequence c_n (cf. Sect. 13.8) is given by

$$c_0 = p^m - 2 \tag{25.30}$$

$$c_n = -1 - \exp[-2\pi i n/(p^m - 1)].$$

This follows from the definition of c_n (13.40), and the (easy to prove) fact that, for $n \not\equiv 0$, the difference $z(k+n) - z(k)$ attains all values from 1 to $p^m - 2$ exactly once, except the value n. The corresponding power spectrum is

$$|A_0|^2 = |A_1|^2 = 1, \tag{25.31}$$

$$|A_k|^2 = p^m, \quad k = 2, 3, \dots, p^m - 2.$$

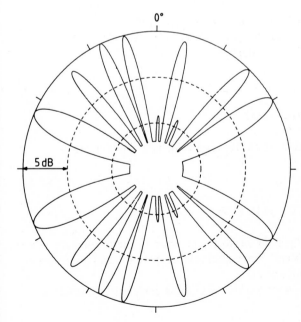

Fig. 25.2. Radiation pattern of constant-amplitude phase array with phase angles based on Zech logarithm for $GF(3^2)$. Array has two periods of period length 8 (total number of nonzero array elements: 14). Spacing between individual antenna elements: 9/16 wavelengths. Note small lobes in the broadside (0°) and second order (26°, corresponding to $r = 2$) directions. The attenuation of the small lobes corresponds closely to the theoretical value $1/p^m \triangleq -9.5$ dB for infinite arrays. By choosing other values for the parameter r any of the other nonbroadside lobes can be made small. (Prepared by M. Rollwage.)

Applied to antenna arrays, operated at a wavelength λ, this result means that a periodic array with element spacing $\lambda/2$ in which all individual elements are driven with equal amplitudes and with phase angles according to (25.29), will radiate equal energies into $p^m - 3$ distinct directions (corresponding to the index values $k = 2$ to $k = p^m - 2$ in (25.31). The "broadside" direction ($k = 0$) and the first "side lobe" ($k = 1$) will receive only $1/p^m$ of the energy going into the other directions.

If the phase angles $2\pi z(n)/(p^m - 1)$ in (25.29) are multiplied by an integer factor r the "undernourished" lobe (other than the broadside) will be the rth lobe. This follows from (25.34), see Sect. 25.9. Figure 25.2 shows the radiation pattern of such an array with element spacing of 9/16 wavelengths, based on $p = 3$, $m = 2$ and $r = 2$. (Note that for this application r does not have to be coprime to $p^m - 1$, so that $rz(n)$ is no longer a perfect permutation obeying (25.26).)

25.9 Spread-Spectrum Communication and Zech Logarithms

Another interesting application of the Zech logarithms is to spread-spectrum communication (see Sect. 15.9). Consider the periodic sequences $a_n^{(r)}$, $r = 1,2, \ldots p^m - 2$, with period $p^m - 1$,

$$a_0^{(r)} = 0 \tag{25.32}$$

$$a_n^{(r)} = \exp[2\pi i r z(n)/(p^m - 1)] \quad \text{for} \quad n = 1,2, \ldots ,p^m - 2 .$$

The autocorrelation sequences are similar to (25.30):

$$c_0 = p^m - 2 \tag{25.33}$$

$$c_n = -1 - \exp[-2\pi i r n/(p^m - 1)] .$$

The corresponding power spectra are flat except for two components:

$$|A_0|^2 = |A_r|^2 = 1 \tag{25.34}$$

$$|A_k|^2 = p^m \quad \text{for} \quad k \not\equiv 0 \quad \text{and} \quad k \not\equiv r .$$

Thus, these sequences too, like the original one ($r = 1$), have constant power in time (except for one "silent" time interval per period) and in frequency (except for two spectral components).

In addition, and most importantly, the cross-correlation coefficient c_{rs} between two such sequences

$$c_{rs} := \sum_{n=0}^{p^m-2} a_n^{(r)} \, a_n^{(s)*} \tag{25.35}$$

equals -1 for $r \not\equiv s$. For $p^m \gg 1$, the cross-correlation is thus small compared to the power of these sequences.

These three properties (nearly constant power in time and frequency and small cross-correlation) make these sequences ideal as "carrier waves" or signature sequences in the design of spread-spectrum communication systems where all channels occupy the same frequency band and yet have small mutual interference.

26. Spectral Properties of Galois Sequences

Certain periodic sequences with elements from the Galois field $GF(p)$, formed with the help of primitive polynomials over $GF(p^m)$, have unique and much sought-after correlation and Fourier transform properties. These *Galois sequences,* as I shall call them, have found ingenious applications in error-correcting codes, interplanetary satellite picture transmission, precision measurements in physiology and general relativity, and even concert hall acoustics. Other applications are in radar and sonar camouflage, and in noise abatement, because Galois sequences permit the design of surfaces that scatter incoming waves very broadly, thereby making reflected energy "invisible" or "inaudible." A similar application occurs in work with coherent light, where a "roughening" of wavefronts (phase randomization) is often desired (for example, to avoid "speckles" in holograms). Excellent structures for this purpose are light diffusers whose design is based on Galois arrays — in a sense the ultimate in frosted (milk) glass. Finally, Galois sequences allow the design of loudspeaker and antenna arrays with very broad radiation characteristics.

26.1 Circular Correlation

Some of the most interesting properties of Galois sequences $\{a_k\}$ for $p = 2$ emerge when we consider their (circular) correlation. But first we shall convert the a_k into b_k by replacing 0's by 1's and 1's by -1's according to the rule

$$b_k = (-1)^{a_k} , \tag{26.1}$$

or, equivalently,

$$b_k = 1 - 2a_k . \tag{26.2}$$

The circular correlation of the sequence b_k is defined, as before, by

$$c_n = \sum_{k=0}^{2^m-2} b_k b_{k+n} , \tag{26.3}$$

where the indices are reduced modulo the period of the b_k, i.e., 2^m-1.

For $n \equiv 0 \ (mod \ 2^m - 1)$, we get

$$c_0 = 2^m - 1 \ . \tag{26.4}$$

For $n \not\equiv 0$, we shall consider the special case $GF(2^4)$. The recursion for the b_k is multiplicative;

$$b_{k+4} = b_{k+1} b_k \ ,$$

or

$$b_k = b_{k-3} b_{k-4} \ , \tag{26.5}$$

which is isomorphic with the corresponding recursion for the a_k (25.18). Now consider the terms in the above correlation sum

$$d_k := b_k b_{k+n} \ . \tag{26.6}$$

It is easy to see that the product sequence d_k has the same recursion as the b_k:

$$d_k = b_k b_{k+n} = b_{k-3} b_{k-4} b_{k+n-3} b_{k+n-4} \ . \tag{26.7}$$

By definition, the product of the first and third factors on the right equals d_{k-3}, and the product of the remaining two factors equals d_{k-4}. Thus,

$$d_k = d_{k-3} d_{k-4} \ , \tag{26.8}$$

as claimed. Since the recursion together with the initial condition determines the sequence uniquely, except for a shift in the index, we have (for $n \not\equiv 0$)

$$d_k = b_{k+s} \tag{26.9}$$

for some shift s.

Using this result in the above correlation, we find

$$c_n = \sum_{k=0}^{2^m-2} b_{k+s} = \sum_{k=0}^{2^m-2} b_k = -1 \ , \tag{26.10}$$

since exactly $2^m/2$ of b_k equal -1 and the remaining $(2^m-2)/2$ equal $+1$. This is so because in the m-tuple representation of $GF(2^m)$ all possible 2^m m-tuples appear exactly once. However, the Galois sequence $\{a_k\}$ was constructed without the 0000-tuple, so that one period of $\{a_k\}$ has one extra 1, i.e., in the sequence $\{b_k\}$ there is one extra -1. Hence, the sum over one period of the b_k equals -1.

Most importantly, this result is true for any $n \not\equiv 0$. Thus, our correlation function is two-valued:

$$
c_n = \begin{cases} 2^m - 1 & \text{for } n \equiv 0 \quad \mathrm{mod}(2^m - 1) \\ \\ -1 & \text{otherwise} \end{cases} , \tag{26.11}
$$

meaning that the spectrum of the b_k is flat [26.1]:

$$
|B_k|^2 = \begin{cases} 2^m & \text{for } k \not\equiv 0 \quad \mathrm{mod}(2^m - 1) \\ \\ 1 & \text{otherwise} \end{cases} . \tag{26.12}
$$

Thus, we have discovered another class of binary-valued periodic sequences with a flat spectrum. But in contrast to the Legendre sequences (Chap. 15), which are not easily generated, our Galois sequences derived from $GF(2^m)$ are generated by a simple *linear* recursion as illustrated in Fig. 25.1. The linearity of the recursion has other important consequences, for example in applications to error-correcting codes (Sect. 26.2).

The periods that are achievable for binary-valued sequences have length $2^m - 1$ and require for implementation an m-stage shift register and a feedback connection based on an primitive polynomial $\pi(x)$ over $GF(2)$ of degree m.

Examples:

$$m = 2 : \pi(x) = 1 + x + x^2$$

$$m = 3 : \pi(x) = 1 + x + x^3$$

$$m = 4 : \pi(x) = 1 + x + x^4$$

$$m = 5 : \pi(x) = 1 + x^2 + x^5$$

$$m = 6 : \pi(x) = 1 + x + x^6$$

Note: $1 + x + x^5$ is not irreducible over $GF(2)$; it factors into $(1 + x^2 + x^3)(1 + x + x^2)$.

Primitive polynomials over $GF(2)$ up to high degrees m have been published [26.2]. The period length of the Galois sequence with $m = 168$ equals $2^{168} - 1$, a 51-digit number. With one clock pulse per picosecond, the sequence will repeat after 10^{22} times the age of the universe — long enough for most purposes.

26.2 Application to Error-Correcting Codes and Speech Recognition

Galois sequences are useful for correcting multiple errors in digital representations, for example in transmission and storage [26.3]. In the simplest case, the m-tuples corresponding to the initial condition are determined by the information to be transmitted, while the recursion generates $2^m - 1 - m$ *check bits*. Thus, putting our example of $GF(2^4)$ to work, with the initial condition 1000, the recursion appends 11 check bits for a 15-bit codeword:

$$1\ 0\ 0\ 0,\ 1\ 0\ 0\ 1\ 1\ 0\ 1\ 0\ 1\ 1\ 1.$$

The efficiency of this code is, of course, only 4/15, or, to put it differently: of the 2^{15} possible binary-valued codewords of length 15, the code uses only 2^4. But as a reward for this frugality, the code can correct up to three errors.

To see this, we note that the *Hamming distance* between any two distinct codewords is $d = 8$. (The Hamming distance is the number of places in which two codewords differ. Since for the code at hand, the difference (or sum) modulo 2 of two codewords is again a codeword, the above result for d follows immediately.) Now, for an even Hamming distance d, the number of errors that can be corrected is $d/2 - 1$ (by choosing the nearest possible codeword). The occurrence of one additional error can be *detected* but it can no longer be corrected, because there are then two possible codewords equidistant from the erroneous one. This so-called Simplex Code based on $GF(2^4)$ can correct three errors and detect a fourth one [26.4].

The general result is that a code based on $GF(2^m)$ in the manner described has efficiency $m/(2^m - 1)$ and corrects $2^{m-2} - 1$ errors per codeword of length $2^m - 1$. Thus, these codes can cope with error rates around 25%! Not bad at all. The fact that the efficiency is low is of little concern in some applications, such as image transmission from interplanetary space vehicles, which take months or years to reach their destinations and have long periods of time available to transmit their pictures of, say, Jupiter's moon Io. But these space communication links need all the error-correcting capabilities they can get, because the typical space probe's transmitter has low power and has to be "heard" over vast distances. In fact, the first good pictures of the "canals" of Mars were obtained with an error-correcting code very similar to the one sketched here. And who wants to confuse a digital error with a new moon of Neptune?

Another advantage of the code described here is that the error correction is quite simple: the received code, converted to the ± 1 representation, is cross-correlated with the above codeword $(1\ 0\ 0 \cdots 0\ 1\ 1\ 1)$, which is also converted to ± 1. If there was no error in transmission, all values of the (circular) cross-correlation will equal -1, except one, which has value 15 and which by its position will indicate which of the 15 allowed codewords was sent.

If there was one error in transmission, the peak value of the correlation will be reduced from 15 to 13 and the other correlation values will be increased from −1 to +1. A second error will give correlation values 11 and up to 3, respectively. Three errors will result in correlations 9 and up to 5, respectively. Thus, the peak value (9) is still distinct from the other values (⩽5), and proper detection is assured.

Four errors will produce correlation values of 7 in the proper location and also 7 in some off-locations, so that correction is no longer possible. But the absence of correlation values larger than 7 is an indication that there are four or more errors.

Error-correcting codes are not only useful for cleansing digital data; they are applicable also to fault finding in other fields. Let us look at (rather than listen to) speech synthesis from text ("talking computers") and *automatic speech recognition* [26.17]. These are the acoustic "bridges" between man and machine

Speech recognition is a particularly difficult problem for the computer, even for small vocabularies enunciated clearly by "his master's voice." But the *talkwriter* (voice-operated typewriter) is waiting in the wings to replace the irreplaceable human secretary. And one of the advances that will speed the entry of the bloodless transcriber will be the proper exploitation of syntax and semantics.

To illustrate, consider a vocabulary of only 2 words and sentences with a length of 15 words. Without constraints, this strange "language" allows the combination of 2^{15} different sentences. Now suppose that good grammar and meaningful statements limit the number of *useful* sentences to $2048 = 2^{11}$. Then 4 words in each sentence are redundant "check words", allowing $2^4 = 16$ different cases to be distinguished, indicating either no error or which of the 15 words in the sentence was recognized incorrectly.

This is a single-errorcorrecting Hamming Code [26.4] of length 15, using four check bits. It is the *dual* of the Simplex Code discussed above, which has also length 15 but $15 - 4 = 11$ check bits. The challenge in the application of finite-field theory to speech recognition is to translate linguistic constraints into algebraic structures that permit fast codebook searches. For artificial languages, such as company telephone numbers or stock item labels, one can often start the language design from scratch, matching it to known good codes.

26.3 Application to Precision Measurements

As we saw in the chapter on error-correcting codes, the Galois sequence $\{b_k\}$ with period $2^m - 1$, derived from an irreducible polynomial $\pi(x)$ of degree m

over $GF(2)$, has the following amazing properties:

1)   Its energy, b_k^2, is distributed perfectly uniformly in "time" (time corresponding to the index k): each "time slot" $k = 0,1, ..., 2^m - 2$ has exactly the same energy, namely 1.

2)   The energy of the sequence is also uniformly distributed over all $2^m - 2$ nonzero frequencies.

Property (1) means that a device limited in peak power, such as a transmitter, can output a maximum amount of energy in a given time interval. This makes it possible to measure at extremely low signal-to-noise ratios.

Of course, a sinewave also has its energy stretched out rather uniformly in time, but it contains only a *single* frequency and therefore has poor time resolution. By contrast, because of Property (2), a Galois sequence contains many frequency components, all having the same power. This means that measurements made with Galois sequences as a test signal — such as measurements of interplanetary distances to check out space-time curvature — will have very high time resolution. In fact, the time resolution will be the same as that obtained with a short pulse having the same broad frequency spectrum. The high temporal precision inherent in Galois sequences is brought out by correlation receivers that cross-correlate the received signal (a radar echo from the planet Mercury, for example, or a faint echo in a concert hall) with the outgoing sequence to form sharp peaks at the delays corresponding to pulse travel times.

Thus, these two properties — the full, optimal use of both Fourier domains (time and frequency) — make the Galois sequences, also known as maximum-length sequences or pseudorandom shift-register sequences, ideal candidates for precision measurements with extremely low energies in the presence of strong interfering noises.

26.4 Concert Hall Measurements

Most measurements of concert halls are done in *empty* halls: no live audience wants to listen to the pops, bangs, hisses and howls produced by acousticians to measure the acoustic characteristics of a hall under scrutiny. But measurements made in an empty hall are often irrelevant, because a sizable audience has a noticeable effect on the acoustics of an enclosure. Can one make measurements during an ongoing concert without interfering with the musical enjoyment of the audience?

The answer to this question is, most surprisingly, yes. A Galois sequence can be radiated (from a loudspeaker) into a concert hall (or any other "noisy" environment) at such extremely low sound levels as to be *inaudible* by the audience (owing to an auditory phenomenon called *masking*). The required detection process is cross-correlation — as in the case of the error-correcting code described above — between the received acoustic signal (at some audience location) and the radiated Galois signal (from the stage, for example). By integrating the correlation over not just one fundamental period, but over many periods, namely the duration of the concert (one hour, say), the required signal-to-noise ratios can be realized [26.5].

For concert hall measurements, the fundamental period of the Galois sequence is chosen to equal approximately the reverberation time, say 2 seconds. The clock frequency chosen must be higher than twice the upper audio cutoff frequency, or about 32 kHz. The sequence must therefore have a period length of at least 2 seconds times 32 kHz = 64000. The smallest $2^m - 1$ that is large enough is obtained for $m = 16$, leading to a period of $2^{16} - 1 = 65535$, which is, in fact, a preferred choice.

The author demonstrates this method of measuring acoustic responses in one of his lecture courses at the University of Göttingen by measuring the impulse response between a point near the lectern and some point in the back of the auditorium *while continuing his lecture.* As long as what he says is sufficiently incoherent (with the radiated Galois sequence), his speech signal averages out in the correlation process, and a "noise-free" response builds up on a TV screen watched by the students. (For didactic reasons, the Galois sequence is played at a soft but *audible* level.)

Galois sequences have also been used in physiological measurements of neuronal systems [26.6].

26.5 The Fourth Effect of General Relativity

General Relativity Theory, the theory of gravitation propounded by Einstein in 1915, passed three important experimental tests during Einstein's lifetime. The following effects predicted by him were confirmed:

1)   The motion of the perihelion (the point closest to the sun) of the orbit of the planet Mercury, already known from observations in the last century.

2)   The bending of light waves near the sun, first observed during a total eclipse in 1919.

3)   The *gravitational* red shift, first seen in the light from massive stars, but now measured even on the earth itself using the ultrasensitive Mössbauer effect.

A fourth effect predicted by Einstein's theory (although not considered by Albert himself) was not confirmed until fairly recently [26.7]: the *slowing* of electromagnetic radiation in a gravitational field — as opposed to the *acceleration* of matter as it approaches a heavenly body.[1] The fourth effect was observed by means of radar echoes from the planets Venus and Mercury as they disappeared behind the sun as seen from the earth ("superior conjunction"). In that position, both the outgoing and returning radar waves have to travel very near (indeed around) the sun. Even after taking plasma effects near the sun's surface and other factors into account, physicists found an extra delay of 200 μs — very close to the prediction of general relativity.

Why was this measurement not done long ago? The reason is that the echo energy from Mercury — exceedingly weak even when visible — drops to 10^{-27} (!) of the outgoing energy as the planet slips behind the sun; in other words, a radar echo would not even suffice to raise the potential of a *single* electron by a thousandth of a volt.

The astounding fact that reliable results have been obtained in spite of these miniscule energies is due mainly to a proper choice of the transmitted sequence of radar pulses, based on irreducible polynomials over finite number fields (called "Galois fields" after the 18th-century French mathematician who died at the age of 20 after a duel over a woman, but not without jotting down some highly ingenious ideas the night before the fatal encounter [26.8]).

26.6 Toward Better Concert Hall Acoustics

Consider the following problem in concert hall acoustics. Recent research, based on a subjective evaluation of the acoustics of 20 major European concert halls [26.9], has shown that many modern halls have poor acoustics because their ceilings are low relative to their widths. Such halls do not provide the listener with enough *laterally* travelling sound waves — as opposed to sound travelling in front/back direction and arriving at the listener's head in his "median" plane (the symmetry plane through his head). Such median-plane sound, of course, gives rise to two very similar acoustic

---

[1] In the long struggle to put his principle of general equivalence of different reference frames into proper mathematical clothing (Riemannian geometry), Einstein discovered — as early as 1909 — that the speed of light could not be constant (as in special relativity) but must depend on the gravitational potential ϕ. Although he had no general theory then, Einstein found that — to first order — $c(\phi) = c + \phi/c$, where c is the usual vacuum velocity of light in field-free space. (Note: $\phi \leqslant 0$.)

Ironically, the slowing of radiation in gravitational fields, although appreciated very early, was not considered a *testable* proposition until the perfection of radar technology, using Galois sequences, in the second half of this century. The reason for this delay in testing the extra delay was, of course, that no one could picture himself (or anyone else, for that matter) floating next to the sun, stopwatch in hand, clocking the passing photons.

signals at the listener's two ears, and it is thought that the resulting excessive "binaural similarity" is responsible for the poor acoustical quality [26.10].

How can one correct this shortcoming? One component of median-plane sound, the direct sound from the stage, cannot be suppressed because it is needed to localize properly the different instruments on the stage. In principle, the ceiling could be raised so that lateral sound would dominate again, as in the preferred (but unprofitable) old-style high and narrow halls. But modern air-conditioning has made the low ceiling possible — and that is where mounting building costs will keep it.

Still another escape would be to *absorb* the sound at the ceiling and thereby curtail is deleterious effect. But although room acoustics is not usually thought of as an "energy problem," we cannot afford to waste precious musical sound energy, especially in a large modern enclosure, where every "phonon" is needed.

There remains only one alternative, to *disperse* the sound from the ceiling into a lateral pattern. In other words, the ceiling should be what the physicist would call a *reflection phase-grating* — a grating that scatters sound waves comprising many different frequency components into a broad lateral pattern, without absorbing them and with a minimum of specular reflection. What would such a ceiling look like?

The far-field or *Fraunhofer diffraction* of a grating is approximated by the Fourier transform of the complex amplitude of the wave as a function of position as it leaves the grating. Thus, if we want a broad diffraction pattern, we have to look for distributions that have a broad Fourier spectrum. Furthermore, if we want the waves leaving the grating to have uniform magnitude, we need a function of constant magnitude which has a broad Fourier spectrum.

A function which has these properties is given by the Galois sequence $\{b_k\}$ discussed in this chapter; it has a magnitude of 1:

$$|b_k|^2 = 1 \quad \text{for all } k \; ,$$

and the nonzero frequencies of its Fourier spectrum all have the same magnitude:

$$|B_n|^2 = 2^m \quad n \not\equiv 0 \mod(2^m-1) \; .$$

Here, in the spatial diffraction-pattern application, each frequency component corresponds to a given diffraction order or *spatial frequency* (to use the modern expression from the field of Fourier optics [26.11]). The diffraction angle α_n is given by

$$\sin\alpha_n = n\lambda/L \; , \quad |n| \leqslant \lfloor \frac{L}{\lambda} \rfloor \; , \tag{26.13}$$

where λ is the wavelength, L is the period of the grating:

$$L = (2^m - 1)w ,$$ (26.14)

and w is the "grating constant," i.e., the step size from one element of the grating to the next [26.12].

The next question is: how do we impart amplitudes distributed like Galois sequence b_k to a wave? The answer is by a hard corrugated surface, such as that shown in Fig. 26.1. A normally incident wave is reflected from a hard surface with a *reflection factor* of $+1$. However, in places where the surface is set back by a quarter wavelength, its phase at the reference plane, after having travelled an extra distance of half a wavelength, is shifted by π, i.e., its complex amplitude is $e^{i\pi} = -1$.

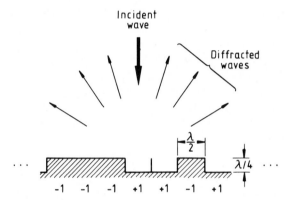

Fig. 26.1. A corrugated surface acting as a reflection phase-grating. If the design is based on a Galois sequence, then the reflected energy is broadly scattered as shown in Fig. 26.2

Of course, the wave reflected from the corrugated structure shown in Fig. 26.1 will not have discontinuous jumps in its amplitude from $+1$ to -1; that would violate the wave equation. We have taken here a kind of Kirchhoff view of diffraction: we consider the diffracted wave to equal the incident wave as modified only in the most obvious manner, without regard to the self-consistency of such assumptions [26.12]. Nevertheless, it is well known that diffraction treated according to Kirchhoff, generally gives results in good agreement with reality, and our case, as proved by subsequent measurements, is no exception.

Figure 26.2 shows the diffraction patterns measured from the corrugated surface shown in Fig. 26.1. We see that, indeed, about equal energies are scattered into the different diffraction orders.

The only disadvantage of a ceiling based on a Galois sequence $\{b_k\}$ from $GF(2^m)$ is that it is binary-valued, $b_k = \pm 1$, so that the corrugated structure only has indentations of one size, namely a quarter wavelength. If we consider an incident wave one octave higher in frequency, i.e., having half the original wavelength, the phase shift it suffers upon reflection by one of the indentations is 2π (instead of π). In other words, there is no phase shift at

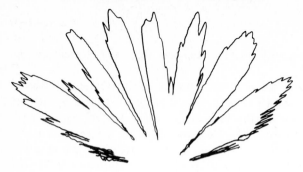

Fig. 26.2. Diffraction pattern from Galois reflection phase-grating shown in Fig. 26.1. Note nearly equal energies being scattered into the (seven) diffraction orders

all, and measurements confirm that there is an almost specular reflection (Fig. 26.3). In fact, Fig. 26.4 shows the reflection from a plane (uncorrugated) surface, and there is hardly any difference between it and the diffraction pattern from the corrugated surface at the octave frequency.

How do we extend the broad diffraction pattern to a broad frequency band comprising several musical octaves? One way would be to make the indentation depths correspond to a quadratic-residue sequence as already discussed (Sect. 15.6). But then we lose the advantage of low specular reflection. Has Galois nothing to offer that works at more than one frequency?

Fig. 26.3

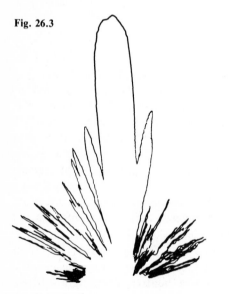

Fig. 26.4

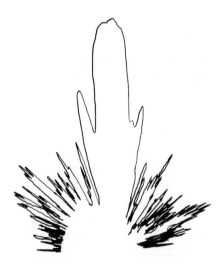

Fig. 26.3. Same as Fig. 26.2 but at half the original wavelength. Note strong specular reflection despite corrugation of reflecting surface

Fig. 26.4. Reflection from plane surface. Note similarity with reflection from corrugated surface shown in Fig. 26.3

How about considering a finite number field $GF(p^m)$ with $p > 2$, and perhaps just $m = 1$? For $p = 11$, a primitive element is 2. Using it to construct $GF(11)$, we get the recursion [see the analogous construction of $GF(2^4)$, Sect. 24.5]:

$$a_{n+1} = 2a_n \pmod{11} . \tag{26.15}$$

Thus, the Galois sequence for $GF(11)$, beginning with 1, is

$$\{a_n\} = 1,2,4,8,5,10,9,7,3,6 . \tag{26.16}$$

However, this sequence is precisely the same as the primitive-root sequence for $p = 11$ and the primitive root $g = 2$ that we studied in Sect. 13.8.

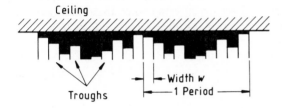

Fig. 26.5. Design of reflection phase-grating based on primitive roots

A scale model of a sound-diffusing structure based on $GF(11)$ is shown in Fig. 26.5. It consists of adjacent "wells" of different depths, the depth being proportional to the following sequence of integers:

$$s_n = 2, 4, 8, 5, 10, 9, 7, 3, 6, 1 , \tag{26.17}$$

and so on (repeated periodically). It is obvious that the sequence of 10 integers above is a permutation of the integers 1 to 10, i.e., each integer appears exactly once. But what is so special about this particular arrangement among the 3,628,800 possible permutations? A little further inspection will reveal that each number in the sequence is twice its predecessor — except when the doubling exceeds 10, in which case 11 is subtracted from it.

In mathematical terms, the sequence s_n is described by the following formula

$$s_n = g^n \pmod{p} , \tag{26.18}$$

where p is a prime number ($p = 11$ in our case) and g is one of its "primitive roots" (here $g = 2$). A primitive root of a prime p is distinguished by the property that the least positive remainder of g^n runs through all the integers from 1 to $p - 1$ (in some order) as n goes from 1 to $p - 1$.

The prime number 11 has 4 primitive roots (6,7,8 are the three others besides 2). Thus, only *one* in nearly a *hundred thousand* permutations (not counting cyclic permutations) is generated by primitive roots.

What distinguishes the permutation generated by a primitive root from the thousands of others? The remarkable fact is that the periodic *Fourier transform* of the sequence

$$r_n = e^{2\pi i s_n/p} \tag{26.19}$$

has components of *equal* magnitude — except the zeroth, which is smaller. It is this fact that results in the desired wave scattering when these numbers are used in the design of a surface structure as illustrated in Fig. 26.5. The depths of the individual wells are

$$d_n = s_n \lambda_1/2p , \tag{26.20}$$

where λ_1 is the fundamental wavelength.

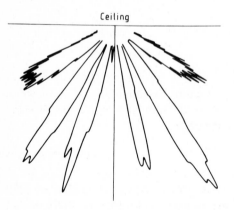

Fig. 26.6. Diffraction pattern from primitive-root design for $p = 7$, $g = 3$

Figure 26.6 shows the energy scattering from a ceiling structured according to the primitive-root sequence in Fig. 26.5.

What happens if the frequency of the waves is 2,3,4 ... or 10 times higher than the design frequency? Whatever the frequency factor, the structure acts essentially as if the above sequence were multiplied by the frequency factor. Thus, for a frequency 5 times higher, for example, the effective sequence is

$$10,9,7,3,6,1,2,4,8,5 \text{ and so on },$$

which is exactly the old sequence shifted cyclically by 4 places to the left. But, as is well known, the magnitude of the Fourier transform is unaffected by such a shift. Quite a trick for a *primitive* root! Only if we multiply by a multiple of 11 does this method fail. Then all remainders modulo 11 are zero and the structure shown in Fig. 26.5 will act very much like a mirror.

While 2,6,7 and 8 are primitive roots of 11, the number 3, for example, is *not*. To see this, we need only evaluate (26.17) for $g = 3$, which gives

$$3, 9, 5, 4, 1, 3, 9, 5, 4, 1,$$

a sequence that has "forgotten" half the numbers between 1 and 10 (2,6,7,8 and 10 are missing!). A ceiling built on the number 3 would lead to poor scattering of sound and lower acoustic quality.

26.7 Higher-Dimensional Diffusors

The wave-scattering principles described in Sect. 26.6 can be applied to two or more dimensions. Suppose the period length of the flat-spectrum sequence has at least two coprime factors m and n, as in $2^4 - 1 = 15 = 3 \cdot 5$ or $31^1 - 1 = 30 = 2 \cdot 15 = 3 \cdot 10 = 5 \cdot 6$. Then the sequence can be converted into a two-dimensional array of length m and width n as illustrated in Fig. 16.1. Such arrays then have the desired spatial correlation and spectrum properties in two dimensions and are useful in noise abatement (to disperse sound waves) or in radar and sonar camouflage.

If the period length can be factored into three or more coprime factors, correspondingly higher-dimensional designs can be realized.

Other interesting generalizations are described by J. H. van Lint, F. J. MacWilliams, and N. J. A. Sloane ("On Pseudo-Random Arrays", SIAM, J. Appl. Math. **36**, 62-72, 1979).

For a Galois sequence whose period length $2^m - 1$ is a Mersenne prime, there are of course no coprime factors and higher-dimensional arrays, based on factoring the period, are impossible. Hence, Mersenne primes, which give us perfect numbers (see Sect. 3.7), levy a certain toll when it comes to the construction of higher dimensional arrays — just like the Fermat primes (see Sect. 13.10).

26.8 Active Array Applications

The principles discussed in Sects. 26.6 and 26.7 are also applicable to active arrays (loud-speaker columns, radio and television antennas) in one, two or more dimensions. In fact, since no Kirchhoff approximation is involved, flat-spectrum sequences lead to arrays with highly uniform directional characteristics. Because of the time reversal invariance (T invariance) of mechanics and electromagnetism (also referred to as "reciprocity"), both transmitting *and* receiving arrays with the desired characteristics can be realized [26.13].

One application is to loud-speaker columns that do not focus sound undesirably or whose directional characteristics can be switched from "narrow" to "broad."

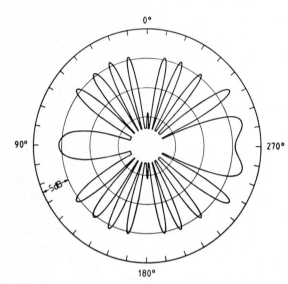

Fig. 26.7. Radiation pattern of 20-element collinear antenna array with half-wavelength spacing. Signals supplied to individual dipoles have constant amplitude across the array but phase angles vary according to primitive-root sequence ($g = 2$, $p = 11$). Note uniformity of radiation in different major lobes, except broadside ($\phi = 0°$ and 180°) where it is low

Figure 26.7 shows the directional characteristics of $2 \cdot 10 = 20$ elementary radiators with a spacing of half a wavelength, whose phases have been selected according to the primitive root $g = 2$ of the prime $p = 11$. The uniformity of the radiated energies into the different "side lobes" is impressive. Only the broadside radiation ($\phi = 0°$ and $\phi = 180°$) is weak — as expected for a primitive-root design. Other antenna applications of number theory are discussed in Chap. 28.

Further facts and applications of binary sequences, flat-spectrum or otherwise, can be found in [26.14-16].

27. Random Number Generators

"Anyone who considers arithmetical methods
of producing random digits
is, of course, in a state of sin."
— *John von Neumann*

In contemporary computation there is an almost unquenchable thirst for random numbers. One particularly intemperate class of customers is comprised of the diverse *Monte Carlo* methods.[1] Or one may want to study a problem (or control a process) that depends on several parameters which one doesn't know how to choose. In such cases random choices are often preferred, or simply convenient. Finally, in system analysis (including biological systems) random "noise" is often a preferred test signal. And, of course, random numbers are useful — to say the least — in cryptography.

In using arithmetical methods for generating "random" numbers great care must be exercised to avoid falling into deterministic traps. Such algorithms never produce truly random events (such as the clicks of a Geiger counter near a radioactive source), but give only *pseudo*random effects. This is nicely illustrated by the following (true) story.

An associate of the author, for a study in human vision, wanted to generate visual noise or "snow" (as seen on the screen of a broken TV set). The unwary researcher selected a widely used random number routine available from a renowned software source. The numbers produced by this algorithm had passed a battery of sophisticated tests with random colors flying. Yet when the researcher selected alternative output samples from the "random source" to determine abscissae and ordinates of white dots on a TV screen, the result was not the expected random "snow" but just a few diagonal streaks!

In the following we will sketch some fundamental facts about both random and *pseudo*random number generators. We will be interested in both discrete and (quasi-)continuous distributions of random numbers. (Remember that nothing is *really* continuous in digital computation.) Our main tools will turn out to be congruences and recursions.

---

[1] The reader may recall that many analytically intractable problems can be solved (albeit with a statistical error) by simulation with a random process and "tabulating" the results of one or more simulation runs.

27.1 Pseudorandom Galois Sequences

In Chap. 25 we became acquainted with a method of generating pseudorandom sequences with elements from GF(p) and period length $p^m - 1$ [27.1]. For $p = 2$, we obtain binary-valued sequences [27.2] with elements 0 and 1 or, as often preferred, 1 and -1.

These sequences a_n are generated by linear recursions [27.3]:

$$a_n = \langle f_1 \, a_{n-1} + \dots + f_m \, a_{n-m} \rangle_p , \tag{27.1}$$

where addition is modulo p, as indicated by the acute brackets, which signify least positive (or nonnegative) remainders. The coefficients f_k are determined by primitive polynomials in $GF(p^m)$ (Chap. 26).

These sequences are, of course, not truly random. Each period contains precisely all m-tuples but one (the all-zero tuple). Also, their correlation is not that of a truly random sequence (like, say, the pulses from a Geiger counter near a radioactive source). In fact, the periodic autocorrelation has only two possible values (a highly valued property in numerous applications; see Chap. 26). Nevertheless, by choosing m very large, say $m = 168$, and selecting a relatively small excerpt from a full period, pseudorandomness can approach true randomness under many statistical tests. For $m = 168$ and $p = 2$, for example, the period length exceeds 10^{50} samples, and even as many as 10^{10} samples from such a sequence constitute a very small portion of the entire sequence.

If used in applications where higher-order correlations are important, however, trouble can still occur. As an illustration, for $m = 5$ and $p = 2$, and the recursion

$$a_n = \langle a_{n-4} + a_{n-5} \rangle_2 , \tag{27.2}$$

or, for $b_n = 1 - 2a_n$,

$$b_n = b_{n-4} \cdot b_{n-5} , \tag{27.3}$$

the third-order correlation coefficient becomes

$$c_{4,5} = \sum b_n \cdot b_{n-4} \cdot b_{n-5} = \sum 1 . \tag{27.4}$$

Thus, $c_{4,5}$, instead of being appropriately randomly small, becomes catastrophically large.[2]

---

[2] Overlooking the importance of higher-order correlations in analyzing nonlinear systems has led to some very misleading results in the animal neurophysiology of hearing when binary pseudorandom sequences of the kind described here have been used as an acoustic input to the animal's ear.

The problem with higher-order correlations can be somewhat alleviated by basing the recursion on primitive polynomials with a maximum number of terms. In our example ($m = 5$, $p = 2$), this would mean replacing (27.3) by a recursion involving four part-terms instead of only two:

$$b_n = b_{n-1} \cdot b_{n-2} \cdot b_{n-3} \cdot b_{n-5} .$$ (27.5)

Now, of course, there is trouble with a fifth-order correlation. But for large m, the recursion can be given so many terms that the nonrandom correlations are usually of an irrelevantly high order.

27.2 Randomness from Congruences

Congruence relations are a preferred device for generating random numbers on a computer [27.4]. A simple congruence

$$x_{n+1} = \langle a\, x_n + b \rangle_m , \quad x_0 = c ,$$ (27.6)

has four parameters: a multiplier a, an additive constant b, a modulus ("base") m, and an initial value ("seed") c.

The integer range of the computer is often chosen as a modulus m, because then if we interpret the acute brackets in (27.6) as meaning "take least absolute residue," the residue reduction is automatic. For example, if the computer is a 32-bit machine, then its integer range is typically from -2^{16} to $2^{16} - 1$ (or from $-2^{16} + 1$ to 2^{16}), in other words, a least absolute residue system of the modulus $m = 2^{32}$. Since integer "overflow" (an integer exceeding the computer's range) is often automatically handled by taking least absolute residue, the remainder operation in (27.6) therefore does not require additional computation. (Never mind that in computerese residue reduction goes by such, possibly unfamiliar terms as "twos-complement," etc.)

The longest period of the x_n generated according to (27.6) is, of course, m. (Once an earlier value of x_n reappears, the whole sequence of x_n will be repeated; (27.6) represents a first-order (pseudo-) Markov process with no memory beyond its immediate past.) If m is power of 2, $m = 2^k$, then to realize a period length of m, the following conditions must be fulfilled [27.4]:

$$a \equiv 1 \quad (\text{mod } 4)$$ (27.7)

and

$$(b, m) = 1 ,$$ (27.8)

i.e., it suffices that b be odd.

With binary computers, a preferred choice for a is $a = 2^5 + 1$, because then multiplication by a is a simple shift operation and an addition.

If the period length is maximum, then since x_n will eventually assume all m possible different values, the choice of the initial value, x_0, is not crucial.

The random numbers generated by a simple congruence with maximum period length show shorter periods in their least significant digits.[3] If this is troublesome, and the least significant digits are actually needed (otherwise they could simply be discarded), it is best to adopt another algorithm such as the following.

A more sophisticated random number generator uses r random initial values instead of only one:

$$x_1, x_2, ..., x_r , \qquad (27.9)$$

which are read out successively. At the same time that an x_n is read out, it is added (modulo m) to the value in the circularly preceding storage location. (In other words, x_r precedes x_1, etc.)

Thus, with the above "seeds," the random sequence will be

$$x_1, x_2, ..., x_{r-1}, \langle x_r + x_1 \rangle_m ,$$
$$\langle x_1 + x_2 \rangle_m, \langle x_2 + x_3 \rangle_m, ...$$
$$\langle x_1 + 2x_2 + x_3 \rangle_m , \quad \text{etc.} \qquad (27.10)$$

27.3 "Continuous" Distributions

If we desire random variants uniformly distributed in the interval $[0,1)$, then simple scaling of the x_n generated by (27.6) or (27.10) will suffice:

$$y_n = \frac{x_m}{m} . \qquad (27.11)$$

Of course, the distribution of the y_n is not continuous — nothing is in a computer (or the world, for that matter) — but for large m, the y_n are often taken as a continuous variable.

Distributions that are not uniform can be obtained from a uniform variable as follows. Say $w(z)$ is the desired frequency distribution and $W(z)$ is the corresponding cumulative distribution:

$$0 \leqslant W(z) \leqslant 1 . \qquad (27.12)$$

---

[3]   Can the reader show why this is so?

Then the random variable z is obtained from y by the inverse W^{-1}, of the (nondecreasing) $W(z)$:

$$z = W^{-1}(y) .$$

(27.13)

For example, for

$$w(z) = \frac{1}{\bar{z}} e^{-z/\bar{z}} , \quad z \geqslant 0 , \quad \text{and}$$

(27.14)

$$W(z) = 1 - e^{-z/\bar{z}} ,$$

(27.15)

inversion of $W(z)$ gives

$$z = -\bar{z} \ln(1 - y) ,$$

(27.16)

or, more simply (because y is uniform):

$$z = -\bar{z} \ln y .$$

(27.17)

Of course, for y near 0, when the logarithm in (27.17) has a singularity, the step size for z is relatively large:

$$\frac{dz}{\bar{z}} = \frac{|dy|}{y} = \frac{1}{my} .$$

(27.18)

27.4 Four Ways to Generate a Gaussian Variable

One of the most often needed distributions is the *normal* or *Gaussian* distribution:

$$w(z) = \frac{1}{\sqrt{2\pi}\sigma} e^{-(z-\bar{z})^2/2\sigma^2} .$$

(27.19)

A Gaussian variable z can be generated according to (27.13), which requires tabulation or computation of the inverse error integral.

In another approach, recourse is had to the Central Limit Theorem of probability theory: an approximation to a Gaussian variable is obtained by summing many uniform variables (at least six, say — but this depends on the application).

A third method uses the fact that a Gaussian variable is obtained by multiplying a Rayleigh variable

$$w(r) = \frac{2r}{\overline{r^2}} e^{-r^2/\overline{r^2}} \qquad (27.20)$$

and the cosine of a uniform variable u, uniform in $[0,2\pi)$. The variable r is obtained with (27.13). Thus, the Gaussian variable g is given by

$$g = \bar{g} + (-2\overline{g^2} \ln y_1)^{\frac{1}{2}} \cdot \cos(2\pi y_2) , \qquad (27.21)$$

where y_1, y_2 are independent uniform variables in $[0,1)$.

A second, uncorrelated Gaussian variable can be generated from the same y_1, y_2 by replacing the cosine in (27.21) by sine. More generally, a Gaussian variable with correlation coefficient c can be generated by adding $\arccos(c)$ to the argument of the cosine in (27.21).

In conclusion, we will give a method for generating two independent Gaussian variables that does not require a trigonometric function as in (27.21). Again, we need two uniform variables, y_1, y_2, which will be "centered":

$$u_1 = 2y_1 - 1 , \quad u_2 = 2y_2 - 1 , \qquad (27.22)$$

and *rejected*, unless

$$r = u_1^2 + u_2^2 \leqslant 1 . \qquad (27.23)$$

With such r, we then compute the two uncorrelated Gaussian variables with mean 0 and variance 1 as follows:

$$g_1 = \frac{\left[-\frac{2}{r} \ln r \right]^{\frac{1}{2}}}{2} \cdot u_1 , \qquad (27.24)$$

and

$$g_2 = \frac{\left[-\frac{2}{r} \ln r \right]^{\frac{1}{2}}}{2} \cdot u_2 . \qquad (27.25)$$

Correlated variables can be obtained by linear combinations of g_1 and g_2.

Of course, none of the methods described here are (or could be!) exact, if only because computation of the logarithm is only approximate. But precision can always be carried as far as needs (and budgets) dictate.

27.5 Pseudorandom Sequences in Cryptography

At first blush, pseudorandom sequences look like a fine tool for scrambling or encrypting secret messages. However, great care must be exercised to avoid some terrible pitfalls when relying on *pseudo*randomness.

As early as 1953, *E. N. Gilbert* of Bell Laboratories wrote a memorandum [27.5] in which he pointed out that binary Galois (linear shift-register) sequences (Chap. 25) of (maximum) period lengths $2^m - 1$ can be distinguished from *truly* random sequences on the basis of *much fewer* than 2^m terms; in other words, long before the sequence betrays its pseudorandomness by periodic repetition.

Specifically, Gilbert showed that the rank of a certain matrix constructed from N consecutive terms of the Galois sequence is less than maximal because of the linear dependence between terms. Conversely, the probability that this will happen for a truly random sequence can be made arbitrarily small.

For example, suppose we consider $N = 50$ successive terms from a Galois sequence with $m = 11$. Although the repetition period is 2047, these 50 "bits" will, of course, fail the randomness test based on maximal matrix rank. But the probability that the randomness test will be failed by a truly random sequence with independently distributed 0's and 1's (each with probability 1/2 is less than 10^{-6} [27.5]. To overcome this inherent weakness of linear shift-register sequences, recent research has concentrated on the problem of increasing the "complexity" of such sequences by taking nonlinear functions of several such sequences [27.6, 7]. This field is presently in considerable flux.

Pronounced biases have also been discovered in pseudorandom (card) shuffles, with interesting number-theoretic implications [27.8].

Good algorithms for pseudorandom permutations [27.9] are of particular interest in cryptography.

New methods for encryption by "random rotations" were described by *Sloane* [27.10] at the Cryptography Workshop held at Burg Feuerstein, Fed. Rep. of Germany, in 1982. Much recent work in cryptography was discussed at that meeting [27.11].

28. Waveforms and Radiation Patterns

In numerous technical applications, as well as some theoretical considerations, an enticing problem is encountered. In its most practical (and real) form the problem may be that of designing a radar (or sonar) transmitter *waveform* with a given spectrum such that its peak factor is a minimum. Peak factor is defined here as the *range* of the waveform values divided by their root-mean-square. This is important in radar and sonar, because one often wants to radiate a maximum amount of signal power, of a prescribed spectral shape, with a given peak power limitation on the transmitter.

In many applications, the prescribed spectral shape is a flat (or "white") spectrum. One waveform that has such a flat spectrum is a sharp impulse; but in this case all the power is concentrated at one point in time, leading to a *maximum* peak power. What we want is a minimum peak power. The problem also occurs in talking computers and electronic speech synthesizers in general: synthetic speech made from sharp impulses sounds harsh and reedy; it suffers from a none-too-pleasant "electronic accent." By contrast, computer speech from low peak-factor waveforms sounds smoother.

How can we avoid sharp pulses[1] without modifying the power spectrum? Obviously, we still have the *phase angles* of the Fourier coefficients to play with.

For a given amplitude spectrum (*magnitudes* of Fourier transform coefficients), how does one choose the *phase angles* of the Fourier coefficients in order to achieve the smallest range of magnitudes in the corresponding inverse Fourier transform?

The results on waveforms described in this chapter are also applicable to antenna directivity problems. Specifically, a low peak factor coresponds to an antenna with a wide radiation or receiving directivity pattern.

Another antenna problem solved by number theory is one of "sparse arrays", namely *minimum-redundancy* antennas. Such antennas play an important role in astrophysics and ocean surveillance, where the individual antenna elements (e.g., steerable parabolic "dishes" or submerged hydrophones) are very expensive or costly to control. In such cases, one

---

[1] Also called *Dirac* functions after the famous theoretical physicist who first formulated quantum electrodynamics (QED), one of the most accurate physical theories. Dirac also discovered, on paper so to speak, as consequences of Lorentz invariance, two very fundamental facts of universal reality: elementary particle spin and antimatter. (While spin had been observed before QED, antimatter was another matter and its existence was even doubted by such supreme judges as Wolfgang Pauli.)

wishes to construct arrays with the smallest number of elements for a given unambiguous target resolution (Sect. 28.5). Minimum redundancy arrays have also become important in real-time diagnostic tomography.

28.1 Special Phases

Early waveforms with low peak factors were found by the author and V. A. Vyssotsky by Monte Carlo computations: several thousand periodic waveforms of a given power spectrum, but different sets of random phase angles, were generated on the computer and sorted according to increasing peak factor. The best waveforms thus found had a peak factor several times smaller than the zero-phase impulse.

Later, the author [28.1] developed a formula for the phase angles α_n of low peak-factor waveforms with a given power spectrum P_n, based on asymptotic spectra of certain frequency-modulated signals:

$$\alpha_n = \alpha_1 - \frac{2\pi}{P} \sum_{k=1}^{n-1} (n-k)P_k , \tag{28.1}$$

where P is the total power:

$$P := \sum_{k=1}^{N} P_k . \tag{28.2}$$

For flat power spectra, $P_k = \text{const}$, (28.1) can be simplified to

$$\alpha_n = \alpha_1 - \frac{\pi}{N} n^2 , \tag{28.3}$$

a *quadratic* dependence of phase on harmonic number (frequency) n.

Figures 28.1 and 28.2 illustrate the reduction in peak factor achieved with formula (28.1) for a nonflat power spectrum.

If the low peak-factor waveform is desired to be symmetric in time, then phase angles are restricted to 0 or π. Formula (28.1) is then replaced by [28.1]:

$$\alpha_n = \pi \left\lfloor \sum_{k=1}^{n-1} (n-k) \frac{P_k}{P} \right\rfloor , \tag{28.4}$$

or, for flat spectra,

$$\alpha_n = \pi \left\lfloor \frac{n^2}{2N} + c \right\rfloor , \tag{28.5}$$

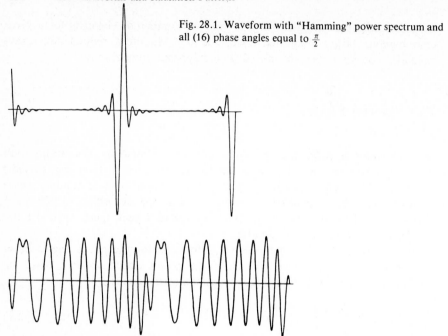

Fig. 28.1. Waveform with "Hamming" power spectrum and all (16) phase angles equal to $\frac{\pi}{2}$

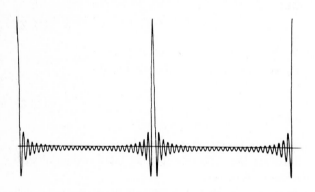

Fig. 28.2. Waveform with same power spectrum as in Fig. 28.1 but phase angles according to (28.1). Note reduction in peak factor (same scale as in Fig. 28.1!)

Fig. 28.3. Waveform corresponding to 31 equal-amplitude harmonics in zero phase

where c is a constant that can be adjusted to minimize rounding effects due to the floor function (Gauss bracket) employed in (28.5).

Figures 28.3 and 28.4 illustrate the 2.64-fold reduction in peak factor achieved with (28.5) for a waveform consisting of 31 harmonics of equal amplitude. The low peak factor is 1.30.

Since this early work on peak factor reduction, several analytical and arithmetic tools have become available which are described in the following sections.

Fig. 28.4. Waveform with same power spectrum as in Fig. 28.3 with phase angles according to (28.5). Reduction in peak factor is by a factor 2.64 (Same scale as in Fig. 28.3)

28.2 The Rudin-Shapiro Polynomials

The so-called Rudin-Shapiro polynomials [28.2] are defined by the following recursion:

$$p_0(t) = q_0(t) = 1 , \tag{28.6}$$

$$p_{n+1}(t) = p_n(t) + e^{it2^n} q_n(t) ,$$

and

$$q_{n+1}(t) = p_n(t) - e^{it2^n} q_n(t) . \tag{28.7}$$

It is clear from these definitions that p_n and q_n have period $2\pi/2^n$. In fact, their Fourier transform consists of 2^n frequency components (beginning at zero frequency) of equal magnitude.

The first low-order Rudin-Shapiro polynomials are (abbreviating $e^{it} = z$):

$$p_1 = 1 + z , \qquad q_1 = 1 - z ,$$

$$p_2 = 1 + z + z^2 - z^3, \quad q_2 = 1 + z - z^2 + z^3 ,$$

$$p_3 = 1 + z + z^2 - z^3 + z^4 + z^5 - z^6 + z^7 ,$$

$$q_3 = 1 + z + z^2 - z^3 - z^4 - z^5 + z^6 - z^7 .$$

It is easy to see from (28.6) and (28.7) that (for $|z| = 1$)

$$|p_n|^2 + |q_n|^2 = 2^{n+1} , \tag{28.8}$$

and therefore

$$|p_n|^2 \leqslant 2^{n+1} ,$$

$$|q_n|^2 \leqslant 2^{n+1} . \tag{28.9}$$

On the other hand, time averaging over one period yields

$$\overline{|p_n|^2} := \frac{1}{2\pi} \int_0^{2\pi} |p_n(t)|^2 \, dt = 2^n \, . \tag{28.10}$$

Thus, the maximum of $|p_n|^2$ exceeds its average value by at most a factor 2 — a property the Rudin-Shapiro polynomials share with a sinusoidal waveform. These polynomials, although perhaps not optimum, are therefore well behaved in terms of their peak factors.

For real waveforms having 2^n frequency components, beginning with frequency m, the real part of $ce^{i2\pi mt}p_{n-1}(t)$ or $ce^{i2\pi mt}q_{n-1}(t)$, where c is a complex constant that determines amplitude and initial phase, is a good choice.

The lowest peak factor achieved for $n = 5$ (31 harmonics after elimination of the zero frequency) was 1.35 for the real part of p_5 and 1.36 for the real part of q_5.

Cyclically shifting the sign sequence of the Fourier components and also allowing their phase angles to be shifted by a constant amount improved the peak factors for p and q to 1.29 and 1.31, respectively. These results are comparable to those achieved with the special phases suggested by the author (Sect. 28.1), but at the price of creating waveforms that are asymmetric in time. For symmetric waveforms the best Rudin-Shapiro peak factor (1.35) is noticeably worse than the peak factor (1.30) of the author's phase combination.

28.3 Gauss Sums and Peak Factors

While the Rudin-Shapiro polynomials are limited to waveforms with 2^n frequency components, waveforms $s(t)$ based on Gauss sums can be expected to be well behaved for a number of components equal to $p-1$, where p is a prime:

$$s(t) = \sum_{k=0}^{p-1} \left[\frac{k}{p}\right] e^{ikt} \, , \tag{28.11}$$

where (k/p) is the Legendre symbol, with $(0/p) = 0$.

The reason why $s(t)$ according to (28.11) can be expected to have a low peak factor is the following. The (periodic) "Legendre" sequence (Chap. 15)

$$a_k := \left[\frac{k}{p}\right] \tag{28.12}$$

has a periodic autocorrelation sequence

$$c_n := \sum_{k=0}^{p-1} a_k a_{k+n} , \tag{28.13}$$

which for zero shift is large:

$$c_n = p - 1 \quad \text{for} \quad n \equiv 0 \pmod{p} , \tag{28.14}$$

and small otherwise:

$$c_n = -1 \quad \text{for} \quad n \equiv 0 \pmod{p} .$$

This follows from the property of the Legendre symbol

$$\left(\frac{k}{p}\right)\left(\frac{k+n}{p}\right) = \left(\frac{k(k+n)}{p}\right) , \tag{28.15}$$

or, after multiplying with

$$\left(\frac{b^2}{p}\right) = 1 , \tag{28.16}$$

where for $n \not\equiv 0$, b can be chosen so that

$$bn \equiv 1 \pmod{p} , \tag{28.17}$$

$$\left(\frac{k}{p}\right)\left(\frac{k+n}{p}\right) = \left(\frac{bk(bk+1)}{p}\right) \quad \text{for} \quad n \not\equiv 0 \pmod{p} . \tag{28.18}$$

In the sum (28.13) we now replace summation over k by $k' = bk$, which does not affect the result, because k' covers the same values as k. Thus

$$c_n = \sum_{k'=0}^{p-1} \left(\frac{k'(k'+1)}{p}\right) = c_1 \quad \text{for} \quad n \not\equiv 0 \pmod{p} ,$$

and, of course,

$$c_n = p - 1 \quad \text{for} \quad n \equiv 0 \pmod{p} . \tag{28.19}$$

Since

$$\sum_{k=0}^{p-1} \left(\frac{k}{p}\right) = 0 , \tag{28.20}$$

we also have

$$\sum_{n=0}^{p-1} c_n = 0 \, , \tag{28.21}$$

and therefore, with (28.19),

$$c_n = -1 \quad \text{for} \quad n \not\equiv 0 \pmod{p} \, . \tag{28.22}$$

Now, if the (k/p) are the amplitude spectrum of a periodic complex waveform as defined in (28.11), the squared magnitude of $s(t)$ has the following Fourier representation:

$$|s(t)|^2 = \sum_{n=0}^{p-1} \tilde{c}_n \, e^{i2\pi nt} \, , \tag{28.23}$$

where the $\tilde{c}_n$ are the *aperiodic* correlation coefficients of the (k/p). Nevertheless, with $|c_n|$ small $(|c_n| = 1$ for $n \not\equiv 0)$, the $|\tilde{c}_n|$ will also be relatively small. Since c_0 is large $(c_0 = p - 1)$, $|s(t)|^2$ will be relatively constant with small fluctuations. As a consequence, the real part of the $s(t)$ is expected to stay within relatively narrow bounds — in other words, the peak factor of $s(t)$ constructed according to (28.11) will tend to be small.

Since (28.11) generates only 30 harmonics, a 31st harmonic was added with either sign. The peak factor for either case is 1.66. When applying cyclic shifts, the resulting smallest peak factor over all 30 cyclic shifts of (k/p) is 1.39 — not bad, but not as good as the result obtained with the quadratic ("chirp") phases (Sect. 28.1).

28.4 Galois Sequences and the Smallest Peak Factors

Another type of sequence with two-valued autocorrelation c_n is that of the Galois sequences g_n (Chap. 26) generated by a linear recursion based on an primitive polynomial over $GF(2^m)$. Their period length is $2^m - 1$ and the periodic correlation is given by

$$c_n = 2^m - 1 \quad \text{for} \quad n \equiv 0 \pmod{2^m-1} \qquad \text{and}$$

$$c_n = -1 \quad \text{for} \quad n \not\equiv 0 \pmod{2^m-1} \, . \tag{28.24}$$

Thus, a periodic waveform

$$s(t) := \sum_{k=1}^{2^m-1} g_k \, e^{i2\pi kt/(2^m-1)} \tag{28.25}$$

constructed with the g_n (or any of their cyclic shifts) will have a relatively constant $|s(t)|^2$, i.e., the peak factor of $s(t)$ will be small (Sect. 28.3).

For $m = 5$ (31 harmonics) there are

$$\frac{\phi(2^5 - 1)}{5} = 6 \tag{28.26}$$

different primitive polynomials over $GF(2)$ of degree 5, namely

$1 + x^2 + x^5$	peak factor $=$	1.259
$1 + x^3 + x^5$		1.252
$1 + x + x^2 + x^3 + x^5$		1.209
$1 + x^2 + x^3 + x^4 + x^5$		1.249
$1 + x + x^2 + x^4 + x^5$		1.240
$1 + x + x^3 + x^4 + x^5$		1.282 .

Here the numbers on the right are the smallest peak factors achieved with each of the six polynomials considering all 31 possible cyclic shifts.[2] As can be seen, all peak factors are smaller than 1.29, and the smallest one, 1.209, is achieved with the third polynomial, giving rise to the recursion

$$g_n = g_{n-1} \cdot g_{n-2} \cdot g_{n-3} \cdot g_{n-5} ,$$

when used with the initial condition

$$g_1 = g_2 = g_3 = g_4 = -1 \quad and \quad g_5 = 1 .$$

This initial condition happens to give the "idempotent":

$$- - - - + - - - + + + - + - + -$$

$$- + - + + + + - - + + - + + - \quad .$$

Idempotent means that the polynomial representing this sequence reproduces itself upon squaring. Applied to the sequence, idempotent means that when only the even-numbered terms of the periodically repeated sequence are taken, the sequence is reproduced. Of all the possibilities investigated, the above sequence is the unique choice that gives the smallest peak factor for 31 harmonics with phase angles restricted to 0 or π. The corresponding waveform is shown in Fig. 28.5. This also holds when the same phase angles are applied to a frequency-shifted waveform with harmonics 51 to 81, which

---

[2] I am grateful to J. L. Hall of Bell Laboratories for calculating these peak factors and for the plots of Figs. 28.5 and 28.6.

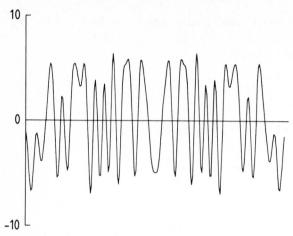

Fig. 28.5. Waveform with 31 constant-amplitude harmonics ($n = 1, 2, \ldots, 31$) and phase angles according to (idempotent) Galois sequence based on primitive polynomial $1 + x + x^2 + x^3 + x^5$ over $GF(2^5)$. Peak factor is lowest so far found for the given power spectrum

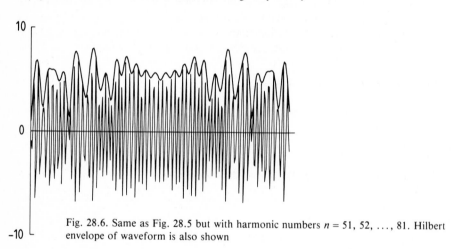

Fig. 28.6. Same as Fig. 28.5 but with harmonic numbers $n = 51, 52, \ldots, 81$. Hilbert envelope of waveform is also shown

has a peak factor of 1.245. In Fig. 28.6, the frequency-shifted waveform is plotted, together with its "Hilbert" envelope. (The Hilbert envelope is defined as the magnitude of the corresponding analytic signal.)

Thus, we see that Galois sequences — already found useful in diffraction problems, precision measurements and error-correction codes — solve yet another practical problem: how to design waveforms with low peak factors for such applications as radar and sonar and computer-synthesized speech. See also [28.3-5].

In the next chapter, still another property of Galois sequences is utilized in the design of necklaces.

28.5 Minimum Redundancy Antennas

The principles of designing waveforms with a low peak factor are also applicable to the design of efficient transmitting or receiving antennas with broad directivity patterns. Given a real or complex sequence having a constant-magnitude Fourier transform, we can construct a linear array with a directivity pattern with constant (or nearly constant) maxima. For the case of a primitive-root sequence (see Sects. 13.8 and 26.6), one resulting directivity pattern is illustrated in Fig. 26.7.

In contrast to such "broadcast" antennas, arrays that favor only *one* direction are often needed. Such antennas have directivity patterns with only one major narrow "lobe." To avoid ambiguity, the spacing between individual elements must be smaller than half the shortest wavelength, λ_{min}, contained in the signal to be received or transmitted. (This requirement follows from the spatial sampling theorem.) For a given directional resolution, the antenna must have a certain minimum size, D. Roughly, for an unambiguous angular resolution $\Delta\alpha$, the antenna size must be

$$D = \frac{\lambda_{min}}{\Delta\alpha} .$$

(28.27)

Thus, for a uniform spacing, smaller than $\lambda_{min}/2$, the number of elements N must be at least

$$N = \frac{2}{\Delta\alpha} + 1 .$$

(28.28)

For the kinds of resolution required in radio astronomy and ocean surveillance, N can be quite large and given the high cost of individual antenna elements (steerable parabolic reflectors or submerged hydrophones), arrays with uniformly spaced elements can often not be economically realized.

Does this mean we have to give up good radio astronomy or affordable sea security? No. For some applications, especially *detection* tasks, it is sufficient if in the array all required spacings occur at least once. Thus, instead of placing four elements on a straight line with constant spacings of 1 unit, giving an array of size 3 units, the three spacings between the four elements can be chosen to be 1, 3 and 2 units, respectively, giving an array of size 6, which has double the angular resolution for the same number of elements. Unambiguous detection requires that all possible element spacings between 1 and 6 occur at least once, which is the case: the spacings 1, 2, 3 occur between adjacent elements and the remaining three spacings are realized as follows:

$$1 + 3 = 4 , \quad 3 + 2 = 5 , \quad 1 + 3 + 2 = 6 .$$

Unfortunately, there are no arrays having more than four elements that give every spacing *exactly* once. For $N > 4$, some spacings will be duplicated resulting in an inefficiency or redundancy. Nevertheless, *minimum redundancy* arrays have been designed that give the greatest unambiguous resolution for a given number of elements [28.6]. For example, for $N = 5$, a minimum redundancy configuration has spacings 1,3,3,2 giving an array size of 9. Except for the spacing 3, all spacings occur only once.

The number theoretical concept behind such antenna designs is that of *difference sets* discussed in greater detail in [28.7], which also gives solutions to other difficult problems in *additive number theory*.

29. Number Theory, Randomness and "Art"

Perfect symmetry is aesthetically boring, and complete randomness is not very appealing either. All works of art show both a degree of regularity and predictability on the one hand, and innovation and "surprise" on the other. In fact, George Birkhoff, of ergodicity fame, once constructed a mathematical formula intended to capture aesthetic value. The formula's two main variables represented regularity and randomness, the aesthetic value function reaching its maximum for ample portions of both ingredients.

While it is probably futile to try to capture artistic achievement in a mathematical formula, there is nevertheless some truth in the basic assumption that both predictability (or a degree of recognizability) and surprise are important in works of art. If we accept this notion, then number theory should be a good store for finding these two ingredients. The distribution of prime numbers, for example, exhibits both some predictability and a certain amount of randomness. Thus, it is not surprising that we can construct, from number-theoretic relationships, some attractive graphic designs and other objects of pleasure. To call them art, however, would be presumptuous.

In this chapter we explore a few of these pleasurable possibilities of number theory.

29.1 Number Theory and Graphic Design

Every third number is divisible by 3, every seventh by 7, etc. In general, the probability that a number picked at random from an infinite range is divisible by the prime number p_i is $1/p_i$, and these probabilities are apparently independent. Thus, the probability that a number is divisible by both 3 and 7 is $1/21$. Similarly, the probability that a given number is *not* divisible by p_i equals $1-1/p_i$.

The probability that *two* numbers picked independently at random are *both* divisible by the *same* prime factor p_i is obviously $1/p_i^2$. And the probability that both are *not* divisible by the same prime factor is consequently $1-1/p_i^2$.

What is the probability P that two randomly chosen numbers have *no* common prime factor, i.e., that they are coprime? If we believe in the independence of prime factors, the answer must be

$$P = \prod_i (1 - 1/p_i^2) \, , \tag{29.1}$$

where, because we are considering an infinite range, the product is extended over *all* primes. Does P have a finite (nonzero) value? For the first 6 primes (2,3,5,7,11,13) its value is 0.618 (no relation to the Golden ratio!). But that does not teach us much — if we recall, for example, how slowly the distances between successive primes diverge. However, the reciprocal of the infinite product (29.1) can be multiplied out, using the formula for geometric series:

$$\frac{1}{P} = \prod_i \frac{1}{1 - 1/p_i^2} = \prod_i \left(1 + \frac{1}{p_i^2} + \frac{1}{p_i^4} + \frac{1}{p_i^6} + \ldots \right) . \tag{29.2}$$

The forbidding-looking infinite product of the infinite sum on the right is nothing but *all* the reciprocal squares $1/n^2$, each square appearing exactly *once* (by the Fundamental Theorem). Thus, as we have seen in earlier chapters:

$$\frac{1}{P} = \sum_{n=1}^{\infty} \frac{1}{n^2} \, , \tag{29.3}$$

a well-known sum (the Riemann zetafunction for $z = 2$). Its value, $\pi^2/6$, can be found in most collections of formulae. Hence $P = 6/\pi^2 = 0.6079\ldots$.

While formally, the result only holds if we pick pairs of integers from an infinite range, the value $6/\pi^2$ is already very close for the range from 2 to 11: exactly 60 of all 100 possible pairs of integers from 2 to 11 have no common divider.

Figure 4.8 shows (as white dots) all the integer pairs in the range 2 to 256 that are coprime, i.e., have no common divider. As expected, the plane lattice is fairly uniformly filled with white dots, and their average density is very close to $6/\pi^2$.

Many well-known number-theoretic relations can be observed in such a graphic display, and who knows — close visual inspection might even suggest new theorems [29.1]. The plot is also related to the topic of "visibility": the white dots are precisely those that can be "seen" from the point with coordinates (1,1), while all other points are "hidden" behind one or several points on the straight line connecting them with the point (1,1). To put it more dramatically, if a pellet of paint exploded at (1,1), the paint would only stick on little pegs at lattice points where the white dots are shown in Fig. 4.8.

As we remarked before, the property of divisibility is a *periodic* one. The Fourier transform, of course, is also "sensitive" to periodicities. Thus, the Fourier transform of the "holy" lattice shown in Fig. 4.8, interpreting a white dot as having value +1 and the other lattice points as having value −1, should be particularly interesting. Figure 4.9 shows the *magnitude* of the two-dimensional Fourier transform of the number-theoretic "coprimality" function

(Fig. 4.8). Not unexpectedly, we get a cluster of "stars" and reflection symmetries around the two diagonals. But there are also near (broken?) symmetries around the horizontal and vertical axes. They are not so easy to explain (how is *near* symmetry defined?), and we will leave it to the reader to explore this symmetry breaking [29.2].

We may also view Fig. 4.9 as the result of x-ray diffraction from a "moth-eaten" two-dimensional crystal with many atoms missing, but with the remaining ones in perfect lattice positions. But no matter how we look at it, number theory has given us quite a pretty picture here.

29.2 The Primes of Gauss and Eisenstein

Prime numbers can be defined in fields other than the integers. In the complex number field **C** we have the *Gauss primes,* a subset of the Gauss integers $n + im$, where $i^2 = -1$. Primes of the form $4k-1$ in **Z** are still primes in **C**, but 2 and primes of the form $4k+1$ can be factored in **C**!

Examples: $2 = (1+i)(1-i)$, $5 = (2+i)(2-i)$, $13 = (2+3i)(2-3i)$, $17 = (4+i)(4-i)$, etc.

The Gauss primes form a pleasing pattern, shown in Fig. 29.1, which has been used in weaving tablecloths and tiling floors.

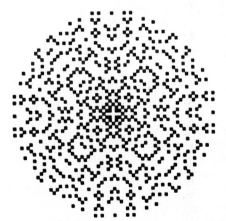

Fig. 29.1. Gauss primes $n+mi$ with norms n^2+m^2 less than 1000. A favorite design for tablecloths

Eisenstein, one of Gauss's favorite mathematicians, defined his own primes based on the complex cube root $\omega = (1 - \sqrt{-3})/2$. (Note: $1 + \omega + \omega^2 = 0$) They are a subset of the Eisenstein integers $n + \omega m$. The Eisenstein primes form an interesting pattern with hexagonal symmetry (Fig. 29.2). Near the

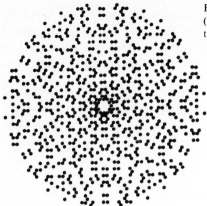

Fig. 29.2. Eisenstein primes $n + m\omega$, where ω is a (primitive) complex root of 3. How would bees react to this pattern?

center of Fig. 29.2, 6 times 6 hexagonal points form 6 hexagons. Are *complete* hexagons formed anywhere else?

The prime 2 and primes of the form $6k-1$ are also Eisenstein primes, but 3 and primes of the form $6k+1$ can be factored.

Examples: $3 = (1-\omega)(1-\omega^2)$, $7 = (2-\omega)(2-\omega^2)$, $13 = (3-\omega)(3-\omega^2)$, $19 = (3-2\omega)(3-2\omega^2)$, etc.

Another pleasing plot is obtained with the set-theoretic *intersection* of the Gauss and Eisenstein *composites*. What are their asymptotic densities?

An even more interesting pattern emerges if the Gauss and Eisenstein primes and their *intersections* are plotted on the same lattice in 3 different colors.

29.3 Galois Fields and Impossible Necklaces

Suppose we inherited a collection of white and black pearls. Could we make a necklace of only 16 pearls (a real choker!) in which all possible variations of 4 adjacent colors appeared? There are $4^2 = 16$ such variations; thus, the necklace would have to have *at least* 16 pearls. But is it really possible?

Figure 29.3 shows the answer, with the 16th pearl, a black one, about to be inserted. With a bit of patience we can find every variation, as promised. For example, the variation of a white pearl followed by a black and two more white ones is found in positions 6 to 9.

On what principle was the necklace designed? Galois field theory! Or, more specifically, the cyclotomic polynomials over the Galois field $GF(2)$ that we encountered in error-correcting codes and in the confirmation of the fourth general relativistic effect with Mercury or Venus in superior conjunction.

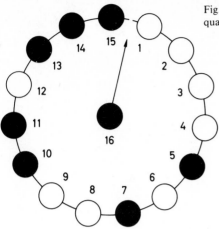

Fig. 29.3. A 15 + 1 pearl necklace with all possible quadruplets of black and white pearls

To design the 16-pearl necklace we need an primitive factor of degree 4 of the polynomial $x^{15} + 1$. Such a factor is

$$x^4 + x^3 + 1 ,\tag{29.4}$$

leading to the recursion

$$a_{n+4} = a_{n+3} + a_n .\tag{29.5}$$

With the initial condition $a_1 = a_2 = a_3 = a_4 = 1$, the recursion gives the following binary sequence of period 15:

$$1\ 1\ 1\ 1\ 0\ 1\ 0\ 1\ 1\ 0\ 0\ 1\ 0\ 0\ 0\ ;\ 1\ 1\ 1\ \text{etc.},\tag{29.6}$$

which contains all variations of length 4 except the quadruplet 0000. This shortcoming can be rectified by adjoining another 0 at the end of the period [(before the semicolon in (29.6)]. If we now identify 1 with a white pearl and 0 with a black pearl and tie one period of the string into a necklace, we obtain the "maximally varied" design shown in Fig. 29.3.

In fact, instead of two colors, we can make such a necklace out of *four* different colors, basing the design on polynomials over $GF(4)$. Of course, now a necklace of 16 pearls can have at most all possible *pairs* of the four colors. That such a necklace is indeed possible is illustrated by Fig. 29.4. The design is based on $GF(4)$ consisting of the four elements $(0,1,\omega,\omega^2)$ with $\omega^3 = 1$ and $1 + \omega + \omega^2 = 0$ and the primitive polynomial $x^2 + x + \omega$. With the initial condition 0, 1 and identifying 0 with red, 1 with blue, ω with yellow and ω^2 with green, we obtain the 16-pearl maximally varied necklace shown in Fig. 29.4 — not, however, before inserting an extra red pearl into the 16th position.

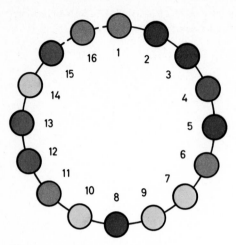

Fig. 29.4. Four-color necklace with 16 pearls exhibiting all possible pairs of colors

Can we go a step further and give each pearl *two* binary properties, such as color and size and have all possible pairs of the four different kinds of pearls (large white, small white, large black, small black) in a necklace of 16 pearls? The answer is yes: we need only identify these four properties with the above four colors to obtain a solution.

What is *not* possible, however, is a necklace with both the stated property for pairs of adjacent pearls and all *quadruplets* of white and black pearls (regardless of size) *and* all quadruplets of large and small sizes (regardless of color).

If we restrict ourselves to 15 pearls, excusing the occasional absence of a small black pearl, then a solution is possible (see Fig. 29.5). For example, we find the quadruplet black/black/white/black in positions 7 to 10, or the quadruplet small/large/small/large in positions 9 to 12. The doublet small white/large black is found in positions 9 to 10. However, the doublets large white/small black and small black/small black are missing unless we sneak in a 16th small black pearl between positions 15 and 1. But even then the doublet small white/small black and the quadruplet black/black/black/black is missing, unless we also allow the 16th pearl to go between positions 5 and 6. In other words, we really need 17 pearls for the *super*-varied necklace. But then some variations occur twice, for example small black/small black — very inelegant! But for that one loose pearl, the arrangement could not be more perfect.

The problem of the loose pearl is deeply related to the fact that 0 has no inverse, not even in a finite field where we expect surprises: for example $3^{-1} = 5$ in $GF(7)$, etc. (Does the reader remember why the inverse of 3 in $GF(7)$ is 5 — and how to calculate it, if it was not so easy to find by trial and error?)

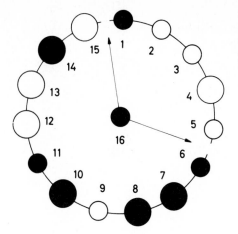

Fig. 29.5. A 16-pearl necklace having pearls of two colors and two sizes. It shows all possible color and size quadruplets and all possible color and size pairs — if pearl number 16 is allowed to move around

If one pearl is always excused, the super-varied necklace can be extended to, say, $63 = 2^6 - 1$ pearls, having the following six binary properties:[1]

$$
\begin{array}{ccc}
\text{large} & \text{or} & \text{small} \\
\text{round} & \text{or} & \text{octagonal} \\
\text{solid} & \text{or} & \text{pierced} \\
\text{polished} & \text{or} & \text{rough} \\
\text{bright} & \text{or} & \text{dark} \\
\text{colored} & \text{or} & \text{grey.}
\end{array}
$$

Except for an occasional missing small-octagonal-pierced-rough-dark-grey pearl, seen in the center of Fig. 29.6, the necklace shown contains all possible $2^6 = 64$ sextuplets of *any* of the 6 binary properties listed above, i.e., $6 \cdot 63 = 378$ prescribed sextuplets.

In addition, if two properties are considered at a time, i.e., $\binom{6}{2} = 15$ paired property choices, all $4^3 - 1 = 63$ triplets can be found, or a total of $15 \cdot 63 = 945$ prescribed triplets.

Further, if *three* properties are considered, in any one of $\binom{6}{3} = 20$ different ways, the necklace has all $8^2 - 1 = 63$ possible doublets, for a total of $20 \cdot 63 = 1260$ doublets.

Finally, considering six properties at a time, we find a total of 63 different singlets in the necklace.

---

[1]   Finite mathematicians Jessie MacWilliams and Andrew Odlyzko acted as assistant jewelers to the author in the design of this necklace.

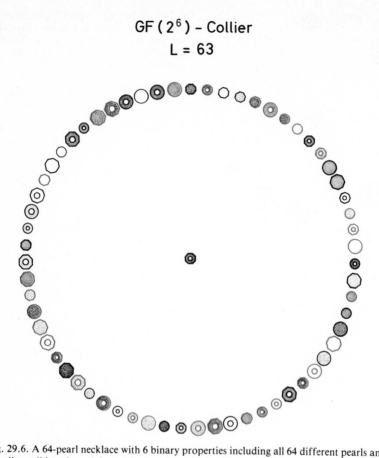

$$GF(2^6) - Collier$$
$$L = 63$$

Fig. 29.6. A 64-pearl necklace with 6 binary properties including all 64 different pearls and showing all possible pairs, triplets and sextuplets (a total of 2646 combinations) — if one pearls is allowed to move around

The total number of constraints that the necklace in Fig. 29.6 obeys is a staggering $\left[\binom{6}{1} + \binom{6}{2} + \binom{6}{3} + \binom{6}{6} \right] \cdot 63 = 42 \cdot 63 = 2646$. Permitting ourselves occasional cheating with the loose small-octagonal-pierced-rough-dark-grey pearl, all singlets and *any* variation of doublets, triplets and sextuplets of pearls occurs [29.3]. Such a super-necklace would not only embellish the discriminating woman (or man) but its design might even benefit some "experimental designs" (as the psychologists are fond of saying).

30. Self-Similarity, Fractals, Deterministic Chaos and a New State of Matter

"O Julia, mir graut's, wenn man bedenkt,
an dünnen Fädchen nur dein Schicksal hängt.
Dein body ist so thin und doch so schön,
wann werde ich dich sweet heart wiederseh'n."

— *Manfred Peschel*

Nature abounds with periodic phenomena: from the motion of a swing to the oscillations of an atom, from the chirping of a grasshopper to the orbits of the heavenly bodies. And our terrestrial bodies, too, participate in this universal minuet — from the heart beat to the circadian rhythm and even longer cycles.

Of course, nothing in nature is *exactly* periodic. All motion has a beginning and an end, so that, in the mathematical sense, strict periodicity does not exist. Nevertheless, periodicity has proved to be a supremely useful concept in elucidating underlying laws and mechanisms.

One reason for the universality of simple harmonic motion is the linearity — or near-linearity — of many physical systems and, more generally, the invariance against displacement in space and time of the laws governing their behavior.

But there are numerous other phenomena in which these fundamental principles break down and, instead of periodicity, we get aperiodic or even chaotic motion: the smooth waves on a well-behaved lake turn to violent turbulence in the mountain brook and the daily sunrise, the paradigm of predictability, is overshadowed by cloud formations, a haven for *chaos* — albeit *deterministic* chaos.

But no matter how chaotic life gets, with all regularity gone to bits, another fundamental bulwark often remains unshaken, rising above the turbulent rubble: *self-similarity,* an invariance with respect to *scaling;* in other words, invariance not against *additive* translations, but invariance against *multiplicative* changes of scale.

Many laws of nature are independent, or nearly so, of a scaling factor. The fact that scaling usually has a limit (Planck's constant, when things get too small, or the speed of light, when objects fly too fast) does no harm to the usefulness of "thinking self-similar," just as the lack (outside mathematics) of strict periodicity is no great impediment in the real world. In fact, self-similarity is nothing but periodicity on a *logarithmic* scale.

What has number theory got to do with all this? Are there any self-similar principles at work behind the integers? If so, are they useful? Can they shed light into dark chaos? Surprisingly, the answers to these questions are affirmative; and, in this final chapter, we shall try to show how.

Take the simplest concept from number theory, ("Kronecker's") integers. Starting from nothing and counting up, we get

0, 1, 2, 3, 4, 5, 6, 7 ...

Next, let us represent these numbers in the binary system:

0, 1, 10, 11, 100, 101, 110, 111, ...

Now take the "digital roots" or parity of each binary number (i.e., the sum of its digits modulo 2):

0 1 1 0 1 0 0 1 ...

The resulting sequence is called the *Morse-Thue sequence*. It has many remarkable properties. For example, it can be constructed recursively by appending to each subsequence the complemented subsequence, as follows:

0
0 1
0 1 1 0
0 1 1 0 1 0 0 1

This kind of construction makes it manifestly aperiodic; it never repeats. But it has a much more interesting property: the infinite sequence is *self-similar*. In fact, striking out every second term reproduces the sequence:

$$0 \; \mathbf{1} \; 1 \; \mathbf{0} \; 1 \; \mathbf{0} \; 0 \; \mathbf{1} \; ... \; = \; 0 \; 1 \; 1 \; 0 \; ... \; . \tag{30.1}$$

The sequence can also be constructed from "nothing" by having each term have a "complementary baby":

```
0
0                   1
0        1          1          0
0    1   1    0     1    0     0    1
```

Such sequences are called self-generating [30.1]. Another nonrandom self-generating sequence is obtained from the signs of the Rudin-Shapiro polynomials, see Sect. 28.2. These sequences mimic, in one dimension, the self-organization of matter, such as crystal growth. However, the Morse-Thue sequence is an *aperiodic* self-generating sequence, thus imitating not the growth of a periodic crystal, but a recently discovered new solidification process.

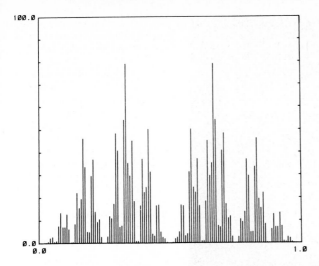

Fig. 30.1. Fourier amplitude spectrum of the self-similar number-theoretic Morse-Thue sequence: 01101001 ..., defined by the *parity* of the integers 0, 1, 2, 3, ... when expressed in the binary notation (0, 1, 10, 11, ...). The (infinite) sequence reproduces itself when, after each element, the complementary element is inserted. Equivalently, the sequence also reproduces itself upon decimation, namely when every other element is struck out. The scaling factor of the self-similarity is thus 2. Although this sequence is aperiodic (it never repeats itself), the amplitude spectrum shows prominent peaks that would be missing in a truly random sequence, such as an aperiodic sequence without the long-range order of the Morse-Thue sequence. (For the discrete Fourier transform shown here, the sequence was truncated to a length of 256.)

Surprisingly, although the Morse-Thue sequence is aperiodic, its Fourier transform or spectrum is not at all noiselike. The self-similarity of the sequence induces *long-range correlations* and causes its spectrum to be highly structured with many high peaks, *as if it were periodic.* In other words, the Morse-Thue sequence, being self-similar, is *both* aperiodic *and* has a spectrum resembling that of a *periodic* sequence, see Fig. 30.1.

This fact would perhaps be only mildly interesting if such sequences did not model something that can actually happen in nature, but was discovered only recently: a *new state of matter!*

We are all familiar with *crystals,* in which the individual atoms are arranged in periodic lattices. And we also know *amorphous* substances, such as ordinary glasses or most liquids, in which the atoms are randomly distributed. Until recently, few if any people suspected that there could be another state of matter sharing important aspects with both crystalline *and* amorphous substances. Yet, this is precisely what *D. Shechtman et al.* [30.2] discovered when they recorded electron diffraction patterns, see Fig. 30.2, of a special aluminum-manganese alloy (Al_6Mn). The diffraction pattern, i.e., the two-dimensional Fourier transform, showed *sharp peaks,* just like those for a

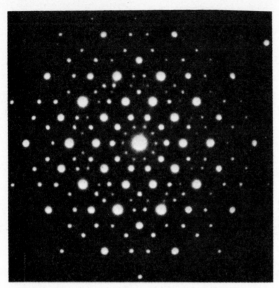

Fig. 30.2. Electron diffraction pattern of an alloy of aluminium and manganese obtained by *Shechtman* et al. [30.2]. The bright spots say that the alloy is not amorphous (like a glass) but "well-ordered". Yet the ten-fold symmetry exhibited by the diffraction pattern cannot occur for any periodic crystal. Thus, the alloy must be both aperiodic and well-ordered: a new state of matter that is closely related in its structure to certain self-similar, but aperiodic, sequences known in number theory (see Fig. 30.3)

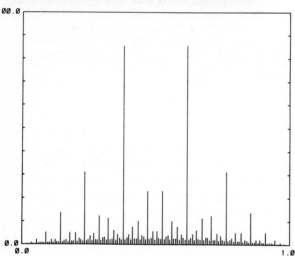

Fig. 30.3. The Fourier amplitude spectrum of the self-similar number-theoretic "rabbit" sequence: 10110101 ..., which can be generated by replacing each 0 (signifying an immature rabbit pair) by a 1 (mature pair) and inserting after each 1 a 0 (birth of a new pair). This generates the next "generation" of the sequence from the previous one. The lengths of successive sequences equal successive Fibonacci numbers, F_n. (In this illustration, the "breeding" process was terminated at $F_{12} = 144$.) The rabbit sequence can also be generated from a Beatty sequence [30.1] $a_k = [ng] = 1, 3, 4, 6, 8, \ldots$, where $g = 1.618 \ldots$ is the Golden ratio, and the a_k specify the indices of the nonzero-terms in the rabbit sequence. (The Beatty sequence $b_k = [ng^2] = 2, 5, 7, \ldots$ gives the indices of the zero-terms.) Note the pronounced spectral peaks near the two frequencies $1/g$ and $1/g^2$ times the sampling frequency, and the self-similar features of the spectrum involving the (approximate) scaling factors g and g^2. This sequence models, in one dimension, the aperiodic long-range order of the new state of matter recently discovered (see Fig. 30.2)

periodic crystal. But the pattern also showed a five-fold symmetry that periodic crystals simply *cannot* have.

However, as we know from the *number-theoretic* Morse-Thue sequence, sharp spectral peaks and aperiodicity are no contradiction, as long as self-similarity prevails.

The number 5, which occurs in the diffraction pattern, points to another self-similar sequence in which perhaps the *Golden ratio* is involved, and this is indeed the case. In fact, it turns out that an aperiodic self-similar sequence derived from the *Fibonacci* numbers is a good model of this new state of matter, showing pronounced peaks in its Fourier spectrum (see Fig. 30.3).

While visiting this newly found Garden of Eden, we will also, in this chapter, make brief calls at an adjoining paradise, that of *Cantor sets* (from which, Hilbert said, we will never be driven away), and at the strangely attractive zoo of *fractal objects,* darlings of a new generation of physical scientists and computer artists.

30.1 Fibonacci, Noble Numbers and a New State of Matter

Let us take another peek at Fibonacci and his rabbits, Fig. 5.9. If we designate a young pair by 0 and a mature pair by 1, Fig. 5.9 (in which this distinction is not shown) becomes this:

$$
\begin{array}{l}
0 \\
1 \\
1\ 0 \\
1\ 0\ 1 \\
1\ 0\ 1\ 1\ 0 \\
1\ 0\ 1\ 1\ 0\ 1\ 0\ 1
\end{array}
\tag{30.2}
$$

where each 1 generates a 1 and a 0 in the row below and each 0 gives rise to a 1. Like the Morse-Thue sequence, this is also a self-generating sequence.

Another law of construction for this bit stream (which I shall call the *rabbit sequence)* is quite apparent: after the first two rows, simply append to each row the previous row to form the next one.

Interestingly, each term in the rabbit sequence r_n can be constructed directly without reference to prior terms or subsequences. For this purpose, the index n has to be decomposed according to the Fibonacci number system (see Sect. 5.5):

$$
n = F_{k_1} + F_{k_2} + \ldots + F_{k_r} .
\tag{30.3}
$$

The parity of the last index, k_r, determines r_n:

$$r_n = \langle k_r + 1 \rangle_2 . \tag{30.4}$$

So, as in the application to the game of Nim (Sect. 5.5), it is the *least* significant Fibonacci number that is important.

By construction, this sequence is aperiodic although arbitrarily long subsequences will appear infinitely often in the sequence. This is of course true even for random sequences. The point is that in the rabbit sequence identical subsequences will be very close by. In fact, for $n > 3$, a given subsequence whose length does not exceed F_n is never farther than F_{n+1} steps away.

But is the rabbit sequence self-similar, and will it have a "peaky" spectrum like that of a periodic sequence? A peaky spectrum is needed in order to model the newly discovered state of matter.

Of course, the self-similarity of the rabbit sequence cannot be as simple as that of the Morse-Thue sequence, where we could strike out every other bit to leave the sequence invariant. The lengths of the rabbit subsequences (30.2) from one row to the next increase in the ratio of successive Fibonacci numbers, tending to the Golden ratio $g = 1.618 ...$

Interestingly, the ratio of 1's to 0's (mature rabbits to young rabbits) is also the ratio of successive Fibonacci numbers. Let us therefore try striking out (rab)bits depending on the bit itself: For a 1, let us hop two places before we strike out, and for a 0 let us hop only 1 place. The result is the following:

1 0 *1* 1 *0̸* 1 0 *1*

i.e., the beginning of the row: 10110. For the infinite sequence our striking rule would simply reproduce the sequence itself: *self-similarity!* In fact, this "self-decimating" rule is simply the inverse of the self-generating rabbit creation rule.

Another way to demonstrate the self-similarity of the rabbit sequence is to replace every bit pair 10 by a 1 and every triplet 110 by a 0. This "condensation" also reproduces the rabbits, as does the mapping $101 \rightarrow 1$ and $10 \rightarrow 0$ and infinitely many other mappings with an increasing degree of condensation.

After our experience with the spectrum of the self-similar Morse-Thue sequence, it is therefore small wonder that the rabbit sequence, too, exhibits pronounced spectral peaks, see Fig. 30.3. These peaks are the one-dimensional analogue of the bright spots in the diffraction pattern (Fig. 30.2). The locations of these peaks, moreover, reflect the Golden ratio (many times) as can easily be observed in Fig. 30.3.

For comparison, Fig. 30.4 shows the spectrum of a *random* binary sequence, calculated, like the Morse-Thue and rabbit spectra, by J. L. Hall of AT&T Bell Laboratories. To make the comparison as close as possible, the random sequence was produced by a three-state Markov process modeled on

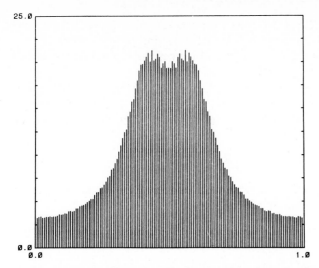

25.0

0.0
0.0   1.0

Fig. 30.4. Average Fourier amplitude spectrum of a three-state Markov process having the same transition probabilities as the rabbit sequence (whose spectrum is shown in Fig. 30.3). Note smooth spectral envelope and absence of pronounced single-harmonic spectral peaks, reflecting the absence of long-range order in the Markov process

the rabbit sequence. The three states correspond to the three possible bit pairs 11, 10, and 01, see (30.2), with deterministic transitions from 11 to 10 (generating a 0) and from 10 to 01 (generating a 1) and a probabilistic transition from 01 to 11 (with probability $\gamma = (\sqrt{5}-1)/2$, the reciprocal of the Golden ratio, generating a 1) and a transition from 01 to 10 (with probability γ^2, generating a 0).

The probabilities of the three states are γ^4, γ^3, and γ, respectively, and the probabilities of 1 and 0 are γ and γ^2, respectively. All these probabilities are in conformity with those of the (infinitely long) rabbit sequence. Yet, although the Markov machine mimics the short-range properties of the rabbit sequence, it lacks the *long-range coherence* of the latter, a fact that is reflected in the absence of sharp peaks in its spectrum, see Fig. 30.4.

The existence of pronounced peaks in Fig. 30.3 means that some long-range order or "phase coherence" exists in the rabbit sequence, extending over its entire length. Although aperiodic, the rabbit sequence, when converted to audible sound, should sound tonal! And the diffraction pattern of a physical substance patterned on such self-similar structure would show bright spots, as if it were a periodic crystal! The new form of matter might also exhibit some rather unorthodox and hitherto unsuspected physical properties.

The two highest peaks in the rabbit spectrum (Fig. 30.3) occur at harmonic numbers ("frequencies") F_{n-1} and F_{n-2} for the subsequence of length F_n. Here, the frequency F_{n-1} is simply the mirror image of the frequency F_{n-2}. (Note: $F_{n-1} = F_n - F_{n-2}$). The ratio F_{n-1}/F_n is approximately $\gamma = 1/1.618\ldots$ and the ratio F_{n-2}/F_n tends toward $\gamma^2 = 1/2.618\ldots$. Why do these high peaks, reflecting periodic behavior, exist and why are they near $1/2.618$ and $1/1.618$ times the "sampling frequency" F_n?

Answer: If we "sample" the rabbit sequence with an average sampling interval of 1.618 samples, we get the same symbol again and again *ad infinitum!* Indeed, the sampled sequences r_k, where k is the largest integer not exceeding n/γ becomes

$$r_k = 1\ 1\ 1\ 1\ 1\ 1\ 1\ \ldots, \quad k = \lfloor \frac{n}{\gamma} \rfloor , \tag{30.5}$$

a fact that follows from the above construction rules. Similarly, sampling with an average interval size of 2.618 yields

$$r_k = 0\ 0\ 0\ 0\ \ldots, \qquad k = \lfloor \frac{n}{\gamma^2} \rfloor . \tag{30.6}$$

The rabbit sequence is also related to the number-theoretic *Beatty sequences* [30.1], defined as follows: if $w < 1$ is a positive irrational, then the two sequences

$$a_m = \lfloor \frac{m}{w} \rfloor , \quad m = 1,2,3, \ldots \tag{30.7}$$

$$b_m = \lfloor \frac{m}{1-w} \rfloor , \quad m = 1,2,3, \ldots , \tag{30.8}$$

together assume every positive integer value exactly once.

This is easy to see, because for the integer k to be a member of the first sequence, a_m, the following inequalities must obtain for some m (with w irrational, equalities need not be considered):

$$\frac{m}{w} - 1 < k = \lfloor \frac{m}{w} \rfloor < \frac{m}{w} , \qquad \text{or}$$

$$m - w < kw < m . \tag{30.9}$$

On the other hand, for k to be a member of the *second* sequence, b_m, the inequalities are the same as (30.9) with $1 - w$ replacing w. Substituting the integer $k - m$ for m, one obtains as the condition for membership in the second sequence that there be some m for which

$$m < kw < m+1 - w . \tag{30.10}$$

Comparing conditions (30.10) and (30.9), one sees that they are precisely complementary: a given integer k fulfills either (30.9) or (30.10) and is therefore a member of either a_m, or b_m. The "densities" of these two sequences are w and $1 - w$, respectively, which add up to 1, as they should if no integer is duplicated or left out.

For a "winding number" (the physicist's term) $w = \gamma = 1/g$, i.e., $1/w = g$ and $1/(1-w) = g^2$, the first sequence becomes

$$a_m = 1,3,4,6,8,9,11, \ldots \quad \text{and the second sequence}$$

$$b_m = 2,5,7,10,13,15,18, \ldots \; .$$

Now a rabbit sequence term r_k equals 1 if k is a member of the Beatty sequence a_m; otherwise (i.e., if k is a member of b_m) it equals 0. — Who can show the equivalence between this rule and (30.3,4)? Hint: use (5.30).

With this kind of long-range order it is small wonder that there are spectral peaks near harmonic numbers $\gamma^2 F_n$ and γF_n. The other peaks visible in the spectrum (Fig. 30.3) result from the self-similarity of the rabbit sequence (scaling factor equal to the Golden ratio 1.618 ...) and the spectral folding ("aliasing") inherent in the Fourier transform of a discrete sequence of finite length.

Beyond the Golden ratio, self-similar sequences and new quasicrystals are generated by the "Silver ratios," $\tau_N^{\pm}$, defined by $\tau_N^{\pm} = N \pm 1/\tau_N^{\pm}$, i.e., all those (quadratic) irrationals that can be expressed by periodic continued fractions with period length one and ± 1 as the numerator. For example, $\tau_2^+ = 1 + \sqrt{2}$ is expected to generate a quasicrystal with the crystallographically forbidden 8-fold symmetry, while $\tau_4^- = 2 + \sqrt{3}$ would underlie a likewise forbidden 12-fold symmetry axis. In addition to the Golden ratio $g = \tau_1^+$, all $\tau_N^{\pm}$, where N is the nth Lucas number (see Sect. 5.8) and the sign superscript equals $(-1)^n$, generate quasicrystals with a 5-fold symmetry!

In another generalization, the *noble numbers,* recently defined as all those numbers whose continued-fraction expansions end in infinitely many 1's, distinguish themselves both in the present case and in the approach to chaos of nonlinear dynamical systems. In this nomenclature the Golden ratio is but the noblest of the noble numbers.

Cassini's divisions in the rings of Saturn are a manifestation of what happens when, instead of noble numbers, simple rationals reign: rocks and ice particles constituting the rings, whose orbital periods are in simple rational relation with the periods of the moons of Saturn, are simply swept out of their paths by the resonance effects between commensurate orbital periods. In fact, the very stability of the solar system depends on the nobility of at least some of the orbital period ratios.

The noble numbers ν, defined by the continued fraction

$$\nu := [a_0; a_1, \ldots, a_n, \overline{1}] , \tag{30.11}$$

can be expressed with the help of the Golden ratio g as follows:

$$\nu = \frac{gA_n + A_{n-1}}{gB_n + B_{n-1}}, \tag{30.12}$$

where A_k and B_k are the numerator and denominator, respectively, of the kth approximating fractions of $[a_0; a_1, ..., a_n]$, see (5.3,4). What is the CF of the noble $(81g + 334)/(73g + 301)$?

A two-dimensional analogue of the new state of matter is the Penrose tiling [30.3] of the plane, see Fig. 30.5. Like the one-dimensional rabbit sequence, it consists of two elements ("arrows" and "kites") that cover the plane without periodic repetition. If we would replace the arrows and kites with 0's and 1's at their centers of gravity, the Fourier transforms of the resulting two-dimensions array would show strong peaks and exhibit the crystallographically forbidden five-fold symmetry apparent in Fig. 30.2. This then is the gist of the paradoxical diffraction pattern shown in Fig. 30.2. For a complete wrap-up of this matter, the three-dimensional case was analyzed by *D. Levine* and *P. J. Steinhardt* [30.4].

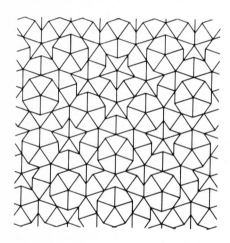

Fig. 30.5. Penrose tiling of the plane with "arrows" and "kites". The five-fold symmetry, forced by the angles of the arrows and kites and their edge-length proportions (equal to the Golden ratio $g = 1.618...$), makes a periodic tiling impossible [30.3]. But there is still a long-range order that is the two-dimensional analog of the recently discovered new state of matter, see [30.2, 4] and Fig. 30.2

30.2 Cantor Sets, Fractals and a Musical Paradox

In the midst of the animated debates on the foundations of mathematics — and the very meaning of the concept *number* — during the last century, Georg Cantor (1845-1918), of set fame, wanted to present his colleagues with a set of numbers between 0 and 1 that has measure zero (i.e., a randomly thrown "dart" would be very unlikely to hit a member) and, at the same time, has *so* many members that it is in fact uncountable (just like *all* the real numbers between 0 and 1).

Many mathematicians, and even Cantor himself for awhile, doubted that such a "crazy" set existed — but it does, and its construction is in fact quite

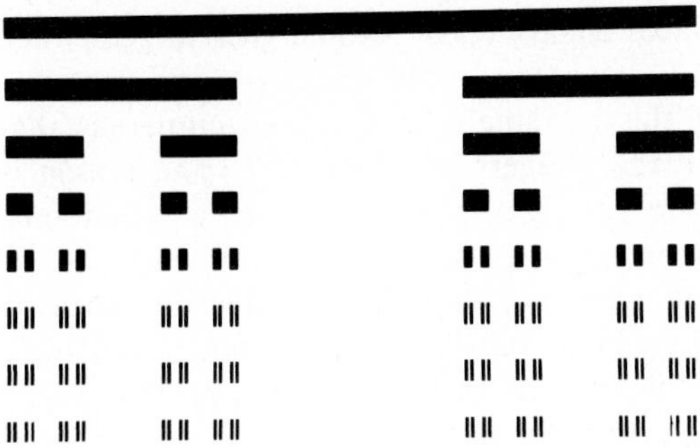

Fig. 30.6. The first steps to the Cantor "dust", the original Cantor set obtained by repeatedly "wiping out" middle thirds of remaining intervals. The resulting set, which can be represented by ternary fractions using only the two digits 0 and 2, has only isolated points and zero measure. Yet its cardinality is as uncountable as that of all the real numbers ("aleph one"). Cantor sets, which are characterized by a (fractal) Hausdorff dimension, play a prominent role in modern nonlinear physics where they describe "strange attractors," i.e., aperiodic orbits in phase space that attract other points from their "basin of attraction". The boundaries of such basins, called Julia sets, are also often fractal (nondifferentiable). The Hausdorff dimension of the original Cantor set equals $d = \log 2/\log 3 = 0.631 \ldots$ [30.5]

straightforward. Imagine the real line between 0 and 1 (drawn with chalk on a blackboard if you will) and wipe out the middle third. Then wipe out each middle third of each remaining third, and so forth *ad infinitum*. The first seven wipings are illustrated in Fig. 30.6, but there is no way to draw the final result, aptly called *Cantor dust* by *B. Mandelbrot* [30.5].

However, there is a neat number-theoretic way to represent the "dust", namely by *ternary* fractions employing the digits 0, 1 and 2. For example, 0.5 corresponds to 0.111 ... in ternary notation. The ternary representation is unique if we forbid periodic fractions ending in a $\bar{1}$ or $1\bar{2}$ and, for example, write 0.0222 ... instead of 0.1 and 0.2 instead of $0.1\bar{2}$.

Now, the numbers in the interval [1/3, 2/3] are precisely all those numbers that in the ternary system have the digit 1 in the first position to the right of the period. Wiping these numbers away on our route to the Cantor dust, we are left with numbers that begin with 0.0 or 0.2.

Similarly, the second wiping (third line in Fig. 30.6) eliminates all numbers with a 1 in the *second* place to the right of the period. In the end, having arrived at the Cantor dust, we are left with all those numbers that have no 1's in *any* place.

Incidentally, the members of the Cantor set form a *self-similar* set: take any line in Fig. 30.6, leave out the left or right half and magnify the remainder three-fold. This results in the line immediately above it. More precisely, the Cantor set is invariant, modulo 1, to scaling by a factor 3. In fact, in the ternary notation, this scaling is nothing but a left-shift of all digits and dropping any 2's that protrude to the left of the period. For example, the Cantor number 0.202202 maps into 0.02202, another Cantor number.

Now it is easy to see why the Cantor set has measure zero: the probability that a random digit in [0,1) has not a single 1 in its ternary expansion is of course zero. More precisely, the expression for the probability that there is no 1 in n ternary places equals $(2/3)^n$, which goes to zero when n goes to infinity.

But why are the members of this extremely "thin" set so numerous that they are not even countable (like the integers or rationals)? The reason is that we can bring the members of the Cantor set in a one-to-one correspondence with *all* the real numbers in the interval between 0 and 1. We simply identify with each Cantor number the *binary* number obtained by changing all 2's to 1's. Thus, for example, 0.020222 corresponds to 0.010111 (=23/64). In this manner, each member of the Cantor set can be mapped into a real number and, conversely, *all* reals between 0 and 1 can be mapped into Cantor numbers, which have thus the same "cardinality" as the reals.

The well-known fact that the reals form an *uncountable* set was proved by Cantor using the "diagonal method" already known to Galileo. The diagonal method is used in an indirect proof that proceeds as follows.

Assume that all the numbers between 0 and 1 form a countable set; they could then be written down, one after another, in the counting sequence. In binary notation, for example, the list might look as follows:

    0.01011 ...
    0.11110 ...
    0.01011 ...
    0.10001 ...
    etc.

Now write down a number whose first digit to the right of the period is the complement of the corresponding digit in the first number in the above list, and whose second digit is the complement of the second digit in the second number, and so forth. The resulting number, which begins

    0.1011

cannot be found anywhere in the above list because it differs from each number in at least one place. Thus, the list cannot be complete and our assumption that the reals form a countable set was false.

But how can we quantify by some *finite* number a point set that has measure zero and cardinality "uncountable"? This we often need to do in physical applications, where Cantor sets may characterize the members of "strange attractors" in nonlinear dynamical systems that revel in deterministic chaos (see Sect. 30.6).

Here the concept of the *Hausdorff dimension,* named after the German mathematician F. Hausdorff (1868-1942), comes to our rescue. For a self-similar set, such as the Cantor set, the Hausdorff dimension d is particularly simple to calculate. We simply look at that part of the set that falls within a "length" one rth the original set, where r is the scaling factor of the self-similarity, and ask what fraction f of the set falls into that portion. Then the Hausdorff dimension is defined as

$$d := \frac{\log(1/f)}{\log r}, \tag{30.13}$$

often a *noninteger* value.

For a straight line f equals of course $1/r$ and $d = 1$, as one would expect. For the Cantor set $f \neq 1/r$; in fact, $r = 3$ and $f = 1/2$. Thus, $d = 0.6309 ...$. The Cantor set having a $d < 1$ reflects its "thinness", its dustlike consistency, compared to a piece of coherent line.

If we calculate the Hausdorff dimension of another self-similar set, the Hilbert curve (Fig. 30.7), then we find for $r = 2$, $f = 1/4$. Thus, $d = \log 4/\log 2 = 2$, which is nice considering the fact that (in the limit) the Hilbert curve goes through every point in the unit square. (But note the monstrous result that a *curve,* something we have always thought of as having just *one* dimension, has now dimension $d = 2$.)

Fig. 30.7. Intermediate stage, on the way to the plane-filling Hilbert curve. The Hausdorff dimension of the Hilbert curve equals $d = \log 4/\log 2 = 2$, as behooves a curve that, in the limit, touches every point in the unit square. The construction of the Hilbert curve accomplishes a *continuous* mapping of the (two-dimensional) unit square onto a (one-dimensional) line, in contrast to the mapping by alternating binary digits, which is discontinuous

The Hausdorff dimension even allows us to salvage a basic Euclidean theorem that otherwise would go to pieces when applied to fractal objects. Look at Fig. 30.8: it shows a hexagon surrounded by 6 identical hexagons. The total area of the figure is 7 times the area of a hexagon, and its perimeter is 3 times that of a hexagon. Note that the total figure is *not* a hexagon, i.e., the figure is not similar to its parts.

But we can make it so! The first step is to break up each straight line in Fig. 30.8 to be similar to one third of the perimeter of the entire figure. For this purpose we break up each straight line by a zig-zag of three straight lines, each having $1/\sqrt{7}$ length, see Fig. 30.8. By repeating this breaking up

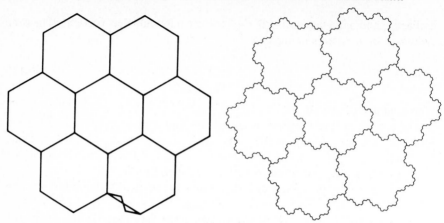

Fig. 30.8. Seven hexagons. Hexagons tile the plane, but they can never form a hexagon again (as opposed to tiling by squares, for example). This tiling defect can be remedied by breaking up each of the 30 straight lines (of unit length) in the figure into 3 straight-line segments, each of length $1/\sqrt{7}$ and repeating the process *ad infinitum*. The result of the first breaking up is illustrated on one edge

Fig. 30.9. The seven hexagons after the first five edge-mutilating steps. The final figure is similar to its seven parts and has a nondifferentiable fractal boundary (Hausdorff dimension $d = \log 3/\log\sqrt{7} = 1.129 \ldots$). The perimeter of the large "hexagon" is 3 times as long as that of a small "hexagon". Yet its area is not $3^2 = 9$ but only 7 times as large: a "non-Euclidean" paradox! (See text.)

process infinitely often, we arrive at a figure, approximated by Fig. 30.9, that is exactly similar to its 7 parts. The area ratio is obviously 7 to 1 and the perimeter is 3 times the perimeter of one of the 7 similar pieces.

Now, following Euclid, areas of similar figures are proportional to their linear dimensions squared. But 3^2 equals 9 and not 7. Where did we go astray? Has Euclid finally been proven wrong??

Well, the old Greeks (with the possible exception of Zeno) can continue to rest in peace. Geometric objects like the one illustrated (imperfectly) in Fig. 30.9 were never on exhibit in Euclid's school. Its boundaries, while everywhere continuous, are nowhere differentiable. The object is a *fractal,* so named by *Mandelbrot* [30.5], with an easily calculable Hausdorff dimension. Since the scaling factor equals $\sqrt{7}$ by construction and f equals 1/3, we obtain with (30.13) for the dimension the following noninteger value:

d = 1.12915 ...

Although noninteger, the appellation "dimension" for d is quite apt, because with it we can reformulate Euclid's theorem about similar areas and obtain a more generally valid result, applicable to fractals and nonfractals alike:

Theorem: FOR   SIMILAR   FIGURES,   THE   RATIOS   OF CORRESPONDING MEASURES ARE EQUAL *WHEN REDUCED TO THE SAME DIMENSION*.

Thus, the area ratio in Fig. 30.9 is 7. Since area has dimension 2, reduction to dimension 1 results in $7^{\frac{1}{2}} = 2.64575\ldots$. and the perimeter ratio 3 when reduced to dimension 1 gives the same number. To wit: $3^{1/d} = 3^{1/1.12915} = 2.64575\ldots$. It is because of properties like this that the Hausdorff dimension is such a useful concept. It is just an extension of the concept of dimension to fractal values, which model many phenomena in the real world.

In Figure 30.9 we have encountered a function, the boundary of the figure, that is everywhere continuous and nowhere differentiable. Such exotic functions were defined by Karl Weierstrass (1815-1897) to show an unbelieving world one of the "impossible" things that could happen with what is innocently called "function". And like Cantor sets, Weierstrass functions are a rich mine of common-sense paradoxes.

For example, a musical chord patterned after a Weierstrass function can have the following weird property. If recorded on magnetic tape and replayed at *twice* the recording speed, the chord will not sound an octave higher in pitch (as every well-behaved recorded sound would) but a semitone *lower*. How is *this* possible?

Let us construct a simple Weierstrass function

$$w(t) = \sum_{k=-\infty}^{\infty} \cos(\beta^k t) . \tag{30.14}$$

If we scale the time dimension t by a factor β, we obtain

$$w(\beta t) = \sum_{k=-\infty}^{\infty} \cos(\beta^{k+1}t) = \sum_{k=-\infty}^{\infty} \cos(\beta^k t) , \tag{30.15}$$

i.e., $w(\beta t)$ *equals* the unscaled function $w(t)$. Thus, $w(t)$ is self-similar. Obviously, such a function cannot have a nonzero derivative anywhere, because derivatives change with scaling. (It is interesting to note that the concept of self-similarity entered mathematics at two independent points, Cantor sets and Weierstrass functions, at about the same time in history and for similar reasons: to elucidate the foundations of mathematics: numbers and functions.)

Now suppose we select $\beta = 2^{13/12}$ and convert $w(t)$ to audible sound:

$$w(t) = \sum_{k} \cos(2^{k\,13/12}t) , \tag{30.16}$$

where k only has to cover the audio range (20 Hz to 20 000 Hz). Playing back $w(t)$ as constructed in (30.16) at twice the speed produces

$$w(2t) = \sum_{k} \cos(2^{k\,13/12+1}t) = \sum_{k'} \cos(2^{k'\,13/12}\,2^{-1/12}t) , \tag{30.17}$$

where $k' = k+1$. Now, if the summations in (30.17) cover the entire audio range, then, as far as the human ear is concerned,

$$w(2t) = w(2^{-1/12}t) .$$ (30.18)

Thus, a doubling of the tape speed will produce a sound with a pitch lowered by a factor $2^{-1/12}$, the chord will sound one semitone *lower* rather than an octave higher!

P. Meyer at the Drittes Physikalisches Institut, University of Göttingen, programmed a desk-top computer to produce a $w(t)$ as in (30.16) with 11 components comprising the frequencies from 10.0 Hz to 18245.6 Hz. By doubling the playback speed the 6th component, for example, will change in frequency from 427.15 Hz to 854.3 Hz. But in comparing the two chords, the human auditory system will identify the doubled 6th component at 854.3 Hz with the *nearest* component of the original chord, namely the 7th at 905.1 Hz. Since 854.3 Hz is a semitone lower than 905.1 Hz, a lowered pitch will be perceived. An astounding paradox, when listened to, but quite easily accounted for.

30.3 The Twin Dragon: a Fractal from a Complex Number System

One of the most widely used number systems in the digital world is the binary system, based on the base 2. In this system, a real number is represented by a sum of powers of 2:

$$x = \sum_k \epsilon_k \, 2^k ,$$ (30.19)

where the binary digits ϵ_k are either 0 or 1; other values are not needed.

To extend the binary system to *complex* numbers z we could represent both real and imaginary parts by sums like (30.19). But there is a more elegant method: we select $1-i$ as a base and write

$$z = \sum_k \delta_k \, (1-i)^k .$$ (30.20)

What values for the digits δ_k do we need to cover all z in a given region of the complex plane? Surprisingly, values of δ_k that are either 0 or 1 suffice, just as in the binary system (30.19) for reals. Thus, (30.20) is a *binary* number system for complex numbers!

What area of the complex plane is occupied by the proper fractions, i.e., all those z for which $\delta_k = 0$ for $k \geqslant 0$? We should not have asked, because the answer is monstrous: the region shown in Fig. 30.10. It has a fractal boundary with surfeit of self-similarities. It is known as the *Twin Dragon*, and its perimeter has Hausdorff dimension $d = 1.5236$, a solution of $2^{3d/2} - 2^d = 2$ [30.5].

Fig. 30.10. A number-theoretic fractal monster: Donald Knuth's "Twin Dragon" [27.4, 30.5], defined as the loci in the complex plane of all proper fractions in the binary number system with the base $1-i$. The Twin Dragon tiles the plane, is similar to its parts, and has a fractal perimeter with Hausdorff dimension $d = 1.5236 \ldots$, but is otherwise harmless

The Twin Dragon decomposes into similar dragons, indicated by different shadings in Fig. 30.10. Conversely, Twin Dragons tile the plane, but rather perversely, considering their fractal skin.

The physically attractive feature of the Twin Dragon is the fact that *strange attractors* in nonlinear dynamical systems often have "basins of attraction" whose boundary is fractal, much like the dragon skin. Attractors characterize stable orbits. They can have a period of 1, in which case they are called fixpoints, or they can be periodic with periods typically equal to 2, 4, 8, 16, etc., depending on the degree of nonlinearity of the system.

As the nonlinearity is increased, the "bifurcation cascade", i.e., the splitting of one orbit into two, continues until the "period" of the attractor is infinite, in which case it is no longer periodic. Initial conditions within the basin of attraction are attracted to a "thin" set of points, a Cantor set (see Sect. 30.6).

30.4 Statistical Fractals

The fractal objects we have encountered so far are strictly self-similar in a *deterministic* way. But the concept of self-similarity can also apply to statistical laws, resulting in statistical fractals. In fact, Brownian motion, misinterpreted for so long, is a statistical fractal *par excellence* (and so are stock market prices). As is well known, the mean-square displacement r^2 of a particle suspended in a warm liquid at rest is proportional to time t:

$$\overline{r^2} = 2Dt \ ,$$

where D is the "diffusion constant". If we sample the motion every t seconds and connect the sample values of $r(t)$ by straight lines, we obtain a rather jagged picture. But this picture has almost nothing to do with the actual motion of the particle, which changes direction upon every impact of the liquid's molecules, perhaps every billionth of a billionth of a second. Indeed, if we sample the path of the Brownian particle 100 times as frequently and magnify the result by 10 diameters, we obtain (statistically) the same picture as before.

This process of more frequent sampling and greater magnification can be repeated many times; it will always give a picture that is statistically indistinguishable from the original. And this is precisely what is meant by *statistical self-similarity*. For the Brownian motion, it spans the range from the mean-free path of the particle to the dimensions of the container — a great many orders of magnitude. The Hausdorff dimension equals, by definition (30.13), $d = 2$. In other words, like the Hilbert curve, see Fig. 30.7, the Brownian motion curve has the same dimension as a two-dimensional surface (which is why a chemical reaction on a surface, governed by a two-dimensional random walk, proceeds much faster than the same reaction in three dimensions — something nature discovered during evolution and still employs to advantage in enzymatic reactions).

For those readers who are disappointed that the dimension of the Brownian motion is an integer ($d = 2$), I will mention the fact that the set of its "zero-crossings" has fractal dimension $d = 1/2$. This "thinness" of zeroes is a well-known fact for random walks ("Wiener processes"), but it is hardly ever mentioned in terms of fractal dimensions and Cantor-like sets. (A numerical model for a Cantor set with $d = 1/2$ is given by the quarternary fractions in which only two nonadjacent digits, 1 and 3 say, are allowed.)

It is interesting to recall that correct analysis of Brownian motion established the existence of atoms and molecules of finite size, the diffusion constant D being critically dependent on *atomic* dimensions. And who laid the foundation stone? One Einstein (in 1905).

Figure 30.11 shows a statistical fractal obtained by computer simulation of a diffusion-limited aggregation process [30.6]. Each dot represents an atom or molecule that attached itself to the existing structure following simple probabilistic rules that depend on occupancies of nearby sites. Such statistical fractals, as we saw in the case of Brownian motion, can be characterized by a fractal Hausdorff dimension. This dimension is often determined numerically by counting the number N of atoms inside spheres of different radii R. In

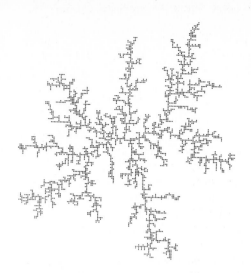

Fig. 30.11. Result of a computer simulation of a two-dimensional diffusion-limited aggregation process, see *Witten* and *Sander* [30.6], a statistical fractal with Hausdorff dimension $d \cong 1.7$, i.e., the number of molecules inside a circle of radius r grows with $r^{1.7}$ (instead of r^2). Two-dimensional electric discharges ("Lichtenberg figures") have similar structures and Hausdorff dimensions. Three-dimensional lightnings (worth looking at) have Hausdorff dimension $d = 2.4$ or greater

two dimensions, for a nonfractal substance, N is proportioned to R^2. But in Fig. 30.11 proportionality of N with $R^{1.7}$ was found. Thus, the fractal has Hausdorff dimension $d = 1.7$.

Electrical discharges on insulators obey laws very similar to diffusion-limited aggregation, and the resulting spark patterns (called Lichtenberg figures, after the 18th-century Göttingen physicist and aphorist who studied them) look indeed very much like Fig. 30.11.

Simulation of such scale-invariant processes in three dimensions give fractal Hausdorff dimensions between 2 and 3. The most beautiful (highly branching) lightnings, for example, have Hausdorff dimensions near 2.4. (Have insurance companies become aware and taken note of this?)

Another important application of fractal dimensionality lies in the analysis of time series of a wide spectrum of natural phenomena. Are long-time temperature variations [30.7] on the earth or other aspects of climate evolution describable as fractal phenomena? Can we analyze brain waves (electro-encephalograms, EEGs) and deduce insights into the state of mind of a patient [30.8] from the Hausdorff dimension of the EEG? Perhaps such fractal evaluation will exceed in diagnostic value even linear predictive analysis [30.9].

30.5 Some Crazy Mappings

A long time ago, Isaac Newton suggested finding the zeroes of a function $f(z)$ by an iteration based on tangent extrapolations. Given an approximate value of z_n to the solution of $f(z) = 0$, Newton finds the next approximation by calculating

$$z_{n+1} = z_n - \frac{f(z_n)}{f'(z_n)} \; , \tag{30.21}$$

where $f'(z_n) \neq 0$ is the derivative of $f(z)$.
   For $f(z) = z^2 - 1$, (30.21) reads

$$z_{n+1} = \frac{1}{2} \left[z_n + \frac{1}{z_n} \right] . \tag{30.22}$$

Not surprisingly, for an initial value z_0, with a positive real part, z_n converges toward the positive root of $z^2 - 1 = 0$, namely $z = 1$. Similarly, for a negative real part of z_0, the solution converges to the negative root, $z = -1$.
   But what happens for purely imaginary $z_0 = i r_0$, where $r_0 \neq 0$ is real? Interestingly, it does not converge at all; the iteration (30.22) cannot make up its mind, so to speak, and hops all over the imaginary axis, according to the mapping

$$r_{n+1} = \frac{1}{2} \left[r_n - \frac{1}{r_n} \right] . \tag{30.23}$$

For example, $g = 1.618 \ldots$ maps into 0.5, -0.5, $0.291\overline{6}$, $-1.56845 \ldots$, etc. But some r_0 behave differently, such as $r_0 = 1 + \sqrt{2}$, which maps into 1, 0, ∞, a kind of fixed point, if rather distant.
   How can we bring order into this chaotic mapping? A trigonometric substitution will do:

$$r = -\cotan(\alpha \pi) \; , \tag{30.24}$$

which turns (30.23) into the simple

$$\alpha_{n+1} = 2\alpha_n \quad \mathrm{mod} \; 1 \; . \tag{30.25}$$

Now everything becomes transparent. If we express α_n as a binary fraction, then α_{n+1} will be simply the fractional part of α_n shifted one place to the left. Thus, periodic binary α_0 will lead to periodic orbits. For example, $\alpha_0 = 1/3 = 0.\overline{01}$ will map into $0.\overline{10} = 2/3$, which will map right back into $\alpha_2 = 0.\overline{01} = \alpha_0$. Indeed, $r_0 = -\cotan(\pi/3) = 1/\sqrt{3}$ maps into $r_1 = \sqrt{3}$, which maps back to $r_2 = 1/\sqrt{3} = r_0$.
   We also see by this analysis that any irrational α_0 will lead to an aperiodic "orbit" along the imaginary axis in the z-plane. This leaves us with the nonperiodic rationals. Where will they end up? Continued left-shifting and taking fractional parts will, sooner or later, produce 0 which maps into ∞. For example, $\alpha_0 = 7/8 = 0.111$ will map into $0.11 = 3/4$, then $0.1 = 1/2$,

which maps into 0. In fact, the corresponding $r_0 = -\cot(7\pi/8) = 1 + \sqrt{2}$ maps into 1, 0, ∞, as we already saw.

Thus, the simple mapping given by Newton's iteration for the function $z^2 - 1$ has rather strange consequences for initial values on the imaginary axis: numbers are classified into

  aperiodic binary rationals,
  periodic binary rationals,

and

  irrationals.

While aperiodic binary rationals "converge" on a fixed point (∞), periodic binary rationals lead to periodic orbits. By contrast, irrational numbers α_0, an uncountable set, give aperiodic orbits: the same value never occurs twice, nor is α_n ever rational.

Thus, we have the curious fact that the tiniest distinction — the difference between rational and irrational numbers — makes a decisive difference in the final fate of a situation. Ordinarily, one should think that physics, and certainly the *tangible* world at large, are untouched by the purely mathematical dichotomy between rationals and irrationals. But in reality this just is not so. While, true enough, everything in the real world can be adequately described by rational numbers, it so happens that a *modeling* that distinguishes between different kinds of numbers is not only extremely useful but catches the true (perhaps hidden) spirit of a physical phenomenon.

More specifically, two different initial conditions that are completely indistinguishable by any finite measurement precision will sooner or later lead to a total divergence as the system evolves in time or space. The essential condition for this to happen is that the corresponding Poincaré map (see Sect. 30.6) be sufficiently *nonmonotonic,* such as (30.25), which has a sawtooth-like nonmonotonicity. The rate of divergence is measured by the so-called *Lyapunov exponent* λ.

In our bare-bones example (30.25), $\lambda = \log(\alpha_{n+1}/\alpha_n)$ and taking natural logarithms gives $\lambda = \log 2 \approx 0.693$.

Newton's method applied to $f(t) = z^3 - 1$ gives

$$z_{n+1} = \frac{2z_n^3 + 1}{3z_n^2} \qquad (30.26)$$

Now the mapping is even crazier. For one thing, the conjecture that all z_0 will converge toward the closest of the three roots is false. For example, $z_0 = -1$ will converge on 1, the root most distant from it.

Fig. 30.12. The iteration $z_{n+1} = (2z_n^3 + 1)/3z_n^2$ (Newton's method for finding the roots of $z^3 = 1$). The three different basins of attraction are indicated by 3 different colors. Whenever two colors meet, the third is also present − a seemingly impossible feat. The meeting points form a Cantor-like, fractal, Julia set, consisting of the origin ($z = 0$) and all of its preimages. The different "annular" zones indicate the number of iteration steps necessary to land in one of the (half black) disks surrounding each of the three attractors. (The black portion of each "ring" lands in the black half of the disk.) Each ring has twice as many black parts as its image ring, illustrating the increasingly complex structure of the iteration near the Julia boundary. Because the map is continuous for $z \neq 0$, the figure shows a great deal of near self-similarity around $z = 0$ and its three preimages, $z = (-0.5)^{1/3}$, their nine preimages, and so forth, *ad infinitum*. While periodic repetition was (and is) one of the artistic staples of artistic expression, periodicity on a *logarithmic* scale, i.e., *self-similarity* is now coming into its own − thanks to modern computer graphics which produced this figure entitled *Orchis Tertiarum* by Forschungsgruppe Komplexe Dynamic (H.-O. Peitgen, P. H. Richter), Universität Bremen

Figure 30.12 shows the intimately intertwined "basins of attraction" of the three roots — a real, or rather imaginary, crazy quilt. In fact, it can be shown that for the mapping (30.26) two basins (colors) can never meet without the third one being present also. This may sound impossible, and in fact it would be — if it were not for fractal boundaries as intimated in Fig. 30.12 and already encountered in the dragon skin (Fig. 30.10).

One way to begin to understand the basins of attraction in Fig. 30.12 is to consider the neighborhood of $z = 0$. By (30.26) the point $z = 0$ maps of course to ∞. But points near 0 participate in a devil's dance before approaching one of the three attractors (or being mapped into 0). In fact, for $|z_0| \ll 1$, $z_1 \cong 1/3z_0^2$, a large number. The next iterations make z smaller again: $z_2 \cong 2/9z_0^2$, etc.; it is almost unpredictable what z_n will do as n goes to infinity.

Other "dangerous" points are the three points $z_0 = (-1/2)^{1/3}$, which land on 0. Since the mapping (30.26) is continuous, the devil's dance around 0 will be mimicked around these three points, as can be seen in Fig. 30.12. In fact, their 9 "preimages", one of which is real ($z_{-1} = -1.1388 \ldots$), show again the same devilish behavior. And the same dance continues for *pre*preimages and so forth *ad infinitum*. Several "generations" of these prepreimages, which form a fractal *Julia set,* are readily visible in Fig. 30.12. The orbits of any points near them become totally irrational (in the common sense of the word). And since all three basins of attraction meet at 0, they also meet at all the preimages — which is everywhere where different basins meet: two basins never meet without the third.

More precisely, the Julia set that forms the boundary between the different basins is the set of all those z that map eventually into 0; it corresponds to the nonperiodic rationals for the map $f(z) = z^2 - 1$ discussed above. And any point z_0 near a member of the Julia set will be mapped in one or several steps into a point near zero, where the *real* dance begins.

The Julia set is self-similar, as is apparent in Fig. 30.12 from the similarity of the neighborhood of 0 with its preimages. The Julia set is also uncountable and has a fractal Hausdorff dimension between 1 and 2, which in the plane is pretty thin dust — but the dust is there; in fact it holds Fig. 30.12 together.

30.6 The Logistic Parabola and Strange Attractors

Newton's map is not the only map with pathological consequences. Even such a simple mapping law as the "logistic parabola"

$$x_{n+1} = rx_n(1-x_n) \tag{30.27}$$

show very strange behavior indeed. Equation (30.27) models the growth (or death) of a population. Without the factor $(1-x_n)$ and for $r > 1$, the population would grow geometrically. But if resources (think of food) are limited, so is the growth, as modeled by the factor $(1-x_n)$.

For $1 < r < 3$, (30.27) has a unique nontrivial limit x^*. Any $x_0 \neq 0$ or 1 will be attracted to x^*. For example, for $r = 2$ the attractor x^* equals 0.5. With $x_0 = 0.8$ (30.27) will yield 0.32, 0.435, 0.492, 0.4998 ... which converges on the attractor $x^* = 0.5$.

But for $r > 3$, the attractor is no longer stable. For example, with $r = 3.2$ and $x_0 = 0.8$, (30.27) gives 0.512, 0.7995, 0.51288,...,0.799455, 0.513044. Now there is no longer a single-valued attractor; x^* has *bifurcated* into an orbit of period 2, also called a periodic attractor characterized by two values of x: 0.513... and 0.799... . Other nontrivial values of x_0 are attracted to this pair of values.

If we keep increasing the "nonlinearity parameter" r the attractor will bifurcate again and have a period of 4. Further increases of r will lead to periods 8, 16, 32, etc. until, at a critical value r_∞, the period has become infinite. In other words, the orbit is now *aperiodic,* comprising a point set of infinitely many values of x that never repeat. However, other values of x are attracted to this point set, which is in fact a Cantor set with Hausdorff dimension d a little over 0.5. Such attractors have been called *strange,* although once one knows about Cantor sets they are really not so strange after all.

In fact, we even encounter the self-similarity again that we first met (Sect. 30.2) when discussing the original Cantor set with $d \cong 0.631$. The reason is that for the logistic parabola (and many other mapping laws) the iterated map $f^{(2)}(x) := f(f(x))$, where $f(x) = rx(1-x)$, has again parabolic regions similar to that of $f(x)$ itself. And the same is true for $f^{(4)}(x)$, $f^{(8)}(x)$, etc. In fact, the important regions of $f^{(2^k)}(x)$ become increasingly similar to each other as k increases. This leads to a self-similar scaling law for the parameter values r_k at which bifurcation takes place: the ratios of successive parameter intervals

$$\frac{r_k - r_{k-1}}{r_{k+1} - r_k}$$

converge to a *universal* constant, the famous (and probably transcendental) *Feigenbaum constant* 4.6692... (originally found by S. Grossmann and S. Thomae). This magic number has earned the epithet "universal" because it applies to many different nonlinear laws, independent of the details of the mapping, as long as the mapping has a parabolic maximum [30.10].

The logistic parabola becomes even more exciting if we allow the parameter r in (30.27) to assume complex values. If we now plot in black all those values in the r-plane for which (30.27) shows a simple periodic behavior and the rest of the r-plane in different shades (depending on the rapidity of divergence in the iteration), we obtain a fractal image with many self-similarities, dominated by a funny looking, rotund little being called "Apfelmännchen" in German, and Mandelbrot set more formally.

Magnifying ever smaller portions of the r-plane, we will discover our model "Apple Manikin" again and again, hiding in every nook and cranny of the complex plane. One such blowup is shown in our last illustration, Fig. 30.13. We shall have to leave the Apfelmännchen and a detailed discussion of the Julia world it (he?) inhabits for another book, in which the many uses (including artistic) of self-similarity might be the main theme.

It is interesting to note that the simple logistic parabola models the behavior of many nonlinear physical systems, including some with rather complex dynamics. The reason is that the often inconceivably complicated orbits in phase space of such systems are adequately described for many purposes by a "cut" through phase space, known as a *Poincaré section* after the famous French mathematician Henri Poincaré (1854-1912). This ingenious trick converts the dynamics into a *mapping* that is much easier to study than the original problem. Especially the chaotic behavior shown by such complex systems is faithfully modeled by the aperiodic behavior observable in such simple equations as the logistic parabola.

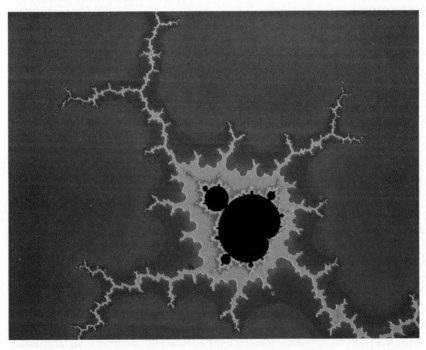

Fig. 30.13. One of the uncountably many *Apfelmännchen* inhabiting the complex plane (and sprouting smaller "Apple Manikins" on its skin). Computer graphic by Forschungsgruppe Komplexe Dynamic (H.-O. Peitgen, P. H. Richter), Universität Bremen

Just as for Newton's method applied to the solution of $z^2 - 1 = 0$, see Sect. 30.5, the logistic parabola, too, shows aperiodic behavior that can be understood as left-shifts of binary digits and the taking of fractional parts, as expressed by equation (30.25). In fact, the *deterministic chaos*, which dominates many a physical discussion these days, is ultimately understandable in these simple terms: the evolution of a complex system, in the final analysis, is akin to the left-shifting of digits of a real irrational number. No matter how precisely the initial conditions are known, temporal evolution will sooner or later bring the originally unknowable, far-right, digits (representing the exact initial conditions) into the neighborhood of the "decimal" point, where they will dominate the system's behavior, which will then seem chaotic.

This has nothing to do with quantum uncertainty or Planck's constant; unpredictability or "chaos" prevails even for completely deterministic systems, which is why it is called *deterministic* chaos. It reflects the fact that initial conditions, no matter how precisely specified, can never be given with an infinite precision, as represented by a real irrational number with infinitely many nonperiodic digits.

This, in a nutshell, is what modern chaos is all about. (The mythical primeval Chaos has not yet been studied in any great detail and may be quite another matter.)

30.7 Conclusion

We live in an age of increasing overlap between the traditional scientific disciplines. Few are the advances in modern technology — and even basic scientific understanding — that rely on a single field of learning. To give a few examples, our understanding of the physical universe, which itself impinges on fundamental philosophical questions of the beginning and the end of being, depends on progress in astronomy, which in turn results from innovations in computer processing, image enhancement, and a host of other new technologies. Our comprehension of the genetic code is influenced by information theory as well as by physical chemistry, which in turn is based on quantum mechanics, including the exclusion principle. Elementary particles and the even more elementary quarks, not to mention atoms, molecules and crystals, are best described by universal principles of symmetry as embodied in mathematical group theory.

The applications of the theory of numbers and finite fields touched upon in this volume — the design of concert halls, precision measurements, error-correcting codes, secret communication, special waveforms and radiation patterns, efficient diffraction gratings, pseudorandom numbers, deterministic chaos and even artistic designs and more playful things — can only provide a flavor of the interdisciplinary endeavors already undertaken, and of those yet to come.

Appendix

A. A Calculator Program for Exponentiation and Residue Reduction

We list here a program for calculating the least nonnegative remainder of a^n modulo m on a Hewlett-Packard 41C or 41CV pocket calculator. In other words, in our acute bracket notation, the calculator calculates

$$\langle a^n \rangle_m \; .$$

The 3 variables, the base a, the exponent n and the modulus m are entered into the calculator in that order.

To call the program, which is labeled "AN", from storage, one presses

GTO   "AN" .

To calculate, say,

$$\langle 2^{340} \rangle_{341} \; ,$$

one proceeds by pressing the following buttons:

2

ENTER

340

ENTER

341

To start the program, one presses

R/S

After 2 seconds the HP 41 display shows in rapid succession

8.
6.
4.
2.

This is the binary decomposition of the exponent 340.

Check: $2^8 + 2^6 + 2^4 + 2^2 = 340$. Check!

Next, the computer will start the necessary repeated squarings and reductions modulo 341, which will take about 7 seconds. Then the display will show the end result, a 1 (without a decimal point!).
Thus,

$$\langle 2^{340}\rangle_{341} = 1 \ .$$

In other words, 341 is a pseudoprime to the base 2.

Similarly, after pressing 2, ENTER, 170, ENTER, 341, R/S, the display shows in succession:

7.   5.   3.   1.   1

where that last display (the 1 without the decimal point) tells us that

$$\langle 2^{170}\rangle_{341} = 1 \ .$$

Proceeding in the same manner, we find

$$\langle 2^{85}\rangle_{341} = 32 \ ,$$

i.e., 341 is not a *strong* pseudoprime to the base 2. Also, by using 3 as a base we find

$$\langle 3^{340}\rangle_{341} = 56 \ .$$

Thus, 341 is certainly not an *absolute* pseudoprime (Carmichael number).

However, for the modulus 2821, we find

$$\langle 2^{2820}\rangle_{2821} = \langle 3^{2820}\rangle_{2821} = 1 \ ,$$

two of the many steps necessary to show that 2821 is an absolute pseudoprime or a Carmichael number.

Note that our little calculator with a limited accuracy and a 10-digit display, in calculating $\langle 3^{2820}\rangle_{2821}$, has coped with a number having 1346 decimal digits! This has been made possible by the frequent intermediate modulo reduction that the program employs.

Listing for "AN"

Comment	Step	Code
call program	01	LBL "AN"
decimal point	02	SF 29
store modulus	03	STO 18
get exponent	04	RDN
store exponent	05	STO 17
get base	06	RDN
store base	07	STO 16
constant	08	2
store 2	09	STO 01
take logarithm	10	LN
store log	11	STO 15
display no fractions	12	FIX 0
constant	13	1
store 1	14	STO 14
constant	15	0
store 0	16	STO 02
subroutine for calculating	17	LBL 10
binary representation of n	18	RCL 17
	19	LN
	20	RCL 15
	21	/
add small constant to	22	0.000000001
avoid inaccurate rounding	23	+
	24	INT
	25	ENTER
	26	ST − IND 01
display the binary	27	PSE
exponents of n	28	1
	29	ST+ 01
	30	RDN
	31	STO IND 01
	32	2
	33	$x < > y$
	34	y^x
	35	ST- 17
	36	RCL 17
binary representation	37	$x = 0?$
of n completed?	38	GTO 11
	39	GTO 10
subroutine for executing	40	LBL 11

Comment	Step	Code
repeated squaring	41	RCL IND 01
	42	$x = 0$?
	43	GTO 12
	44	RCL 16
	45	RCL 18
	46	2
	47	/
	48	$x < > y$
	49	$x > y$?
	50	XEQ 16
	51	x^2
	52	RCL 18
take remainder	53	MOD
modulo n	54	STO 16
	55	1
	56	ST − IND 01
	57	GTO 11
subroutine for cal-	58	LBL 12
culating intermediate	59	RCL 16
squaring results	60	ST∗ 14
	61	RCL 14
	62	RCL 18
	63	MOD
	64	STO 14
	65	1
	66	ST- 01
	67	RCL 01
	68	2
	69	$x = y$?
	70	GTO 13
	71	GTO 11
	72	LBL 16
	73	RCL 18
	74	−
	75	RTN
Subroutine for	76	LBL 13
recalling and	77	RCL 14
displaying residue	78	CF 29
display residue	79	STOP
ready to start over	80	GTO "AN"
	81	RTN
	82	END

B. A Calculator Program for Calculating Fibonacci and Lucas Numbers

To call the program, which is labeled "FN", from storage, press

GTO "FN"

To calculate, for example, the 12th Fibonacci number, F_{12}, press

12
R/S .

After about 1 second one sees the answer:

144,

with a comma at the end. Pressing

R/S

again gives the 12th Lucas number, L_{12}:

322 .

with a period at the end (to distinguish it from the Fibonacci number and to tell the user that the calculator is ready to accept the next index). The calculator is accurate up to F_{39} and L_{37}.

Pressing the multiplication sign yields

$$F_{2n} = F_n \cdot L_n = 46368 \quad (\text{for } n = 12) .$$

This rule comes in handy to calculate F_n for large even n.

The odd-index F_n are calculated by

$$F_{2n+1} = gF_{2n} = 75025 \quad (\text{for } n = 12) ,$$

where $g = 1.618 \ldots$ is the Golden ratio, stored in register 02.

Listing for "FN"

Comment	Step	Code
initialize by calculat-ing two constants	01	LBL "FN"
	02	5
	03	SQRT

Comment	Step	Code
$\sqrt{5}$	04	STO 01
	05	1
	06	+
	07	0.5
	08	*
Golden ratio	09	STO 02
	10	RDN
subroutine for calculat-	11	LBL 11
ing F_n and L_n	12	RCL 02
	13	$x < > y$
	14	y^x
	15	STO 03
	16	RCL 01
	17	/
	18	0.5
	19	+
	20	INT
	21	FIX 0
	22	CF 28
display F_n	23	STOP
	24	RCL 03
	25	0.5
	26	+
	27	INT
	28	SF 28
display L_n	29	STOP
ready to start over	30	GTO 11
	31	END

C. A Calculator Program for Decomposing an Integer According to the Fibonacci Number System

To call the program, press

GTO "FNS"

and enter the integer to be decomposed, for example,

1000 .

Then press

R/S

to start the calculation. The display will show in rapid succession the Fibonacci decomposition of 1000 and the least significant term once more with two zeroes after the decimal point (because this is important in a winning strategy for nim-like games; see Chap. 5):

987.   13.   13.00

More specifically, in such a parlor game, when confronted with a pile of 1000 chips, take 13 away to start your winning strategy. (Rules of the game: No player can take more than twice the number of chips his predecessor just took, but each player must take at least one chip, and the first player may not take all the chips. He who takes the last chip(s) wins.)

Listing for "FNS"

Comment	Step	Code
	01	LBL "FNS"
initialize by calculating	02	5
two constants	03	SQRT
$\sqrt{5}$	04	STO 01
	05	1
	06	+
	07	0.5
	08	*
Golden ratio	09	STO 02
	10	LN
	11	STO 03
	12	RDN
subroutine for	13	LBL 14
storing new integer		
subroutine for	14	STO 04
Fibonacci decomposition		
	15	LBL 15
	16	0
	17	RCL 04
decomposition completed?	18	$x \leqslant y$?
	19	GTO 13
	20	RCL 01
	21	*
	22	LN

Comment	Step	Code
hyperbolic cosine	23	COSH
from MATH 1	24	2
	25	*
	26	LN
	27	RCL 03
	28	/
	29	INT
	30	RCL 02
	31	$x < > y$
	32	y^x
	33	RCL 01
	34	/
	35	0.5
	36	+
	37	INT
	38	FIX 0
display Fibonacci decomposition	39	PSE
	40	ST- 04
	41	STO 05
	42	GTO 12
subroutine for recalling and	43	LBL 13
displaying least significant term	44	RCL 05
	45	FIX 2
display least significant term	46	STOP
ready to start over	47	GTO 14
	48	END

Glossary of Symbols

$a_1 \equiv a_2 \pmod{b}$	a_1 congruent to a_2, modulo b; $a_1 - a_2$ divisible by b
$\mathbf{C}$	the field of complex number
$d(n)$	the number of (positive) divisors of n; $\sigma_o(n)$
$d \mid n$	d divides n; n is a multiple of d; there is an integer q such that $dq = n$
$d \nmid n$	d does not divide n
e	base of natural logarithms; 2.718281828459045 ...
exp{ }	exponential function
F_n	Fermat numbers: $2^{2^n} + 1$. Or Fibonacci numbers
$f(x) = 0(g(x))$	$f(x)/g(x) \to 0$ as $x \to \infty$
$f(x) = 0(g(x))$	there is a constant c such that $\lvert f(x) \rvert < cg(x)$
i	square root of -1; $i^2 = -1$
ln x	natural logarithm of x
(m,n)	GCD (greatest common divisor) of m and n; highest common factor of m and n
$[m,n]$	LCM (least common multiple) of m and n. Also, the block of consecutive integers, m, $m+1$,..,n

M_p Mersenne numbers: $2^p - 1$

$n!$ factorial n;
                                   $1 \times 2 \times 3 \times \ldots \times n$

$\begin{pmatrix} n \\ k \end{pmatrix}$ n choose k; the binomial
                                   coefficient $n!/k!(n-k)!$

$\begin{pmatrix} p \\ q \end{pmatrix}$ or (p/q) Legendre symbol, also fraction

$p^a \| n$ p^a divides n, but p^{a+1}
                                   does not divide n

p_n the nth prime, $p_1 = 2$,
                                   $p_2 = 3$, $p_3 = 5, \ldots$

$\mathbf{Q}$ the field of rational numbers

$r_k(n)$ least number of numbers not exceeding n, which
                                   must contain a k-term arithmetic progression

$\lfloor x \rfloor$ Gauss bracket or floor of x; greatest integer
                                   not greater than x

$\lceil x \rceil$ ceiling of x; least integer
                                   not less than x

$\langle x \rangle_n$ least positive (or nonnegative) remainder
                                   of x modulo n

$\mathbf{Z}$ the ring of integers

$\mathbf{Z}_n$ the ring of integers, 0, 1,
                                   $2, \ldots, n-1$ (modulo n)

γ Euler's constant;
                                   0.577215664901532...

π ratio of circumference of
                                   circle to diameter;
                                   3.141592653589793...

$\pi(x)$	number of primes not exceeding x, also primitive polynomial
$\pi(x;a,b)$	number of primes not exceeding x and congruent to a, modulo b
Π	product
$\sigma(n)$	sum of divisors of n; $\sigma_1(n)$
$\sigma_k(n)$	sum of kth powers of divisors of n
Σ	sum
$\phi(n)$	Euler's totient function; number of positive integers not exceeding n and prime to n
ω	complex cube root of 1, $\omega^3 = 1$, $\omega \neq 1$, $\omega^2 + \omega + 1 = 0$
$\omega(n)$	number of distinct prime factors of n
$\Omega(n)$	number of prime factors of n, counting repetitions

References

Chapter 1

1.1   T. M. Apostol: *Introduction to Analytic Number Theory* (Springer, Berlin, Heidelberg, New York 1976)

1.2   I. Asimov: *Asimov on Numbers* (Doubleday, Garden City, NY, 1977)

1.3   A. O. L. Atkin, B. J. Birch (eds.): *Computers in Number Theory* (Academic, London 1971)

1.4   E. R. Berlekamp, J. H. Conway, R. K. Guy: *Winning Ways* (Academic, London 1981)

1.5   W. Kaufmann-Bühler: *Gauss. A Biographical Study* (Springer, Berlin, Heidelberg, New York 1981)

1.6   P. J. Davis: *The Lore of Large Numbers* (Random House, New York 1961)

1.7   L. E. Dickson: *History of the Theory of Numbers*, Vols. 1-3 (Chelsea, New York 1952)

1.8   U. Dudley: *Elementary Number Theory* (Freeman, San Francisco 1969)

1.9   C. F. Gauss: *Disquisitiones Arithmeticae* [English transl. by A. A. Clarke, Yale University Press, New Haven 1966]

1.10   W. Gellert, H. Küstner, M. Hellwich, H. Kästner (eds.): *The VNR Concise Encyclopedia of Mathematics* (Van Nostrand Reinhold, New York 1977)

1.11   R. K. Guy: *Unsolved Problems in Intuitive Mathematics*, Vol. I, Number Theory (Springer, Berlin, Heidelberg, New York 1981)

1.12   H. Halberstam, C. Hooley (eds.): *Progress in Analytic Number Theory*, Vol. I (Academic, London 1981)

1.13   G. H. Hardy: *A Mathematician's Apology* (Cambridge University Press, Cambridge 1967)

1.14   G. H. Hardy, E. M. Wright: *An Introduction to the Theory of Numbers*, 4th ed. (Clarendon, Oxford 1960)

1.15   L. H. Hua: *Introduction to Number Theory* (Springer, Berlin, Heidelberg, New York 1982)

1.16   K.-H. Indlekofer: *Zahlentheorie*, Uni-Taschenbücher 688 (Birkäuser, Basel 1978)

1.17   K. Ireland, M. Rosen: *Elements of Number Theory* (Bogden and Quigley, New York 1976)

1.18   H. Minkowski: *Diophantische Approximationen* (Teubner, Leipzig 1907; reprinted by Physica, Würzburg 1961)

1.19   T. Nagell: *Introduction to Number Theory* (Wiley, New York 1951)

1.20   C. S. Ogilvy: *Tomorrow's Math* (Oxford University Press, Oxford 1962)

1.21   O. Ore: *Number Theory and Its History* (McGraw-Hill, New York 1948)

1.22   H. Rademacher: *Lectures on Elementary Number Theory* (Blaisdell, New York 1964)

1.23   H. Rademacher, O. Toeplitz: *The Equipment of Mathematics* (Princeton University Press, Princeton 1957)

1.24   A. Scholz, B. Schoenberg: *Einführung in die Zahlentheorie*, Sammlung Göschen 5131 (Walter de Gruyter, Berlin 1973)

1.25   C. E. Shannon: Communication theory of secrecy systems. Bell Syst. Tech. J. *28*, 656-715 (1949)

1.26   W. Sierpiński: *250 Problems in Elementary Number Theory* (American Elsevier, New York 1970)
1.27   J. V. Uspensky, M. A. Heaslet: *Elementary Number Theory* (McGraw-Hill, New York 1939)
1.28   D. J. Winter: *The Structure of Fields*, Graduate Texts in Mathematics, Vol. 16 (Springer, Berlin, Heidelberg, New York 1974)

Chapter 2

2.1   L. H. Hua: *Introduction to Number Theory* (Springer, Berlin, Heidelberg, New York 1982)
2.2   R. Plomp, G. F. Smoorenburg: *Frequency Analysis and Periodicity Detection in Hearing* (A. W. Sijthoff, Leiden 1970)

Chapter 3

3.1   C. Pomerance: The search for prime numbers. Sci. Am. *247*, No. 6, 136-147 (1982)
3.2   W. H. Mills: A prime representing function. Bull. Am. Math. Soc. *53*, 604 (1947)
3.3   T. Nagell: *Introduction to Number Theory* (Wiley, New York 1951)
3.4   D. Slowinski: Searching for the 27th Mersenne prime. J. Recreational Math. *11*, 258-261 (1978-79)
3.5   D. B. Gillies: Three new Mersenne primes and a statistical theory. Math. Comp. *18*, 93-97 (1963)
3.6   G. H. Hardy, E. M. Wright: *An Introduction to the Theory of Numbers*, 5th ed., Sect. 2.5 (Clarendon, Oxford 1984)
3.7   W. Kaufmann-Bühler: *Gauss. A Biographical Study* (Springer, Berlin, Heidelberg, New York (1981)
3.8   C. Chant, J. Fauvel (eds.): *Science and Belief* (Longman, Essex 1981).

Chapter 4

4.1   P. Erdös, M. Kac: The Gaussian law of errors in the theory of additive number theoretic functions. Am. J. Math. *62*, 738-742 (1945)
4.2   P. D. T. A. Elliot: *Probabilistic Number Theory*, Vols. 1-2 (Springer, Berlin, Heidelberg, New York 1980)
4.3   G. Kolata: Does Gödel's theorem matter to mathematics? Science *218*, 779-780 (1982)
4.4   P. Erdös: On a new method in elementary numbers theory which leads to an elementary proof of the prime number theorem. Proc. Nat. Acad. Sci. U.S.A. *35*, 374-384 (1949)
4.5   H. M. Edwards: *Riemann's Zeta Function* (Academic, New York 1974)
4.6   Z. Füredi, J. Komlos: The eigenvalues of random symmetric matrices. Combinatorica *1*, 233-241 (1981)
4.7   M. R. Schroeder: A simple function and its Fourier transform. Math. Intelligencer *4*, 158-161 (1982)
4.8   U. Dudley: *Elementary Number Theory* (Freeman, San Francisco 1969)
4.9   G. H. Hardy, E. M. Wright: *An Introduction to the Theory of Numbers*, 5th ed. Sect. 22.3 (Clarendon, Oxford 1984)
4.10   D. Zagier: "Die ersten 50 Millionen Primzahlen" in *Lebendige Zahlen,* ed. by F. Hirzebruch (Birkhäuser, Basel 1981)

Chapter 5

5.1   C. D. Olds: *Continued Fractions* (Random House, New York 1963)
5.2   H. S. Wall: *Analytic Theory of Continued Fractions* (Van Nostrand, Princeton 1948)
5.3   A. N. Khovanskii: *The Application of Continued Fractions and Their Generalizations to Problems in Approximation Theory* (Noordhoff, Groningen 1963)
5.4   A. Y. Khinchin: *Continued Fractions* (University of Chicago Press, Chicago 1964)
5.5   C. J. Bouwkamp, A. J. Duijvestijn, P. Medema: Tables relating to simple squared rectangles (Dept. of Mathematics and Mechanics, Technische Hogeschool, Eindhoven 1960)
5.6   V. E. Hoggatt: *Fibonacci and Lucas Numbers* (Houghton Mifflin, Boston 1969)
5.7   P. H. Richter, R. Schranner: Leaf arrangement. Naturwissenschaften *65*, 319-327 (1978)
5.8   M. Eigen: "Goethe und das Gestaltproblem in der modernen Biologie," in H. Rössner (ed.): *Rückblick in die Zukunft* (Severin und Siedler, Berlin 1981)
5.9   O. Ore: *Number Theory and Its History* (McGraw-Hill, New York 1948)
5.10   A. Koenig (personal communication)
5.11   W. Gellert, H. Küstner, M. Hellwich, H. Kästner (eds.): *The VNR Concise Encyclopedia of Mathematics* (Van Nostrand Reinhold, New York 1977)
5.12   L. K. Hua, Y. Wang: *Applications of Number Theory to Numerical Analysis IX.* (Springer, Berlin, Heidelberg, New York 1981)
5.13   J. C. Lagarias; A. M. Odlyzko: Solving "low-density" subset sum problems. (to be published)
5.14   R. L. Graham (personal communication)
5.15   R. K. Guy: *Unsolved Problems in Intuitive Mathematics*, Vol. I, Number Theory (Springer, Berlin, Heidelberg, New York 1981)
5.16   M. Gardner: Mathematical games. Sci. Am. *239*, No. 4, 22-26 (1978),
5.17   R. L. Graham: A theorem on partitions. J. Austral. Math. *4*, 435-441 (1963)
5.18   E. Landau: *Elementary Number Theory* (Chelsea, New York 1958)
5.19   E. H. Neville: *The Farey Series of Order 1025* (Cambridge University Press, Cambridge 1950)
5.20   C. M. Rader: Recovery of undersampled periodic waveforms. IEEE Trans. ASSP-*25*, 242-249 (1977)
5.21   M. Gardner: *Wheels, Life and Other Mathematical Amusements* (Freeman, New York 1983)
5.22   R. T. Gregory and E. V. Krishnamurthy: *Methods and Applications of Error-Free Computation* (New York Springer-Verlag 1984)

Chapter 6

6.1   G. H. Hardy, E. M. Wright: *An Introduction to the Theory of Numbers*, 5th ed., Sect. 5.2 (Clarendon, Oxford 1984)
6.2   P. J. Davis: *The Lore of Large Numbers* (Random House, New York 1961)
6.3   L. E. Dickson: *History of the Theory of Numbers*, 1-3 Vols. (Chelsea, New York 1952)

Chapter 7

7.1   G. H. Hardy, E. M. Wright: *An Introduction to the Theory of Numbers*, 5th ed., Sect. 5.4 (Clarendon, Oxford 1984)
7.2   T. L. Heath: *Diophantus of Alexandria* (Dover, New York 1964)

7.3   C. F. Gauss: *Disquisitiones Arithmeticae* [English transl. by A. A. Clarke, Yale University Press, New Haven 1966]

7.4   R. Tijdeman: On the equation of Catalan. Acta Arith. *29*, 197-209 (1976); "Exponential Diophantine Equations," in Proc. Int. Congr. Math., Helsinki (1978)

7.5   W. Kaufmann-Bühler: *Gauss. A Biographical Study* (Springer, Berlin, Heidelberg, New York (1981)

7.6   M. Abramowitz, I. A. Stegun: *Handbook of Mathematical Functions* (Dover, New York 1965)

7.7   M. D. Hirschhorn: A simple proof of Jacobi's four-square theorem. J. Austral. Math. Soc. *32*, 61-67 (1981)

7.8   H. Minkowski: Peter Gustav Lejeune Dirichlet und seine Bedeutung fur die heutige Mathematik. Jahresbericht der Deutschen Mathematiker-Vereinigung *14*, 149-163 (1905)

7.9   M. R. Schroeder: Eigenfrequenzstatistik und Anregungsstatistik in Räumen. Acustica *4*, 45-68 (1954)

Chapter 8

8.1   H. Halberstam, C. Hooley (eds.): *Progress in Analytic Number Theory*, Vol. I (Academic, London 1981)

8.2   R. H. Hudson: A common combinatorial principle underlies Riemann's formula, the Chebyshev phenomenon, and other subtle effects in comparative prime number theory. J. reine angew. Math. *313*, 133-150 (1980)

Chapter 9

9.1   N. J. A. Sloane: "Error-Correcting Codes and Cryptography," in *The Mathematical Gardener*, ed. by D. Klarner (Prindle Weber Schmidt, Boston 1981) pp. 347-382. Republished in Cryptologia *6*, 128-153, 258-278 (1982)

9.2   A. G. Konheim: *Cryptography: A Primer* (Wiley, New York 1981)

9.3   W. Diffie, M. E. Hellman: Privacy and authentication: An introduction to cryptography. Proc. IEEE *67*, 397-427 (1979)

9.4   J. J. Quisquater, C. Couvreur: Fast decipherment algorithm for RSA public-key cryptosystem. Electron. Lett. *18*, 905-907 (1982)

9.5   G. Brassard: A note on the complexity of cryptography. IEEE Trans. IT-*25*, 232-233 (1979)

9.6   C. S. Ogilvy: *Through the Mathescope* (Oxford University Press, London 1956), cited in M. Dekking, M. Mendés France, A. van der Poorten: Folds!. Math. Intelligencer *4*, 130-138 (1982)

9.7   C. H. Meyer, S. Matyas: *Cryptography: A New Dimension in Computer Data Security* (Wiley, New York 1982)

Chapter 10

10.1   G. H. Hardy, E. M. Wright: *An Introduction to the Theory of Numbers*, 5th ed., Sect. 18.2 (Clarendon, Oxford 1984)

Chapter 11

11.1   G. H. Hardy, E. M. Wright: *An Introduction to the Theory of Numbers*, 5th ed., Sect. 22.8 (Clarendon, Oxford 1984)

11.2   M. Abramowitz, I. A. Stegun (eds.): *Handbook of Mathematical Functions* (Dover, New York 1970)

11.3   J. Kubilius: *Probabilistic Methods in the Theory of Numbers* Translations of Mathematical Monographs *11* (Auger. Math. Soc., Providence 1964)

11.4   S. W. Graham: The greatest prime factor of the integers in an interval. J. London Math. Soc. (2) *24*, 427-440 (1981)

11.5   C. Couvreur, J. J. Quisquater: An introduction to fast generation of large prime numbers. Philips J. Res. *37*, 231-264 (1982)

Chapter 12

12.1   R. L. Rivest, A. Shamir, L. A. Adleman: A method for obtaining digital signatures and public key cryptosystems. Comm. ACM *21*, 120-126 (1978)

12.2   A. G. Konheim: *Cryptography: A Primer* (Wiley, New York 1981) pp. 331-347

12.3   E. N. Gilbert, F. J. MacWilliams, N. J. A. Sloane: Codes which detect deception. Bell Syst. Tech. J. *53*, 405-424 (1974)

12.4   V. Fak: Repeated use of codes which detect deception. IEEE Trans. IT-*25*, 233-234 (1979)

Chapter 13

13.1   C. F. Gauss: *Disquisitiones Arithmeticae* [English transl. by A. A. Clarke, Yale University Press, New Haven 1966]

13.2   M. Abramowitz, I. A. Stegun (eds.): *Handbook of Mathematical Functions* (Dover, New York 1970)

13.3   C. S. Ogilvy: *Tomorrow's Math* (Oxford University Press, Oxford 1962)

13.4   L. J. Alex: Solving exponential diophantine equations. Math. Mag. *54*, 259-260 (1981)

13.5   S. W. Golomb: *Shift Register Sequences* (Holden-Day, San Francisco 1967)

13.6   H. E. Rowe: *Signals and Noise in Communication Systems* (Van Nostrand Reinhold, New York 1965)

13.7   M. R. Schroeder, D. Gottlob, K. F. Siebrasse: Comparative study of European concert halls. J. Acoust. Soc. Am. *56*, 1195-1201 (1974)

13.8   M. R. Schroeder: Binaural dissimilarity and optimum ceilings for concert halls: More lateral sound diffusion. J. Acoust. Soc. Am. *65*, 958-963 (1979)

13.9   M. R. Schroeder: Toward better acoustics for concert halls. Phys. Today *33*, No. 10, 24-30, October (1979)

13.10   H. W. Strube: More on the diffraction theory of Schroeder diffusors. J. Acoust. Soc. Am. *70*, 633-635 (1981)

13.11   H. P. Lawther, Jr.: An application of number theory to the splicing of telephone cables. Am. Math. Monthly *42*, 81-91 (1935)

Chapter 14

14.1   N. J. A. Sloane: "Error-Correcting Codes and Cryptography," in *The Mathematical Gardener*, ed. by D. Klarner (Prindle Weber Schmidt, Boston 1981) pp. 347-382. Republished in Cryptologia *6*, 128-153, 258-278 (1982)

14.2   R. C. Merkle, M. E. Hellman: Hiding information and signatures in trapdoor knapsacks. IEEE Trans. IT-*24*, 525-530 (1978)

14.3   A. Shamir: "A Polyominal Time Algorithm for Breaking Merkle-Hellman Cryptosystems," Internal Report Applied Mathematics, The Weizmann Institute, Rehovot, Israel

14.4   Y. Desmedt, J. Vandewalle, R. Govaerts: "Critical Analysis of the Security of Knapsack Public Key Algorithms," in *Proceedings of the IEEE International Symposium on Information Theory* (IEEE, New York 1982) pp. 115-116
14.5   J. C. Lagarias (personal communication)

Chapter 15

15.1   G. H. Hardy, E. M. Wright: *An Introduction to the Theory of Numbers*, 5th ed., Sect. 6.5-7 (Clarendon, Oxford 1984)
15.2   E. Jahnke, R. Emde: *Tables of Functions* (Dover, New York 1945)
15.3   M. Born, E. Wolf: *Principles of Optics* (Pergamon, Oxford 1970)
15.4   M. R. Schroeder, R. E. Gerlach, A. Steingrube, H. W. Strube: Response to "Theory of Optimal Plane Diffusors." J. Acoust. Soc. Am. *66*, 1647-1652 (1979)
15.5   M. R. Schroeder: Constant-amplitude antenna arrays with beam patterns whose lobes have equal magnitudes. Archiv für Elektronik und Übertragungstechnik (Electronics and Communication) *34*, 165-168 (1980)
15.6   J. E. Mazo: Some theoretical observations on spread-spectrum communications. Bell Syst. Tech. J. *58*, 2013-2023 (1979)
15.7   I. F. Blake, J. W. Mark: A note on complex sequences with low correlations. IEEE Trans. IT-*28*, 814-816 (1982)
15.8   R. A. Scholtz: The origins of spread-spectrum communication. IEEE Trans. Communications, *30, 822-852* (1982); see also other articles in the same issue (May 1982)
15.9   R. M. Lerner: Signals having good Correlation Functions. Western Electronic Show and Convention, San Francisco, August 1961. Paper 9/3
15.10   S. W. Golomb: *Shift Register Sequences* (Holden-Day, San Francisco 1967)

Chapter 16

16.1   L. H. Hua: *Introduction to Number Theory* (Springer, Berlin, Heidelberg, New York 1982)
16.2   C. E. Shannon (personal communication)
16.3   J. H. McClellan, C. M. Rader: *Number Theory in Digital Signal Processing* (Prentice-Hall, Englewood Cliffs, NJ 1979)
16.4   H. J. Nussbaumer: *Fast Fourier Transform and Convolution Algorithms* (Springer, Berlin, Heidelberg, New York 1981)
16.5   B. Gold, C. M. Rader, A. V. Oppenheim, T. G. Stockham: *Digital Processing of Signals* (McGraw-Hill, New York 1969)
16.6   J. W. Goodman: *Introduction to Fourier Optics* (McGraw-Hill, New York 1968)
16.7   C. M. Rader: Discrete Fourier transforms when the number of data samples is prime. Proc. IEEE *56*, 1107-1108 (1976)

Chapter 17

17.1   A. Hedayat, W. D. Wallis: Hadamard matrices and their applications. Ann. Statistics *6*, 1184-1238 (1978)
17.2   M. Harwit, N. J. A. Sloane: *Hadamard Transform Optics* (Academic, New York 1979)
17.3   H. J. Nussbaumer: *Fast Fourier Transform and Convolution Algorithms* (Springer, Berlin, Heidelberg, New York 1981)

Chapter 18

18.1   W. Kaufmann-Bühler: *Gauss. A Biographical Study* (Springer, Berlin, Heidelberg, New York 1981)
18.2   G. H. Hardy, E. M. Wright: *An Introduction to the Theory of Numbers*, 5th ed., Sect. 8.1 (Clarendon, Oxford 1984)

Chapter 19

19.1   R. Morris: On the converse of Fermat's theorem. Comp. Sci. Tech. Rep. to be published.
19.2   C. Pomerance: Recent developments in primality testing. Math. Intelligencer *3*, 97-105 (1981)
19.3   A. J. van der Poorten, A. Rotkiewicz: On strong pseudoprimes in arithmetic progressions. J. Austral. Math. Soc. *29*, 316-321 (1980)
19.4   A. Rotkiewicz: On Euler Lehmer pseudoprimes and strong pseudoprimes with parameters L, Q in arithmetic progressions. Math. Comp. *39*, 239-247 (1982)
19.5   S. Goldwasser, S. Micali: "Probabilistic Encryption and How To Play Mental Poker," in *Proceedings of the 4th ACM Symposium on the Theory of Computing* (Assoc. Comp. Machinery, New York 1982) pp. 365-377
19.6   S. Micali (personal communication)
19.7   A. Shamir, R. L. Rivest, L. M. Adleman: "Mental Poker," in *The Mathematical Gardener*, ed. by D. Klarner (Prindle Weber Schmidt, Boston 1981) pp. 37-43
19.8   S. S. Wagstaff: Large Carmichael numbers. Math. J. Okayama Univ. *22*, 33-41 (1980)
19.9   M. O. Rabin: Probabilistic algorithm for testing primality. J. Number Theory *12*, 128-138 (1980)
19.10   C. F. Gauss: *Disquisitiones Arithmeticae* [English transl. by A. A. Clarke, Yale University Press, New Haven 1966]
19.11   J. D. Dixon: Asymptotically fast factorization of integers. Math. Comp. *36*, 255-260 (1981)
19.12   E. Lucas: *Théorie des Nombres* (Blanchard, Paris 1961)
19.13   J. Brillhart, D. H. Lehmer, J. L. Selfridge: New primality criteria and factorizations of $2^m \pm 1$. Math. Comp. *29*, 620-647 (1975)
19.14   L. M. Adleman, C. Pomerance, R. S. Rumely: On distinguishing prime numbers from composite numbers (to appear)

Chapter 20

20.1   L. H. Hua: *Introduction to Number Theory* (Springer, Berlin, Heidelberg, New York 1982)
20.2   T. M. Apostol: *Introduction to Analytic Number Theory* (Springer, Berlin, Heidelberg, New York 1976)
20.3   G. H. Hardy, E. M. Wright: *An Introduction to the Theory of Numbers*, 5th ed., Sect. 17.7 (Clarendon, Oxford 1984)

Chapter 21

21.1   L. H. Hua: *Introduction to Number Theory* (Springer, Berlin, Heidelberg, New York 1982)
21.2   G. H. Hardy, E. M. Wright: *An Introduction to the Theory of Numbers*, 5th ed., Sect. 17.1-5, 19.3-15 (Clarendon, Oxford 1984)

21.3   L. R. Rabiner, B. Gold: *Theory and Application of Digital Signal Processing* (Prentice-Hall, Englewood Cliffs, NJ 1975)

21.4   W. Feller: *An Introduction to Probability Theory with Its Applications* (Wiley, New York 1968)

21.5   M. Abramowitz, I. A. Stegun (eds.): *Handbook of Mathematical Functions* (Dover, New York 1970)

21.6   S. Ramanujan: *Notebooks* (Tata Institute of Fundamental Research, Bombay 1956)

21.7   G. H. Hardy: *Ramanujan* (Chelsea, New York 1978)

21.8   D. M. Bressoud: Analytical and combinational generalization of the Rogers-Ramanujan identities. Memoirs Am. Math. Soc. *24*, No. 227, 1-54 (1980)

Chapter 22

22.1   C. F. Gauss: *Disquisitiones Arithmeticae* [English transl. by A. A. Clarke, Yale University Press, New Haven 1966]

22.2   W. Gellert, H. Küstner, M. Hellwich, H. Kästner (eds.): *The VNR Concise Encyclopedia of Mathematics* (Van Nostrand Reinhold, New York 1977)

22.3   H. Rademacher: *Lectures on Elementary Number Theory* (Blaisdell, New York 1964)

22.4   J. H. McClellan, C. M. Rader: *Number Theory in Digital Signal Processing* (Prentice-Hall, Englewood Cliffs, NJ 1979)

Chapter 23

23.1   A. Papoulis: *Signal Analysis* (McGraw-Hill, New York 1977)

23.2   J. H. McClellan, C. M. Rader: *Number Theory in Digital Signal Processing* (Prentice-Hall, Englewood Cliffs, NJ 1979)

23.3   S. Winograd: On computing the discrete Fourier transform. Math. Comp. *32*, 175-199 (1978)

Chapter 24

24.1   L. H. Hua: *Introduction to Number Theory* (Springer, Berlin, Heidelberg, New York 1982)

24.2   D. Hilbert: Über die Irreduzibilität ganzer rationaler Funktionen mit ganzzahligen Koeffizienten. J. reine angew. Math. *110*, 104-129 (1982)

24.3   J. H. McClellan, C. M. Rader: *Number Theory in Digital Signal Processing* (Prentice-Hall, Englewood Cliffs, NJ 1979)

Chapter 25

25.1   F. J. Mac Williams, N. J. A. Sloane: *The Theory of Error-Correcting Codes* (North-Holland, Amsterdam 1978)

25.2   S. W. Golomb and H. Taylor: Constructions and properties of Costas Arrays. Proc. IEEE *72*, 1143-1163 (1984)

Chapter 26

26.1   S. W. Golomb: *Shift Register Sequences* (Holden-Day, San Francisco 1967)

26.2   W. Stahnke: Primitive binary polynomials. Math. Comp. *27*, 977-980 (1973)

26.3   E. R. Berlekamp: *Algebraic Coding Theory* (McGraw-Hill, New York 1968)

26.4   F. J. Mac Williams, N. J. A. Sloane: *The Theory of Error-Correcting Codes* (North-Holland, Amsterdam 1978)

26.5   M. R. Schroeder: Integrated-impulse method of measuring sound decay without using impulses. J. Acoust. Soc. Am. *66*, 497-500 (1979)

26.6   A. R. Møller: Use of stochastic signals in evaluation of the dynamic properties of a neuronal system. Scand. J. Rehab. Med., Suppl. 3, 37-44 (1974)

26.7   I. I. Shapiro, G. H. Pettengill, M. E. Ash M. L. Stone, W. B. Smith, R. P. Ingalls, R. A. Brockelman: Fourth test of general relativity. Phys. Rev. Lett. *20*, 1265-1269 (1968)

26.8   T. Rothman: The short life of Évariste Galois. Sci. Am. *246*, No. 4, 136-149 (1982)

26.9   M. R. Schroeder, D. Gottlob, K. F. Siebrasse: Comparative study of European concert halls. J. Acoust. Soc. Am. *56*, 1195-1201 (1974)

26.10   M. R. Schroeder: Binaural dissimilarity and optimum ceilings for concert halls: More lateral sound diffusion. J. Acoust. Soc. Am. *65*, 958-963 (1979)

26.11   J. W. Goodman: *Introduction to Fourier Optics* (McGraw-Hill, New York 1968)

26.12   M. Born, E. Wolf: *Principles of Optics* (Pergamon, Oxford 1970)

26.13   M. R. Schroeder: Constant-amplitude antenna arrays with beam patterns whose lobes have equal magnitudes. Archiv für Elektronik und Übertragungstechnik (Electronics and Communication) *34*, 165-168 (1980)

26.14   G. Hoffmann de Visme: *Binary Sequences* (The English University Press, London 1971)

26.15   F. J. MacWilliams: The structure and properties of binary cyclic alphabets. Bell Syst. Tech. J. *44*, 303-332 (1965)

26.16   F. J. MacWilliams: A table of primitive binary idempotents of odd length. IEEE Trans. IT-*25* 118-121 (1979)

26.17   M. R. Schroeder (ed.): *Speech and Speaker Recognition* (S. Karger, Basel 1985)

Chapter 27

27.1   F. J. Mac Williams, N. J. A. Sloane: *The Theory of Error-Correcting Codes* (North-Holland, Amsterdam 1978)

27.2   G. Hoffmann de Visme: *Binary Sequences* (The English University Press, London 1971)

27.3   S. W. Golomb: *Shift Register Sequences* (Holden-Day, San Francisco 1967)

27.4   D. E. Knuth: *The Art of Computer Programming*, Vol. 2, Seminumerical Algorithms (Addison-Wesley, Reading, MA 1969)

27.5   E. N. Gilbert: Unpublished notes (1953)

27.6   T. Herlestam: "On the Complexity of Functions of Linear Shift Register Sequences," in *Proceedings of the International Symposium on Information Theory* (IEEE, New York 1982) p. 166

27.7   H. J. Baker, F. C. Piper: Communications security, a survey of cryptography. IEE Proc. *A 129*, No. 6, 357-376 (1982)

27.8   D. P. Robbins, E. D. Bolker: The bias of three pseudo-random shuffles. Aequationes Math. *22*, 268-292 (1981)

27.9   P. Diaconis, M. Shahshahani: Generating a random permutation with random transpositions. Z. Wahrscheinlichkeitstheorie *57*, 159-179 (1981)

27.10   N. J. A. Sloane: "Encrypting by Random Rotations," in [Ref. 27.11] pp. 71-128

27.11   T. Beth (ed.): *Cryptography*, Proc. Workshop, Burg Feuerstein, March 29-April 2, 1982, Lecture Notes in Computer Science, Vol. 149 (Springer, Berlin, Heidelberg, New York 1983)

Chapter 28

28.1   M. R. Schroeder: Synthesis of low peak-factor signals and binary sequences with low autocorrelation. IEEE Trans. IT-*13*, 85-89 (1970)

28.2   W. Rudin: Some theorems on Fourier coefficients. Proc. Am. Math. Soc. *10*, 855-859 (1959)

28.3   D. A. Shedd, D. V. Sarwate: Construction of sequences with good correlation properties. IEEE Trans. IT-*25*, 94-97 (1979)

28.4   A. Gersho, B. Gopinath, A. M. Odlyzko: Coefficient inaccuracy in transversal filtering. Bell Syst. Tech. J. *58*, 2301-2316 (1979)

28.5   I. F. Blake, J. W. Mark: A note on complex sequences with low correlations. IEEE Trans. IT-*28*, 814-816 (1982)

28.6   A. T. Moffet: Minimum redundancy linear arrays. IEEE Trans. Antennas and Propagation AP-*16*, 172-175 (1968)

28.7   A. Rosa, G. Sabadussi, J. Turgeon (eds.): *Theory and Practice of Combinatorics* (North-Holland, Amsterdam 1982)

Chapter 29

29.1   M. R. Schroeder: Images from computers and microfilm plotters. Comm. ACM *12*, 95-101 (1969). See also IEEE Spectrum, March 1969, pp. 66-78

29.2   M. R. Schroeder: A simple function and its Fourier transform. Math. Intelligencer *4*, 158-161 (1982)

29.3   M. R. Schroeder: Number theory in physics, engineering and art. Interdisciplinary Sci. Rev. *6*, No. 3, 239-248 (1980)

Chapter 30

30.1   N. J. A. Sloane: *A Handbook of Integer Sequences* (Academic Press, Orlando, FL, 1973)

30.2   D. Shechtman, I. Blech, D. Gratias and J. W. Cahn: Metallic phase with long-range orientational order and no translational symmetry. Phys. Rev. Lett. *53*, 1951-1953 (1984)

30.3   M. Gardner: Extraordinary nonperiodic tiling that enriches the theory of tiles. Scientific American *236*, 110-121 (Jan. 1977)

30.4   D. Levine and P. J. Steinhardt: Quasicrystals: A new class of ordered structures. Phys. Rev. Lett. *53*, 2477-2480 (1984)

30.5   B. Mandelbrot: *The Fractal Geometry of Nature* (Freeman, San Francisco 1983)

30.6   T. A. Witten and L. M. Sander: Phys. Rev. Lett. *47*, 1400-1403 (1981); Phys. Rev. B *7*, 5686-5697 (1983)

30.7   C. Nicolis und G. Nicolis: Gibt es einen Klima-Attraktor? Phys. Blätter *41*, 5-9 (1985)

30.8   E. Basar: Toward a physical approach to integrative physiology. I. Brain dynamics and physical causality. Am. J. Physiol. *245* (Regulatory Integrative Comp. Physiol. *14*), R510 – R533 (1983); see also A. Abraham, A. Mandel and D. Farmer, in *Proceedings Nonlinear Functions of the Brain* (Santa Barbara 1982)

30.9   M. R. Schroeder: Linear prediction, entropy and signal analysis. IEEE ASSP Magazine *1*, 3-11 (July 1984)

30.10   M. J. Feigenbaum: Universal behavior in nonlinear systems. Los Alamos Science *1*, 4-27 (1981); see also M. J. Feigenbaum: Quantitative universality for a class of nonlinear transformation. J. Statistical Physics *19*, 25-52 (1978)

Name Index

Subject Index